WOOD ADHESIVES

CHEMISTRY AND TECHNOLOGY
Volume 2

edited by

A. Pizzi

Division of Processing and
 Chemical Manufacturing Technology
Council for Scientific and Industrial Research
Pretoria, South Africa

CRC Press
Taylor & Francis Group
Boca Raton London New York

CRC Press is an imprint of the
Taylor & Francis Group, an informa business

First published 1989 by Marcel Dekker

Published 2019 by CRC Press
Taylor & Francis Group
6000 Broken Sound Parkway NW, Suite 300
Boca Raton, FL 33487-2742

© 1989 by Taylor & Francis Group, LLC
CRC Press is an imprint of Taylor & Francis Group, an Informa business

First issued in paperback 2019

No claim to original U.S. Government works

ISBN 13: 978-0-367-45112-7 (pbk)
ISBN 13: 978-0-8247-8052-4 (hbk)

Visit the Taylor & Francis Web site at
http://www.taylorandfrancis.com

and the CRC Press Web site at
http://www.crcpress.com

Library of Congress Cataloging in Publication Data

Main entry under title:

(Revised for vol. 2)

Wood adhesives.

    Includes bibliographical references and indexes.
      1. Adhesives. 2. Wood-Bonding. I. Pizzi,
A. (Antonio)
TP968.W66 1983   568'3   83-20870
ISBN 0-8247-1579-9 (v.1)
ISBN 0-8247-8052-3 (v.2)

Preface

Wood Adhesives: Chemistry and Technology, Volume 1 was written as a basic textbook for adhesives chemists and wood technologists training in a fascinating, important, and sometimes obscure—but never dull—field. Volume 2 is a natural follow-up. It contains a host of interesting and important concepts and information that could not be included or expanded upon in the previous volume. It is by no means exhaustive, since some breakthroughs and concepts, such as in carbohydrates-based adhesives for wood, are still in such rapid flux that a report on them now would be incomplete and perhaps misleading.

The material presented, however, is of such interest that no self-respecting wood technologist or chemist in this field can afford to be unaware of it. The various chapter authors are considered specialists in their particular fields or have participated in and been greatly involved in the history of the reported developments. The book could be divided broadly into two sections, concepts and adhesives, but by necessity, the borders of such a division are fictitious; the two are often intertwined and meld into each other.

In the preface to the first volume, protein adhesives were described as a dying line. They are included in this volume because they are still used extensively in some countries and have resisted the onslaught of synthetic resins in the last half a century, indicating an unusually resilient product. Possibilities definitely exist for their modification, upgrading, and use in countries where synthetic adhesives are not produced. Hence, in this field, the accumulated knowledge of perhaps centuries must not be allowed to be forgotten. They are masterfully described in the chapter by Alan L. Lambuth. In another chapter, William E. Johns describes the exciting develop-ment and intriguing concept of "gluing without glue," of which he was so much at the forefront during its recent development. Lawrence Gollob's chapter on structure/property relationships in phenol-formaldehyde adhesives is a clear indication of things to come, of the direction in which adhesive technologists must go and are bound to go in the future. Kenneth R. Geddes fills the vacuum on the chemistry of polyvinyl-acetate that was so obvious in the first volume. He also describes the development of the continuous recycle reactor, in the commercialization of which he played such a major role—a few are already in operation in several countries. Rainer Marutzky's chapter on formaldehyde emission makes the point once again with his usual sharpness and incisiveness on this very important topic of recent times.

My own contribution consists of chapters covering three topics that have, or are likely to have, some impact on wood adhesives. The first, "honeymoon" fast-set adhesives for glulam and finger-jointing, is a concept born in North America less than two decades ago that matured to extensive modifications and commercialization in Africa. It is a true success story, both technically and commercially. The second, the approach to phenolformaldehyde cellulose adhesion forces by conformational analysis, a theoretical technique borrowed from the research on the spatial structure of macromolecules, is of interest because so many other similar problems in wood adhesives could possibly be approached fruitfully by the same technique. The third, the "blue glue," is a success story in the making. A vast step forward in phenol/resorcinol/formaldehyde resins, it already indicates what the impact of true resin engineering on the adhesives world is likely to be.

The other South African writers are less well known in the wood adhesives world, but judging from the topics addressed, it is quite likely that some of their developments will be known better in a few years. Of particular mention is the work of Gerrit H. van der Klashorst on bagasse lignin-formaldehyde reactions. I have also participated extensively enough in this area of adhesives polymerization, formulation, and application to judge its importance in this, as well as in other fields. The lignin-based cold-set reported is a world first, and both the thermosets and the cold-sets are now on the verge of commercial exploitation.

My sincere thanks go to the authors who have contributed these very provocative and stimulating chapters.

A. Pizzi

Contributors

F. A. Cameron* Division of Processing and Chemical Manufacturing Technology, Council for Scientific and Industrial Research, Pretoria, South Africa

Gérard Elbez Centre Technique du Bois et de L'Ameublement, Paris, France

Kenneth R. Geddes† Research Department, Crown Decorative Products Ltd., Darwen, Lancashire, England

Lawrence Gollob Chemical Division, Georgia-Pacific Resins, Inc., Decatur, Georgia

William E. Johns Department of Mechanical and Materials Engineering, Washington State University, Pullman Washington

Alan L. Lambuth Timber and Wood Products Group, Boise Cascade Corporation, Boise, Idaho

Rainer Marutzky Fraunhofer-Institute for Wood Research, Braunschweig, Federal Republic of Germany

A. Pizzi Division of Processing and Chemical Manufacturing Technology, Council for Scientific and Industrial Research, Pretoria, South Africa

Neville E. Quixley‡ Expandite (PTY) Ltd., Isando, South Africa

---

Present affiliations:
*Industrial Laminates (PTY) Ltd., Alrode, South Africa
†Crown Berger Licensing Group, Crown Berger Europe Limited, Darwen, Lancashire, England
‡T.A.C. National (PTY) Ltd., Benoni South, South Africa

R. Smit* Division of Processing and Chemical Manufacturing Technology, Council for Scientific and Industrial Research, Pretoria, South Africa

Gerrit H. van der Klashorst Division of Processing and Chemical Manufacturing Technology, Council for Scientific and Industrial Research, Pretoria, South Africa

Present affiliation: Department of Chemistry, Rand Afrikanse Universiteit, Johannesburg, South Africa.

Contents

PREFACE   iii

CONTRIBUTORS   v

1. PROTEIN ADHESIVES FOR WOOD   1
   Alan L. Lambuth

   I.   Introduction   1
   II. Soybean Adhesives   2
   III. Blood Glues   13
   IV. Casein Glues   22
   V.   Other Proteins   26
        References   27

2. THE CHEMISTRY OF PVA   31
   Kenneth R. Geddes

   I.   Introduction   31
   II.   Process Outline   31
   III. Vinyl Acetate Monomer   33
   IV.   Internal Plasticization and Copolymerization   33
   V.   Water   36
   VI.   Initiators   36
   VII. Surface-Active Agents   38
   VIII. Colloids: Polyvinyl Alcohol   41
   IX.   Other Colloids   42
   X.   Buffers and pH Adjusters   43
   XI.   External Plasticizers, Coalescents, Tackifiers
         and Crosslinking Agents   43
   XII. Fillers and Pigments   44
   XIII. Fungicides and Preservatives   44
   XIV. Polymerization Reactions and Kinetics   45
   XV.   Copolymerization Kinetics   49
   XVI. Glass Transition Temperature of Copolymers   51
   XVII. Emulsion Polymerization by the Batch Process
         in the Presence of Surfactant   51

XVIII.   Emulsion Polymerization of Vinyl Acetate
         Stabilized by Polyvinyl Alcohol Alone   54
  XIX.   Continuous Emulsion Polymerization   55
   XX.   High Pressure Polymerization   60
  XXI.   Practical Aspects of Production of PVA
         Adhesives   62
 XXII.   Film Formation and Film Properties   65
         References   66

3.   THE CHEMICAL BONDING OF WOOD   75
     William E. Johns

     I.   Introduction   75
    II.   Varieties of Chemical Bonding   77
   III.   Discussion   87
    IV.   Summary   92
          References   92

4.   PHENOL-FORMALDEHYDE STRUCTURES IN
     RELATION TO THEIR ADHESION TO WOOD CELLULOSE   97
     A. Pizzi and R. Smit

     I.   Introduction   97
    II.   The Structure of Some Phenol-Formaldehyde
          Condensates for Wood Adhesives   98
   III.   Adhesion to Wood Cellulose   112
    IV.   Conclusions   118
          References   119

5.   THE CORRELATION BETWEEN PREPARATION
     AND PROPERTIES IN PHENOLIC RESINS   121
     Lawrence Gollob

     I.   Introduction   121
    II.   Concepts of Synthesis/Structure/Property
          Relationships   123
   III.   Outlook   150
          References   150

6.   LIGNIN FORMALDEHYDE WOOD ADHESIVES   155
     Gerrit H. van der Klashorst

     I.   Introduction   155
    II.   Polymerization Reactions for Alkali Lignin   157

III. The Utilization of Alkali Lignin in Phenol-
Formaldehyde Wood Adhesives 160
IV. Industrial Soda Bagasse Lignin-Based Cold-
Setting Wood Adhesives 164
V. Lignin-Based "Honeymoon" Adhesives 166
VI. Soda Bagasse Thermosetting Wood Adhesives 166
VII. Meta-Modified Lignin-Based Wood Adhesives 173
VIII. Reaction of Industrial Lignin with Formalde-
hyde at the Meta Positions 181
IX. Preparation of Cold-Setting Wood Adhesive
Based on the Metamodification of Industrial
Pine Kraft Lignin 184
X. Conclusion 186
Reference 187

7. LOW-RESORCINOL PRF COLD-SET ADHESIVES—
THE BRANCHING PRINCIPLE 191
A. Pizzi

I. Introduction 191
II. Chemistry of PRF Resins and Branching 193
III. Different Branching Molecules 195
IV. Branched PRF Characteristics 201
V. Base- and Acid-Catalyzed Formulations;
Pot-Life and Shelf-Life 205
VI. Formulation 208
VII. Conclusion 209
References 210

8. HOTMELTS FOR WOOD PRODUCTS 211
Neville E. Quixley

I. Introduction 211
II. Background 211
III. How Do Hotmelts Work? 212
IV. Application Areas 212
V. Ethylene-Vinyl Acetate Hotmelts for Edging 215
VI. Polyamide Hotmelts for Edging 219
VII. Edge-Bonding Techniques 220
VIII. Health and Safety 227

9.   FAST-SETTING ADHESIVES FOR FINGER-
     JOINTING AND GLULAM   229
     A. Pizzi and F. A. Cameron

        I.   Introduction   229
       II.   Components A and B   230
      III.   Adhesive Resins Preparation   234
       IV.   Finger-Jointing   236
        V.   Testing   240
       VI.   Statistical Analysis   240
      VII.   Initial Industrial Application   244
     VIII.   Variations in Paraformaldehyde Content
             and pH   257
       IX.   Factory Trials   262
        X.   Variation in the Proportion of A and B Components   263
       XI.   Minimum Curing Temperatures   273
      XII.   Laminated Beams (Glulam) Application:
             Laboratory Requirements   284
     XIII.   Industrial Plant Applications   287
      XIV.   Conclusion   302
             References   304

10. RELEASE OF FORMALDEHYDE BY WOOD PRODUCTS   307
     Rainer Marutzky

        I.   Introduction   307
       II.   Formaldehyde: An Interesting Compound   309
      III.   Formaldehyde: Limits, Regulations, and
             Guidelines   314
       IV.   Reasons for the Release of Formaldehyde
             by Wood Products   317
        V.   Analysis of Adhesives with Respect to
             Formaldehyde   323
       VI.   Analytical Methods for Detecting and
             Quantifying Formaldehyde   327
      VII.   Factors Influencing the Release of Formalde-
             hyde in Wood Products   334
     VIII.   Methods for Determination of Formaldehyde
             Release in Wood Products   339
       IX.   Differences in Emission Behavior Among
             Wood Products   357
        X.   Reduction of Formaldehyde Release in
             Wood Products   362
       XI.   Future Developments   372
             References   373

11. EXPANDING RESORCINOL COLD-SETS FOR
    GAP-FILLING ADHESIVES FOR WOOD   389
    Gérard Elbez

        I. Introduction   389
       II. Development of Low Temperature Formulations 390
      III. Industrial Application Tests   393
       IV. Economic Aspects   403
        V. Conclusions   406
            References   407

INDEX   409

Contents of Volume 1

PREFACE   iii

CONTRIBUTORS   v

1. PROPERTIES AND PERFORMANCE OF WOOD ADHESIVES   1

    John M. Dinwoodie

    I.   Introduction   2
    II.   Properties of Wood Adhesives   4
    III.   Performance of Wood Adhesives   12
    IV.   Performance of Bonded Wood Products   16
    V.   Selected Relevant Standards   51
          References   53

2. AMINORESIN WOOD ADHESIVES   59

    A. Pizzi

    I.   Introduction   59
    II.   Chemistry of Aminoresins   60
    III.   Chemistry and Technology of Application of
          Aminoresin Adhesives for Wood   80
          References   102

3. PHENOLIC RESIN WOOD ADHESIVES   105

    A. Pizzi

    I.   Introduction   106
    II.   Chemistry of Phenol-Formaldehyde Condensations   107

III.   Chemistry and Technology of Application of Phenolic
       Resin Adhesives for Wood   144
       References   173

4.   TANNIN-BASED WOOD ADHESIVES   177

     A. Pizzi

     I.   Introduction   178
     II.   Chemistry of Condensed Tannins   179
     III. Reactivity of Tannins as Macromolecules   192
     IV.   Chemistry and Technology of Industrial Tannin
           Adhesive Formulations   213
     V.   Conclusions   242
           References   242

5.   LIGNIN-BASED WOOD ADHESIVES   247

     Horst H. Nimz

     I.   Introduction   248
     II.   Chemical Backgrounds of the Curing Reaction of
           Lignin   249
     III. Application of Lignin as an Adhesive for Particle-
           board, Plywood, and Fiberboard   254
     IV.   Lignin in Combination with Phenol-Formaldehyde
           Adhesives   270
     V.   Lignin in Combination with Urea-Formaldehyde
           Resins   279
     VI.   Outlook   285
           References   287

6.   DIISOCYANATES AS WOOD ADHESIVES   289

     K. C. Frisch, L. P. Rumao, and A. Pizzi

     I.   Introduction   289
     II.   Chemistry of Diisocyanates   290
     III. Technology of Diisocyanate Adhesives for Wood   306
           References   316

7.   POLYVINYL ACETATE WOOD ADHESIVES   319

     Terence Martin Goulding

     I.   Introduction   320
     II.   Background   320
     III. Chemistry of Polyvinyl Acetates   321
     IV.   Formulating a PVA-Based Adhesive   325
     V.   Aspects of Application   340

Contents   *xv*

    VI.   Performance of PVA Adhesives   344
   VII.   Conclusion   349
         References   350

AUTHOR INDEX   351
SUBJECT INDEX   359

1

Protein Adhesives for Wood

ALAN L. LAMBUTH / *Boise Cascade Corporation*
Boise, Idaho

I. INTRODUCTION

Natural adhesives made from starch, collagen, and blood represent
ancient technology, dating back at least to the Egyptians, for bonding wood [1]. During later but still historical times, the proteins
from milk curd, fish skins, and legumes were added to the roster
of adhesive substances used for gluing wood and paper. These
included both decorative and utilitarian products. The bonds obtained
were strong and durable for centuries when kept dry. Moisture
produced rapid weakening and deterioration. Thus, exterior and
marine service requirements could be met only with natural thermoplastics such as bitumen and tree pitch. Unfortunately, neither
of these produced rigid bonds [2].

For structural gluing, animal and vegetable proteins were the
only real options from ancient times down through the Industrial
Revolution and nearly to the present. They could be made to perform briefly under exterior conditions only with careful surface
protection. During recent decades, mainly since World War II, several
different adhesive polymers based on synthetic organic chemicals
were developed and quickly preempted entire segments of the wood
products industry from protein adhesives. For example, vinyl and
acrylic emulsion polymers largely displaced collagen glues from the
furniture, cabinet and millwork industries. Phenol-formaldehyde
and urea-formaldehyde resin adhesives took over the structural
and decorative plywood industries from blood, soybean and starch
glues. Room temperature curing resorcinol-formaldehyde resin adhesives preempted lumber laminating and other construction special-
ties from casein glues.

In spite of this intense competition from the new synthetics, protein adhesives maintained a strong position in bonding wood products for interior applications until about 1960. They were able to do this on the basis of unique curing properties and low cost. After 1960, however, the very low commodity prices for most petrochemicals enabled the synthetic resin adhesives to take over virtually all remaining major markets still served by protein glues.

As a result, any useful current discussion of animal and vegetable protein glues must necessarily center on the technology and refinements developed during the peak years of protein adhesive consumption, mainly from 1940 to 1960. In addition, it is appropriate to review those continuing adhesive applications involving unique properties of proteinaceous materials. It is also important to remember that despite their displacement by synthetic resin adhesives over the past 20 years or so, protein glues represent very practical and immediately useful adhesive technology for the wood products industry to fall back upon if the supply of synthetic resin raw materials is jeopardized by world events affecting crude oil.

This chapter will address three of the most widely used families of protein-based adhesives for wood; namely, soybean, blood, and casein. The technology of soybean glues will be discussed first because soybean flour is frequently combined with blood or casein to yield adhesives of intermediate properties.

II. SOYBEAN ADHESIVES

A. Raw Material Source and Preparation

Soybeans are legumes, the seeds of a low-growing field vine. These vines are ancient in culture; the written record of their domestication in China dates back almost 3000 years [3]. From that time until now, soybeans have remained a very important agricultural crop for almost every civilization because of their unusually high contents of oil and edible protein.

To process soybeans into these useful products, the beans are dehulled and the oil is removed by crushing at very high pressure or by solvent extraction. If the resulting dry soybean meal is intended for food, it is heated to 70°C or higher to coagulate the proteins and caramelize the carbohydrates, thus improving their nutritional qualities. If the soybean meal is intended for adhesive uses instead, it is carefully processed at temperatures below 70°C to preserve the alkaline solubility of the proteins [4].

The protein content of oil-free soybean meal ranges from about 35 to 55% on a worldwide basis. However, the industrial grades are generally blended to yield a uniform protein content of 44-55%, depending on source. The other principal constituents of soybean meal are carbohydrates, totaling about 30%, and ash at 5 or 6%. The moisture content after processing is quite low, usually less than 10%.

Long experience has shown that adhesive grade "untoasted" soybean meal must be ground to an extremely fine flour in order to perform well as a protein glue [5]. Typically, the dry extracted meal is ground or milled until at least 40% and preferably 60-80% will pass through a 46 μm (325-mesh) screen. For easier quality control with flours of this fineness an alternative "Specific surface" test method is available that determines average particle size in terms of surface area per gram [6]. For the range of mesh sizes recommended above, the corresponding specific surface values are about 3000-6000 cm^2/g.

B. Formulation

Soybean flour will wet and swell in water but will not "disperse" to yield useful adhesive properties. For this purpose, treatment with a soluble alkaline material is necessary. Almost any organic or inorganic alkali will disperse wetted soybean flour to some degree. However, soybean wood glues of maximum bonding efficiency require dispersion with several percent of a strong alkali such as sodium hydroxide or trisodium phosphate [7]. The effect of this strongly alkaline treatment is to break the internal hydrogen bonds of the coiled protein molecules, literally unfolding them and making all their complex polar structure available for adhesion to wood. This alkaline dispersion process, while essential for adhesion, causes the gradual destruction of the protein molecules themselves by hydrolysis. Thus, an alkaline soybean glue has a definite useful life, slowly losing viscosity and adhesive functionality over a storage period of 6-12 hours.

Although these strongly alkaline soybean glues are nearly colorless in an applied film, they cause a reddish-brown stain on wood surfaces as they cure due to "alkali burn" of the cellulose [2]. If a colorless glue line on wood is desired, the wetted soybean flour must be dispersed with a less strongly alkaline material such as hydrated lime or ammonia [8]. However, the adhesive bond strength of these low color, mildly dispersed soybean glues is considerably less than that obtained with fully dispersed, highly alkaline formulations. Typical high alkali and low alkali soybean glue formulations are listed below.

High Alkali Formulation

Ingredient	Amount (kg)
Water at 16–21°C	87.5
"Adhesive-grade" soybean flour	48.5*
Pine oil or diesel oil defoamer: mix 3 minutes or until smooth	1.5*
Water at 16–21°C: mix 2 minutes or until smooth	72.5
Fresh hydrated lime (as a slurry in)	6.0
Water at 16–21°C: mix 1 minute	12.0
50% sodium hydroxide solution: mix 1 minute	7.0
Sodium silicate solution: mix 1 minute	12.5†
Orthophenyl phenol: mix 10 minutes	2.5

*Normally dry-blended for easier handling and dust control.
†8.90% Na_2O, 28.70% SiO_2, 41° Baumé.

The additions of hydrated lime and sodium silicate solution in this high alkali mix accomplish two purposes: (a) they help maintain a level glue viscosity for a longer adhesive working life, and (b) they improve the water resistance of the cured glue film by forming some insoluble proteinates [9].

The starchy constituents of soybean flour also disperse in the presence of strong alkali to become useful adhesive molecules contributing to dry bond strength. However, this starchy fraction also retains its well-known sensitivity to water and is considered mainly responsible for limiting the performance of soybean glues with respect to water resistance [10].

The final addition of preservative shown in this formulation is essential in virtually all protein glues to provide mold resistance in high humidity service. Without this protection, even heat-cured soybean adhesives *will* mold as the moisture content of the bonded wood approaches 20% [11]. Copper-8-quinolinolate, copper naphthenate, and orthophenyl phenol are among the few remaining preservatives permitted to be used in the United States at this time for wood products. Where the use of chlorinated phenols is still permitted, they are also very effective preservatives for protein glues at the addition level shown. In each case, the sodium hydroxide content of the glue formulation converts the water-insoluble fungicide to its soluble sodium salt.

Large quantities of this high alkali soybean glue formulation were used to bond interior grades of softwood plywood between about 1940 and 1960. It was also used to some extent for assembling prefabricated wooden building components [12]. Its primary advantages were very low cost and the capability to bond almost any dry wood surface. It also offered real versatility in pressing; that is, it could be hot pressed or cold pressed to promote cure. Appropriate pressing schedules for each curing mode will be shown later in the chapter.

Low Alkali Formulation

Ingredient	Amount (kg)
Water at 16-21°C	112.5
"Adhesive-grade" soybean flour	48.5*
Pine oil or diesel oil defoamer: mix 3 minutes or until smooth	1.5*
Water at 16-21°C: mix 2 minutes or until smooth	75.0
Fresh hydrated lime (as a slurry in)	15.0
Water at 16-21°C: mix 5 minutes	25.0

*Normally dry-blended for easier handling and dust control.

This glue has been widely used as a briquetting binder for wood and charcoal particles. It is particularly suitable for paper and softboard laminating where a colorless glue line and minimum swelling of the glue film on high humidity exposure are desired [13]. It is not recommended for structural uses such as sheathing plywood because of its lower degree of protein dispersion and thus lower bonding strength. Substituting borax or monosodium phosphate for the hydrated lime dispersing agent will yield similar nonstaining glues.

Over the years a number of "denaturants" or "crosslinkers" have been added to soybean glues to improve their water resistance, working life, and consistency. These may be roughly categorized as formaldehyde donors, sulfur compounds, and inorganic complexing salts. Each of these groups of compounds appears to react with

the starchy constituents of whole soybean flour as well as the
dispersed protein molecules. Formaldehyde itself acts too rapidly
and thus is difficult to control. Instead, such compounds as dialde-
hyde starch, dimethylol urea, sodium formaldehyde bisulfite, and
hexamethylenetetramine have been successfully used to toughen
the cured glue film and improve its water resistance [14]. Similarly,
carbon disulfide, thiourea, and ethylene trithiocarbonate, among
the sulfur compounds, and the soluble salts of cobalt, chromium,
and copper have been used to improve soybean glue working properties and adhesive performance [15]. These modifiers are generally
added last when preparing the glue. The range of addition of
all such denaturants is 0.1-1.0% based on the weight of soybean
flour. Also, 5-20% of an aliphatic epoxy resin has been added to
soybean glues, yielding significantly improved durability, but
the cost is high [16].

C. Mixing, Application, and Pressing

Soybean glues are very easy to mix, provided they are wetted
with plain water as a first step. (If any form of alkali is present
in the first mixing water, the dry soybean flour will form permanent
lumps.) As with all protein glues, the first mix is kept thick in
order to physically break down any lumps of dry powder that
may be present. The division of water additions in both formulations
given earlier demonstrates this mixing procedure. Once the soybean
flour particles have been uniformly wetted, further dilution and
dispersion steps can follow without difficulty. Water additions are
adjusted to yield a mixed glue viscosity in the broad range of
500-25,000 cP at 25°C, depending on purpose. Briquetting and
paper laminating glues would typically be 500-1000 cP, while cold
press plywood glues should be 10,000-20,000 cP for best performance
[17]. Hot press formulations would be midrange. Particularly because
of the heavy first mixing stage and the high final viscosities, an
appropriate soybean glue mixer should have relatively large and
slow-turning blades plus a provision for scraping glue buildup
off the mixer walls continuously.

   Low-viscosity soybean briquetting adhesives are generally applied
by spray. Similarly, paper and softboard laminating glues are
usually applied by curtain coater, knife, or indirect roller. High
viscosity plywood and lumber assembly formulations are fairly well
limited to application by spreader roll or extrusion. For very small
assembly jobs, soybean glue can easily be applied by brush.

   One of the real advantages of protein glues generally is their
ability to be cured (under pressure) hot or cold. Following are
typical commercial schedules for each mode of cure.

Soybean Glue: Hot-Pressing Schedule for Interior-Type Douglas Fir Plywood*

Rough panel thickness (mm)	Number of plies	Panels per press opening	Platen temperature (°C)	Pressing time (min·dt full pressure)
4.8	3	2	110	3
6.4	3	2	116	3
7.9	3	2	121	3-1/2
7.9	3	3	127	5
9.5	3	2	127	4
11.1	3	1	116	3
11.1	3	2	132	4-3/4
12.7	5	1	110	3-1/2
12.7	5	2	121	6
14.3	5	1	110	3-3/4
14.3	5	2	121	6
15.9	5	1	116	4
17.5	5	1	121	4
19.0	5	1	127	4-1/4
20.6	5	1	132	4-1/2
20.6	7	1	127	5
23.8	7 and 9	1	132	5-1/2
27.0	7 and 9	1	132	6
30.2	7 and 9	1	132	7
33.4	9	1	132	7-1/2
36.5	9	1	132	8
39.7	9 and 11	1	132	9

*Total time assembly not to exceed 15 minutes; hydraulic pressure not less than 14 kg/cm^2.

Glue Spread

Core thickness (mm)	Mixed glue per single glue line (g/m^2)
1.59	195
1.54, 2.82	208
3.18	220
3.63, 4.23, 4.76, 6.35	232
>20.6 mm	245

Soybean Glue: Cold-Pressing Schedule ("No-Clamp" Process) for
Interior-Type Douglas Fir Plywood

Core thickness (mm)	Mixed glue per single glue line (g/m^2)*
2.54	305–318
2.82	313–323
3.18	318–330
3.63	325–337
4.23	330–342
4.76	330–342
All 5 ply, 20.6 mm constructions	367
For rough core, add an extra 20 g over these spreads.	

*Use these wet glue spreads according to core thickness.

Special Pressing Instructions

Hold load 5 minutes before applying pressure.

Load must be under pressure within 25 minutes after the first
panel is laid.

Pressing time to be measured after pressure gauge reaches full
pressure.

Use 12-14 kg/cm^2 hydraulic pressure.

Pressure to be retained for 15 minutes.

Do not use veneer over 43°C.

    This cold pressing schedule is the result of an interesting
laboratory observation and subsequent industry-wide patent [17].
Alkaline protein glues, particularly soybean glues, lose water quite
rapidly into adjacent dry wood surfaces. As a result, they gain
sufficient "gel strength" in 15 or 20 minutes to permit removal
of a glued wood assembly from its clamping device without loss
of contact between the glued surfaces. Cure is then completed
over the next 6-12 hours simply by placing the bonded products
in storage at ambient temperatures with minimum handling. This
method of cold pressing, called the no-clamp process, was used

throughout the softwood plywood industry for many years. Prior to its introduction, all protein-bonded cold press plywood was clamped for 6-8 hours with bulky steel beams and turnbuckles.

Because protein glues develop bond strength primarily by water loss over time, roll pressing has proved unsuccessful as a clamping method for wood products. The short, intense period of pressure simply squeezes the still-fluid mix off the glue line without affording sufficient time for water loss and gelation. An exception is the soft roll lamination of paper to paper or paper to wood [18]. This can be accomplished at moderate speeds because of the extreme rapidity with which dry paper removes water from protein glues.

D. Blended Formulations

As mentioned earlier, the single largest commercial use of soybean flour in wood glues during the recent past has been as a blend with other adhesive proteins, mainly blood and casein, for bonding interior grade plywood, doors, and millwork. These blended formulations exploit several unique properties of the soybean glues themselves and incorporate useful characteristics of the other protein materials.

1. Soybean-Blood Glues

For example, a blend of soybean flour with spray-dried soluble animal blood, a fairly expensive but efficient adhesive protein, yields a glue with the best properties of each material [19]. Namely, the cost becomes moderate and the consistency is ideal for assembly time tolerance (slightly granular) because of the soybean flour. The hot press curing time is very short and the cured glue bonds are considerably more water resistant because of the blood content. Fortunately, both proteins require the same neutral wetting procedure and strongly alkaline dispersion steps. They are otherwise compatible in all proportions, yielding a series of cost/performance-related adhesives. Soybean-blood blend glues were by far the most widely used protein hot press adhesives for interior structural plywood from the early 1940s until about 1960 [20]. Also, when the oil embargo of 1973 quickly placed phenolic resin adhesives on allocation through petrochemical restrictions, the plywood industry immediately returned to the use of soybean-blood blend glues for interior structural grades. Following are typical examples of low and high blood content soybean blend glues.

Soybean Blend Glues

Low blood content glue	Amount (kg)
Water at 16–21°C	100.0
"Adhesive-grade" soybean flour	36.0*
Dried soluble animal blood	7.5*
74-μm wood flour†	5.0*
Pine oil or diesel oil defoamer: mix 3 minutes or until smooth	1.5*
Water at 16–21°C: mix 2 minutes or until smooth	110.0
Fresh hydrated lime (as a slurry in)	4.0
Water at 16–21°C: mix 1 minute	8.0
Sodium silicate solution: mix 1 minute	20.0‡
50% sodium hydroxide solution: mix 5 minutes	5.0
Orthophenyl phenol: mix 5 minutes	2.5

High blood content glue	Amount (kg)
Water at 16–21°C	80.0
Dried soluble animal blood	35.0*
"Adhesive-grade" soybean flour	8.5*
74-μm wood flour†	5.0*
Pine oil or diesel oil defoamer: mix 3 minutes or until smooth	1.5*
Water at 16–21°C: mix 2 minutes or until smooth	19.50
Fresh hydrated lime (as a slurry in)	4.0
Water at 16–21°C: mix 1 minute	8.0
Sodium silicate solution: mix 1 minute	22.5‡
50% sodium hydroxide solution: mix 5 minutes	8.0
Powdered hexamethylenetetramine: mix 3 minutes	1.0

*Normally dry-blended for easier handling and dust control.
†0.074 mm (200-mesh) and finer.
‡8.90% Na O, 28.70% SiO , 41° Baumé.

Both glues are ready to use when mixed and have a working life of 6-8 hours at inside temperatures. Several points of difference between these glues should be noted:

The water content of the high blood glue formulation is much larger, which offsets most of the material cost increase. This is possible because the "water requirement" of alkaline-dispersed blood is much higher than that of soybean flour.

The order of addition of alkaline dispersing agents in the high blood mix is partially reversed. This helps impart a more granular consistency to the dispersed blood, which is otherwise very slick and smooth.

The final addition of hexamethylenetetramine illustrates the use of a formaldehyde donor to partially denature the dispersed proteins. This adds some further granular character to the mixed glue and also improves the water and mold resistance of the cured adhesive film.

As with straight soybean glues, the low blood content formulation requires a mold-inhibiting ingredient to meet plywood performance standards while the high blood content glue does not [11].

2. Soybean-Casein Glues

Blends of soybean flour with ground and screened casein also yield a very useful series of protein adhesives; in this case, mostly cured cold [21]. While alkaline-dispersed casein yields strong and water-resistant cold-cured bonds in wood, its sticky dispersed consistency does not permit the rapid water loss needed for quick-clamping procedures. By combining it with soybean flour, the no-clamp process becomes possible. Formulations of this type have proved so successful for bonding plywood faces onto wooden flush door frames in a short cold pressing cycle that the entire industry has employed these protein adhesives from about 1950 to the present [22]. As a special performance property, the bonds of soybean-casein door glues maintain strong adhesion in a fire until the glue lines are literally charred away. Thus, glues of this type are widely used in the assembly of flush design fire doors. They are also excellent adhesives for millwork in general [23]. Some current formulations also contain minor amounts of blood [24].

The formulation of the typical soybean-casein blend glue listed below is quite different from any protein adhesive described thus far in that *all* ingredients, dispersing agents included, are dry-blended into a single packaged composition which requires only the addition of water to prepare. The oiling of all ingredients during the blending step slows down the solution of the alkaline

ingredients long enough for the soybean flour and casein to wet
out under reasonably neutral conditions. Then the alkaline agents
dissolve. Highly alkaline dispersing conditions are provided by
reaction of the sodium salts with lime to yield sodium hydroxide
"in situ" plus insoluble calcium salts [8]. The sodium hydroxide
needed for strongly alkaline dispersion could not be included in
this one-package composition, of course, because of its hygroscopic
behavior. These solution and dispersion reactions require some
time for completion which is the reason for the 15-minute pause
in glue preparation.

Dry Glue Composition

Ingredient	Amount (kg)
"Adhesive-grade" soybean flour	29.0
250-μm lactic acid casein[†]	9.5
Fresh hydrated lime	3.5
74-μm wood flour*	2.5
Granular sodium carbonate	2.5
Granular sodium fluoride	1.0
Granular trisodium phosphate	0.5
Pine oil or diesel oil defoamer	1.5

*0.074 mm (200-mesh) and finer.
[†]0.250 mm (60-mesh).

The dry ingredients are intensively blended in an appropriate
mixer while the defoamer is sprayed in to provide uniform distribution. Mixing directions are as follows.

Ingredient	Amount (kg)
Water at 16-21°C	100
Dry glue: mix until smooth; let stand 15 minutes or until thinning has occurred; mix until smooth	50
Water at 16-21°C: mix 2 minutes or until smooth	25

The second water addition may be increased or decreased to
obtain the desired final viscosity. (A normal range is 4000-8000 cP.)
Working life is 4-6 hours at inside temperatures. Application rates
must be determined by experience but will generally range from
245-345 g of mixed glue per square meter of single glue line.
    While the soybean-casein blend glue can be used according
to the short cycle no-clamp process on dry softwood, it will require
4-6 hours of clamp time to cure to machining strength when used
on dense hardwoods. Water removal from the glue film is simply
too slow on hardwoods to develop adequate early gelation. As a
caution, this formulation represents about the maximum casein
content at which short cycle clamping is possible for flush door
or millwork assembly.

III. BLOOD GLUES

Since soybean-blood blend glues were covered in the preceding
section, this portion of the chapter will deal only with all-blood
adhesive compositions.

A. Raw Material Source and Preparation

Historically, animal blood was mainly used for adhesives in reasonably
fresh liquid form. These glues performed well on wood. However,
the very rapid spoilage rate of liquid blood imposed real limitations
on the general availability and use of this adhesive raw material.
It was not until about 1910 that techniques were developed for
drying whole blood in commercial quantities without denaturing
its protein content, thus maintaining its water solubility [25].
As a result, blood could be collected, processed, and stored indefi-
nitely for later use. The effect of this development was to stimulate
a rapid growth in the technology of blood-based adhesives, especially
for wood.
    Virtually all the proteins in animal blood can be dispersed
into useful adhesive form. These include the serum albumin and
globulin and even the red cell hemoglobin [26]. Only the fibrin
clotting substance must be removed before drying (by agitation
or acidification) because of its instability in solution. Thus, except
for residual moisture content, dried blood is essentially 100% active
adhesive protein.
    The principal North American bloods sold in quantity for adhesive
uses are beef and hog, with lesser amounts from sheep and horses.
Because of its high lysine content, poultry blood is almost exclusively

utilized as a feed additive/binder and is seldom available otherwise. For adhesive purposes, there are significant viscosity differences relating to species among these dried bloods, beef being highest and poultry lowest [27]. Viscosity and water-holding properties are also influenced by animal age, diet, activity, and other factors. As a result, industrial-grade dried soluble blood is generally blended in large quantities to provide average and reproducible properties for adhesive formulation.

The method employed for drying blood is now entirely spray drying. Formerly, a certain amount of vacuum pan-dehydrated blood was also available. Spray-drying conditions relating to temperature, dwell time, and humidity can be adjusted to produce a wide range of blood solubilities [28]. Also, chemical denaturants such as glyoxal can be added to the blood solution prior to drying to further modify its adhesive characteristics. Solubilities from about 20-95% can be prepared with ±5% control. (Dried bloods below 20% solubility are difficult to redissolve without destroying a portion of the adhesive proteins.) This permits the formulation of blood glues with a variety of handling and performance properties.

Generally speaking, the lower the solubility of a dried blood product, the more granular and water-holding is its alkaline-dispersed form [29]. For instance, blood glues of 20-40% solubility make excellent cold press formulations [28]. They also yield the most water- and mold-resistant (near-exterior) glue bonds when cured hot. By comparison, highly soluble bloods in the 85-90% range yield very slick and livery alkaline dispersions of somewhat lower water-holding capacity. Soybean flour is normally blended with these high soluble bloods to produce glues of appropriate granular consistency. If not soybean flour, they must be combined with a particulate filler such as wood flour or nut shell flour to develop an appropriate consistency. Examples of these glue types are listed in the following section.

B. Formulation

As with the soybean glues previously discussed, dried blood adhesives must be initially wetted or redissolved in neutral water and then subjected to one or more alkaline dispersing steps. Unlike vegetable proteins, however, high solubility blood proteins can be adequately dispersed and rendered usefully adhesive by more moderate alkaline agents such as hydrated lime or ammonia [30]. Glues of this type were successfully used during and after World War I to laminate aircraft propellers and to bond various structural components. Especially with a denaturing compound added, these glues represented the most water-resistant adhesives available until

the advent of phenol-formaldehyde resins [2]. An example is as
follows:

Ingredient	Amount (kg)
90% soluble dried animal blood	50.0
Water at 16-21°C: mix 3 minutes or until smooth	40.0
Water at 16-21°C: mix until smooth	30-60
Ammonium hydroxide, sp gr 0.90: mix 3 minutes	3.0
Powdered paraformaldehyde (sift in slowly while mixing)	7.5
Allow mix to stand 30 minutes; mix briefly until glue is fluid and smooth.	

This mix is unique in the quantity of denaturant it employs.
The reaction causes the blood protein to actually gel for a short
period before thinning out again to a working viscosity level.
The useful life is 6-8 hours. This formulation can be cured hot
or cold, but hot pressing yields the most durable bonds.

The next resurgence of blood glue technology came after World
War II. By that time, the highly alkaline multistep dispersing systems
of soybean glues had become well established and were successfully
used with blood glues. Following are two examples utilizing low
solubility blood in typical plywood glue formulations. The second
mix depends on the use of hot water to coagulate the blood and
lower its solubility *during* the mixing procedure [31].

Dry Heat-Treated Blood

Ingredient	Amount (kg)
Water at 16-21°C	150
20% soluble animal blood	37.5*
74-μm wood flour†	10.5*
Pine oil or diesel oil defoamer: mix 3 minutes or until smooth	2.0*
Water at 16-21°C: mix 2 minutes or until smooth	165.0

Ingredient (continued)	Amount (kg)
Fresh hydrated lime (as a slurry in)	5.0
Water at 16–21°C: mix 1 minute	10.0
50% sodium hydroxide solution: mix 10 minutes	8.0
Sodium silicate solution: mix 5 minutes ·	17.5‡

*Normally dry blended for easier handling and dust control.
†0.074 mm (200-mesh) and finer.
‡8.90% Na_2O, 28.70% SiO_2, 41° Baumé.

Hot Water-Coagulated Blood

Ingredient	Amount (kg)
Water at 63°C	100.0
90% soluble dried animal blood	40.0*
74-μm wood flour†	9.0*
Pine oil or diesel oil defoamer: mix 10 minutes	1.0*
Water at 10–16°C	175.0
Pine oil or diesel oil defoamer: mix 2 minutes or until smooth	1.0
Fresh hydrated lime (as a slurry in)	3.5
Water at 10–16°C: mix 2 minutes	7.0
50% sodium hydroxide solution: mix 2 minutes	7.5
Sodium silicate solution: mix 5 minutes	17.5‡

*Normally dry-blended for easier handling and dust control.
†0.074 mm (200-mesh) and finer.
‡8.90% Na_2O, 28.70% SiO_2, 41° Baumé.

Both these glues are excellent adhesives for interior grade plywood when cured either hot or cold. Preservative or denaturant additions are not normally required to meet plywood performance standards. As a point of interest, blood glues are not affected by many of the protein denaturants used with soybean glues, including sulfur compounds and complexing salts [32,33]. However, they *are* very sensitive to aldehyde-acting compounds and these have been employed at levels of 0.1–1.0% to yeild improved consistency and water resistance.

A special class of blood protein denaturants are the alkaline phenol-formaldehyde resins. Low molecular weight, low-alkali PF resins cause granulation of dispersed blood protein without much effect on viscosity, usually a reduction [34]. Highly advanced, high alkali PF resins such as those used as plywood adhesives generally cause rapid thickening and gelation of dispersed blood glues [35]. Resins of intermediate advancement and alkalinity are almost passive to dispersed blood. These interactions have been exploited to formulate blood-resin glues for different applications at almost every level of combination. Two examples at the extremes of the range will suffice.

Low Resin Content Formulation

Ingredient	Amount (kg)
Water at 16-21°C	87.5
20% soluble dried animal blood	25.0*
90% soluble dried animal blood	12.5*
74-μm wood flour†	11.0*
Pine oil or diesel oil defoamer: mix 3 minutes or until smooth	1.5*
Water at 16-21°C: mix 2 minutes or until smooth	200.0
Fresh hydrated lime (as a slurry in)	3.0
Water at 16-21°C: mix 1 minute	6.0
Sodium silicate solution: mix 1 minute	22.5‡
45-50% solids low to intermediate advancement PF resin: mix 3 minutes	13.5§

*Normally dry-blended for easier handling and control.
†0.074 mm (200-mesh) and finer.
‡8.90% Na O, 28.70% SiO , 41° Baume.
§Georgia Pacific 3195, Borden Cascophen WLP-3, Chembond Cerac 118.

The PF resin addition in this case is functioning as a preservative agent and was widely used for the purpose. This formulation is for hot pressing only. The bond durability level approaches "mid-exterior."

High Resin Content Formulation

Ingredient	Amount (kg)
Water at 16–21°C	250.0
74-μm nut shell flour*	75.0
Winter wheat flour	25.0
Dried soluble animal blood	17.5
Defoamer: mix 5 minutes or until smooth	2.5
50% Sodium hydroxide solution: mix 2 minutes	28.0
Granular sodium carbonate: mix 15 minutes	10.5
43% Solids highly advanced PF resin: mix 5 minutes while cooling the glue to 21–27°C	610.0†

*0.074 mm (200-mesh) and finer.
†Georgia-Pacific 5763, Borden Cascophen 335, Chembond Cerac 303.

A partial addition of PF resin can be made just after the initial mix and before the sodium hydroxide addition if more fluidity is needed for propeller-type stirring. This second formulation is more properly termed a blood-fortified exterior PF resin adhesive for general hot press laminating [36]. The effect of the animal blood, even in quantities this small, is to reduce the hot press curing time by 20–30% over phenolic resins used alone. For purposes of adhesive solids calculation, the blood content can be legitimately included with the phenolic resin solids.

A special application for which 80% soluble blood is particularly suited is its use in phenolic resin glues as a foaming agent to produce "air-extended" PF adhesives [37]. These are currently used to manufacture plywood on automated layup lines. For this purpose, the mixed adhesive containing blood is put through a special high speed stirring and air-injection system that lowers the specific gravity of the adhesive from about 1.1 to 0.2 with very fine air bubbles. The low density adhesive foam is then extruded onto passing veneer surfaces, which are assembled and hot pressed to produce exterior grades of plywood. (Recycled glue is defoamed and recirculated.) The primary advantage of this kind of adhesive is lowered cost—for example, savings up to 25% over conventionally applied phenolic adhesives. A typical foamable glue mix is as follows:

Ingredient	Amount (kg)
Water at 16–21°C	170
Industrial wheat flour	50
80% Soluble animal blood: mix 7 minutes	20
PF plywood resin	110*
50% Sodium hydroxide solution: mix 15 minutes	12
PF plywood resin	275*
50% Sodium hydroxide solution: mix 2 minutes	5
Surfactant: mix 2 minutes or until smooth	1†

*43% solids phenolic resin; Borden Cascophen 3136, Chembond Cerac 305, Georgia-Pacific 4922.
†Emersol or equivalent.

C. Mixing, Application, and Pressing

Straight blood and soybean-blood blend glues are mixed in the same sequence and manner shown earlier for soybean glues. Finished glue viscosity ranges are somewhat lower, typically 5000–10,000 cP for hot press formulations and 8000–20,000 cP for the thicker and grainier cold press glues. Glue life at room temperature is 4–8 hours, the cooler the longer.

With respect to application methods, blood glues can be spread on wood surfaces by most conventional means. These include roller, knife, spray, and extrusion but do *not* include curtain coating, for which the glues must be thinned below practical wood-bonding levels.

The major advantage of alkaline-dispersed blood glues over all other wood glues except resorcinol-based synthetic resin adhesives is their sensitivity to heat, resulting in extremely fast hot press curing times [38]. This property is of sufficient importance to warrant reproducing an entire plywood hot pressing schedule for purposes of comparison. The commercial blood glue pressing times shown below are literally half those of current phenolic plywood resin adhesives. They are also significantly faster than those for soybean glues listed earlier.

Blood Glue Hot-Pressing Schedule for Interior-Type Douglas Fir Plywood*

Rough panel thickness (mm)	Number of plies	Panels per press opening	Glue spread per single glue line (g/m²)	Minimum stand time (min)†	Pressing time (min. at full pressure)			
					110°C	116°C	127°C	138°C
6.4	3	2	171	3	3	2-3/4	—	—
7.9	3	1	183	3	1-3/4	1-1/2	1-1/4	1
7.9	3	2	183	3	3-1/2	3	2-3/4	2-1/4
7.9	3	3	195	3	—	7	6	5-1/4
9.5	3	1	195	3	1-3/4	1-3/4	1-1/2	1-1/4
9.5	3	2	195	3	—	4-1/4	3-3/4	3
11.1	3	1	208	3	2	1-3/4	1-3/4	1-1/2
11.1	3	2	208	3	—	4-1/2	4-1/4	4
11.1	5	1	171	3	3	—	—	—
11.1	5	2	171	3	—	6	—	—
12.7	5	1	183	3	2-3/4	2-1/4	2	1-3/4
12.7	5	2	195	3	—	6	5-1/2	5
14.3	5	1	195	3	2-3/4	2-1/2	2-1/4	2
14.3	5	2	208	3	—	7	6	5-1/4
15.9	5	1	195	3	3-1/2	3	2-1/2	2-1/4
15.9	5	2	208	3	—	7-1/2	6-1/2	5-1/2

17.5	5	1	208	3	4	3-1/2	3	2-1/2
17.5	7	1	208	3	4-1/4	4	3-3/4	3-1/4
19.0	5	1	208	3	—	4-1/4	3-3/4	3
19.0	7	1	208	3		5	4-1/4	3-3/4
20.6	5	1	208	3		4-1/2	3-3/4	3-1/4
20.6	7	1	208	3		6	4-1/4	4
22.2	7	1	208	3		—	4-1/4	4
23.8	7 and 9	1	220	4		—	6	5
25.4	7 and 9	1	220	4		—	7	5-1/2
27.0	7 and 9	1	232	4		—	7	6
28.6	7 and 9	1	232	4		—	7-1/2	6-1/4
30.2	7 and 9	1	232	4		—	8	6-1/2
33.4	9	1	244	5		—	9	7
36.5	9	1	244	5		—	10	8
39.7	9 and 11	1	244	5		—	12	9

*Total time assembly limit, 16 minutes. Veneer not to exceed 43°C. Use 12 g extra spread, as minimum, on rough or warm core. Veneer moisture content not to exceed 5%. Not less than 12 kg/cm² hydraulic pressure.

†Time before loading press.

This hot pressing schedule is suitable for all straight blood adhesive formulations and also for soybean-blood blend glues containing at least half blood as the active dry ingredient.

Blood and soybean-blood blend glues of appropriate high viscosity and granular consistency can be pressed cold according to the schedule shown for soybean glues. For this purpose, they must contain a terminal addition of about 5% preservative in order to meet product standards for mold resistance [11].

IV.   CASEIN GLUES

A.   Raw Material Source and Preparation

As with soybean and blood, the adhesive qualities of casein curd from milk were recognized in relatively ancient times. Mixed with a simple alkali such as lime, casein protein became an important adhesive for furniture and paint pigments and the preferred sizing agent for the canvas of Renaissance paintings [39,40].

Medieval furniture assembly was divided between casein glues and animal gelatin adhesives made from boiled extracts of bone and hide. The gelatin glues were applied to joints as a hot solution and gained bond strength almost immediately on cooling. The casein glues required much long clamping times to develop adhesion by water loss and calcium caseinate formation. However, the casein glues had true water resistance while the animal gelatin glues remained forever sensitive to moisture [2]. Thus, casein tended to be used where durability was required. This association with water resistance has remained a favorable performance factor for casein glues down to the present.

Casein protein is recovered from skim milk by acid precipitation to pH 4.5. Mineral acids may be used, or the milk can be cultured with bacteria that convert lactose (milk sugar) to lactic acid, which in turn precipitates the casein. The precipitated protein curd is washed free of acid with hot or cold water and is then dried and ground. The commercial designation for casein often includes its method of acid precipitation (e.g., "lactic acid casein").

Since industrial casein competes directly with the worldwide food uses of milk and its proteins, the price of casein tends to vary widely as the supply/demand economics of milk products rise and fall. In recent years, the cost has remained well over a dollar (U.S.) a pound. Even at this price, however, certain casein blend and specialty glues continue to hold a significant place in current markets.

B. Formulation

For adhesive uses, the particle size of ground casein is normally
controlled within the range of about 250-500 μm [41]. Particles
coarser than 500 μm (30-mesh) may not dissolve and disperse
completely. Those much finer than 250 μm (60-mesh) tend to form
lumps on wetting, even if oiled. For "single-package" casein glues,
preliminary oiling of the dry ingredients is a very important manufacturing step. It helps prevent the pickup of atmospheric moisture
by alkaline salts in the dry composition and then premature attack
on the casein during storage. It slows down the solution of these
salts in water at the time of glue mixing, thus allowing the casein
particles to become wetted and lump-free in reasonably neutral
water.

The lime content of casein glues is similarly important. A high
percentage of lime (above 30% of dry casein weight) ensures maximum
water resistance of the cured glue film but sharply reduces adhesive
working life. A lime content below 10% provides a long working
life and strong dry bonds on wood but significantly reduces moisture
resistance. Most commercial adhesive formulations balance these
properties by utilizing lime additions in the 15-25% range [42].

As with blood and soybean flour, the maximum adhesive capability
of casein is attained only by complete aqueous dispersion of the
folded protein molecules with a strongly alkaline inorganic salt
such as sodium hydroxide [41]. Since sodium hydroxide cannot
be successfully incorporated into a dry adhesive composition, it
is quickly produced on mixing through reaction of calcium hydroxide
with strongly ionized but less alkaline salts such as sodium fluoride,
sodium carbonate, and trisodium phosphate. (The residues from
this reaction are insoluble calcium compounds.)

The viscosity and consistency of casein glues can be substantially
altered by reaction with most of the classic protein denaturants
such as sulfur compounds, formaldehyde donors, and complexing
metal salts. One or more of these are frequently used as manufacturing controls to offset the natural variability of casein and produce
glues of uniform properties. The water resistance of cured casein
glues is also improved by moderate denaturing.

Finally, to provide mold resistance adequate for interior and
covered exterior structural requirements, a fungicide must be added
to casein glues [23]. In this case, there is no excess of sodium
hydroxide in the glue composition to convert the fungicide to its
soluble sodium salt. Therefore, it is added as a prepared soluble
salt in order not to upset the fairly precise alkaline balance in
the dry glue composition needed to fully disperse the casein. Sodium
orthophenylphenate and sodium pentachlorophenate are examples.

The following casein adhesive formulation embodies all the foregoing technology.

Dry Glue Composition

Ingredient	Amount (kg)
500–250 μm lactic acid casein	15.0
500–250 μm sulfuric acid casein	15.0
74 μm wood flour*	5.0
Fresh hydrated lime	6.5
Granular trisodium phosphate	4.0
Granular sodium fluoride	2.0
Powdered dimethylol urea	0.05
Defoamer	1.45
Sodium orthophenylphenate†	1.0

*0.074 mm (200-mesh) or finer.
†Dowicide A, Dow Chemical Co.

The dry ingredients are intensively blended in an appropriate mixer while the defoamer is sprayed in to provide uniform distribution. The dimethylol urea addition, a protein denaturant, is variable for glue viscosity control. The adjustment is made in defoamer. Mixing directions are as follows.

Ingredient	Amount (kg)
Water at 16–21°C	100
Dry glue: mix 2 minutes or until smooth	50
Let stand 15 minutes or until thinning has occurred; mix 2 minutes or until smooth.	

Finished glue viscosity should be in the range of 4000–8000 cP at room temperature, thickening gradually over several hours and attaining a firm gel overnight.

In a totally different area of application, casein adhesives for paper sizing, laminating, and label gluing are more nearly "casein solutions" [43,44]. They are simple dispersions with ammonia or borax at a moderate pH and low viscosity. They are frequently combined with latexes or soluble rosin derivatives for special performance improvements [45].

C. Mixing, Application, and Pressing

Casein glues for wood pass through an early thick-consistency stage that requires fairly strong agitation to reduce them to a uniform and lump-free state. The mixer should be equipped for sidewall scraping to continuously work thickened glue back into the stirred composition. Counterrotating paddle mixers and bread dough mixers have proved ideal for this purpose.

Because of their thick, sticky consistency, casein glues are generally applied to only one of a mating pair of wood surfaces by roller, knife, or extrusion. Adequate adhesive wetting and transfer occur when the wood surfaces are brought together.

The stickiness of alkaline-dispersed casein glue provides one of its best performance attributes; namely, long assembly time tolerance. A film of casein glue on dry lumber, for example, may allow an open/closed assembly time of 1-2 hours before clamping is required. This property is especially useful in the timber laminating industry where it permits many pieces of lumber to be assembled into large and complex laminated beams [46]. This useful assembly tolerance plus the gap-filling capabilities of casein glues made them the outstanding choice for laminated structural wood products from the mid-1930s onward. Today's phenol-resorcinol-formaldehyde laminating adhesives, which ultimately displaced casein glues on the basis of exterior durability, could still use a measure of these performance properties of casein.

While casein glues can be heat cured and were employed in the past to make hot press plywood, most of the high volume bonding applications have involved cold pressing. Casein glue films are adequately cured by water loss and insolubilizing of the proteins through various chemical reactions at room temperature [47]. Heating does not yield significantly improved water resistance. Except for soybean-casein blend glues, which take on the granular consistency of the soybean constituent, the inherent stickiness of straight casein glues dictates a fairly long clamping time to bring about water loss and adhesive hardening. Progressive shear tests have shown these glues develop about half their dry strength in 3 hours and substantially all of it in 6-8 hours at room temperature. However, moisture resistance continues to improve for several days [48].

Another performance attribute of casein glues that recommended their use in structural wood laminates is fire resistance [49]. While all three of the proteins discussed in this chapter burn to a char before losing bond strength, casein adhesives appear particularly durable in this respect. Thus, they remain the adhesives of choice for the economical assembly of wood-based fire doors of flush and panel designs.

V. OTHER PROTEINS

Reference was made to the historical use of collagen glues made from the gelatin extracts of animal bones and hides. This does not properly indicate the true importance to the wood industry of these materials. From ancient times down to the present, animal glues have in fact remained one of the primary assembly adhesives for wooden furniture, cabinets, and musical instruments [50]. Applied as a hot, viscous solution to furniture joints, they rapidly develop gel strength on cooling that permits the prompt removal of clamping pressure. On subsequent drying, these glues cure to resilient, high strength bonds between wood surfaces, especially those involving end-grain. Animal glue bonds are strong and permanent as long as they are kept dry, but are subject to softening and fungicidal attack when moistened. Water resistance can be improved through the incorporation of most of the protein denaturants listed earlier [51,52].

Animal glues are widely used in a variety of ways with paper. For example, they have been the dominant adhesive for rewettable gummed paper tapes, labels, and envelope seals [53]. They are an important coadditive with synthetic wet strength resins and rosin sizes for sheet paper products [54]. They have been a primary binder for the grit that forms sandpaper [55].

In contrast to their widespread use in furniture and paper products, animal glues have not proved useful as structural adhesives for wood. When used alone, their water sensitivity is excessive compared with other available proteins. When combined with soybean, blood, or casein, animal gelatin glues are completely hydrolyzed and destroyed by the strong alkalies used to disperse these proteins. In addition, they soften when severely heated [2].

Although in recent years animal glues have been substantially replaced by the newer synthetic adhesives, particularly the vinyl and acrylic emulsions, large quantities are still sold in dry and stable liquid forms for furniture and paper bonding applications.

Similar comments can be made with respect to fish skin adhesive extracts regarding these and other adhesive applications. Fish skin glues are normally prepared in stable liquid form and are frequently combined with animal glues for improved rewettability and adhesion to glass or metal surfaces [56].

Other vegetable protein sources are occasionally mentioned as substitutes for soybeans. These have included cottonseed meal, peanut flour, Alaska pea, and rapeseed meal, to name just a few. While they do contain 25-35% useful protein substance, they have never made significant inroads on soybean flour for wood gluing applications on the basis of comparable performance. However, they *can* be used and are handled for adhesive purposes in the same manner as soybean flour itself.

REFERENCES

1. Lucas, A. (1926). "Ancient Egyptian Materials." *Analyst*, 51.
2. Truax, T. R. (1929). "The Gluing of Wood." U.S. Department of Agriculture Bulletin No. 1500.
3. Pen Ts'ao Kong Mu (2838 B.C.). (The Records of Chinese Emperor Sheng-Nung).
4. Johnson, O. (1923). U.S. Patent 1,460,757, July 3.
5. Davidson, G. (1929). U.S. Patent 1,724,695, Aug. 13.
6. Lea, F. M., and Nurse, R. W. (1939). "The Specific Surface of Fine Powders." *Trans. J. Soc. Chem. Ind.*, 58.
7. Brother, G. H., Smith, A. K., and Circle, S. J. (1940). "Soybean Protein." U.S. Department of Agriculture, Bureau of Agricultural Chemistry, Washington, DC.
8. Davidson, G., and Laucks, I. F. (1931). U.S. Patent 1,813,387, July 7.
9. Laucks, I. F., and Davidson, G. (1928). U.S. Patents 1,689,732, Oct. 30, and 1,691,661, Nov. 13.
10. Laucks, I. F. (1943). *Chemurgic Digest*, 2.
11. Douglas Fir Plywood Association (1952). "Mold Resistance of Plywood Made with Protein Adhesives." Tacoma, WA, DFPA, Feb. 15.
12. Arneson, G. N. (1946). "Glues and Gluing in Prefabricated House Construction." U.S. Department of Agriculture Forest Products Laboratory Project 575B, Madison, WI.
13. Sheeran, N.J. (1957). U.S. Patent 2,788,305, Apr. 9.
14. French, D., and Edsall, J. T. (1945). *Advances in Protein Chemistry*, Vol. 2. New York, Academic Press.
15. Bjorksten, J. (1951). "Cross Linkages in Protein Chemistry," in *Advances in Protein Chemistry*, Vol. 6. New York, Academic Press.
16. Lambuth, A. L. (1965). U.S. Patent 3,192,171, June 29.

17. Galber, H., and Golick, A. J. (1946). U.S. Patent 2,402,492,
    June 18.
18. Stillinger, J. R., and Williams, W. Jr. (1955). "Production
    of Plyveneer," Timberman, January.
19. Cone, C. N., and Galber, H. (1934). U.S. Patent 1,976,435,
    Oct. 9.
20. U.S. Department of Commerce (1958). U.S. Tariff Commission
    Reports of 1948; Census of Manufactures, 1942 through Prelimi-
    nary 1958. Washington, DC.
21. Bradshaw, L., and Dunham, H. V. (1931). U.S. Patents
    1,829,258-9, Oct. 27.
22. U.S. Department of Agriculture Forest Products Laboratory.
    (1954). "Hollow Core Flush Doors." Report No. 1983, Madison,
    WI, June.
23. Selbo, M. L. (1949). "Durability of Woodworking Glues for
    Dwellings." Forest Products Journal, 3.
24. Dunham, H. V. (1932). U.S. Patent 1,892,486, Dec. 27.
25. Eichholz, W. (1907). German Patent 199,093, Aug. 6.
26. Meyer, K. H. (1950). Natural and Synthetic High Polymers.
    New York, Wiley-Interscience, pp. 572-584.
27. Drugge, C. E., and Hine, J. M. (1960). U.S. Patent 2,963,454,
    Dec. 6.
28. Lambuth, A. L. (1967). U.S. Patent 3,324,103, June 6.
29. Sheeran, N. J. (1959). U.S. Patent 2,870,034, Jan. 20.
30. U.S. Department of Agriculture Forest Products Laboratory
    (1955). "Blood Albumin Glues: Their Manufacture, Preparation
    and Application." Report Nos. 281-282, Madison, WI, revised
    July 1938 and March 1955.
31. U.S. Plywood Corp. Sales Bulletin (1952). "Weldwood LI-R
    Plywood." American Institute of Architects File No. 19-F,
    February.
32. Cohen, A. (1933). U.S. Patent 1,935,434, Nov. 14.
33. Jarvi, R. A. (1955). U.S. Patent 2,705,680.
34. Carmichael, O. C. (1945). U.S. Patent 2,375,195, May 8.
35. Ash, J. R., and Lambuth, A. L. (1957). U.S. Patent 2,817,639,
    Dec. 24.
36. Monsanto Company Product Bulletin (1968). "PF3097 Phenolic
    Resin Adhesive for Exterior Type Softwood Plywood," Schedule
    No. PG1976. Seattle, WA, Dec. 6.
37. Nylund, S. (1987). "Foam Gluing Innovations," Proceedings
    of the Forest Products Research Society Conference on Struc-
    tural Wood Composites, November, Madison, WI.
38. Gossett, J. M., Estep, M. H., Jr., and Perrine, M. J. (1959).
    U.S. Patent 2,874,134, Feb. 17.
39. Spellacy, J. R. (1953). Casein Dried and Condensed Whey.
    San Francisco, Lithotype Process Co.

40. Salzberg, H. K. (1962). "Casein Glues and Adhesives," in *Handbook of Adhesives* (Irving Skeist, ed.), New York, Reinhold.
41. Sutermeister, E., and Brown, F. L. (1939). *Casein and Its Industrial Applications*, 2nd ed. New York, Reinhold.
42. Higgins, H. G., and Plomley, K. F. (1950). *Australian Journal of Applied Science*, 1.
43. Jones, S. (1954). *Adhesion and Adhesives* (S. Clark, J. E. Rutzeler, and R. L. Savage, cds.), New York, Wiley.
44. Lodge, R. J. (1959). "Casein Adhesives." *Mod. Packag.-Encyclopedia Issue.*
45. Mark, J. G. (1942). U.S. Patent 2,279,256, Apr. 7.
46. U.S. General Services Administration (1955). "Adhesives, Casein-Type, Water- and Mold-Resistant." Federal Spec. MMM-A-125, Washington, DC.
47. U.S. Department of Agriculture Forest Products Laboratory (1950). "Casein Glues: Their Manufacture, Preparation and Application." Report No. D280, Madison, WI.
48. U.S. Department of Agriculture Forest Products Laboratory (1950). "Control of Conditions in Gluing with Protein and Starch Glues." Report No. R1340, Madison, WI.
49. U.S. Department of Agriculture Forest Products Laboratory (1960). "Adhesives: Their Use and Performance in Structural Lumber Products." Report No. 2199, Madison, WI.
50. Gill, R. C. (1957). "Animal Glues in the Modern Furniture Plant." *Furniture Manuf.*, 79(2), February.
51. Hubbard, J. R. (1936). U.S. Patent 2,043,324, June 9.
52. Stainsby, G. (1958). In *Recent Advances in Gelatine and Glue Research.* New York, Pergamon Press.
53. Hubbard, J. R. (1962). "Animal Glues," in *Handbook of Adhesives* (Irving Skeist, ed.), New York, Reinhold.
54. TAPPI (Technical Association of the Pulp and Paper Industry (1961). "Protein and Synthetic Adhesives for Paper Coating." Monograph No. 22, Atlanta.
55. Oglesby, N. E. (1943). U.S. Patent 2,322,156, June 15.
56. Walsh, H. C. (1962). "Fish Glue," in *Handbook of Adhesives* (Irving Skeist, ed.), New York, Reinhold.

2

The Chemistry of PVA

KENNETH R. GEDDES* / *Crown Decorative Products Ltd. Darwen, Lancashire, England*

I. INTRODUCTION

PVA adhesives are widely manufactured, water-dilutable products. They consist of a stable suspension of polyvinyl acetate particles that dry to give a continuous film of polymer. The estimated production was 250,000 tons in 1985 in northern Europe [1] and 715,000 tons in the United States in 1986 [2].

The advantages of this type of adhesive over other systems include low cost, low toxicity, low flammability, and simple application. There is no requirement for heat during application or curing, and excellent adhesion is given to cellulosic surfaces such as wood and paper.

The PVA adhesives are most effectively used in simple high volume applications, and they are at a disadvantage only where evaporation of water is difficult, or where faster drying must be achieved.

The preparation and chemistry of these adhesives and associated products is discussed at a mainly qualitative level, but reference to original papers will give scope for much deeper reading if required.

II. PROCESS OUTLINE

The most significant materials present in a PVA adhesive are polyvinyl acetate, stabilizers, plasticizers, and water.

---

*Present affiliation: Crown Berger Europe Limited, Darwen, Lancashire, England

The manufacture of polyvinyl acetate emulsion from vinyl acetate monomer is conducted commercially in both batch and continuous reactors.

The polymerization reaction is initiated by free radicals formed from the breakdown of materials such as hydrogen peroxide, inorganic persulfates, and organic hydroperoxides. The polymerization is purely an addition process, and no elimination of small molecules takes place.

$$\left[\begin{array}{c} CH_2\!=\!CH \\ | \\ R \end{array}\right]_n \rightarrow \begin{array}{c} -CH_2-CH-CH_2-CH-CH_2-CH- \\ \qquad | \qquad\qquad | \qquad\qquad | \\ \qquad R \qquad\qquad R \qquad\qquad R \end{array}$$

If this reaction is conducted in the presence of water and stabilizers, the polyvinyl acetate is formed as microscopic spherical particles.

A very simplified recipe is as follows:

Water	40.7
Polyvinyl alcohol	4.0
Sodium dodecyl sulfate	0.1
Sodium bicarbonate	0.1
Vinyl acetate	50.0
Ammonium persulfate	0.1
Water	5.0

To begin the process, heat the solution of polyvinyl alcohol, sodium dodecyl sulfate, and sodium bicarbonate with agitation to 65°C. Add 2.5 parts of vinyl acetate (5%) together with 2 parts of ammonium persulfate solution (40%). Raise to 80°C and add the balance of the vinyl acetate and ammonium persulfate solution at a steady rate over 4 hours. Hold for a further 30 minutes at 80–85°C before cooling.

Agitation should be just sufficient to keep the reaction mixture homogeneous and to prevent any pooling of monomer on the surface. Excessive shear may cause deformation of the particles, which will cause poor rheology and possibly mechanical breakdown.

Temperature of reaction, which is governed by the need to activate both the vinyl acetate and the free radical initiator, should not exceed about 90°C to avoid boiling of the water, which would disrupt the emulsion. Redox reactions take place at lower temperatures (say 40–65°C) and may pose more problems of cooling as little or no reflux takes place below 66°C.

Total nonvolatiles content can be adjusted readily and is usually in the range of 50–65%, depending on polyvinyl alcohol type and content, particle size, and viscosity requirements or constraints.

III. VINYL ACETATE MONOMER

Vinyl acetate is manufactured from acetylene or ethylene and acetic acid and has the following properties:

Appearance	Clear, colorless, mobile liquid
Molecular weight	86.09
Relative density (20°C)	0.934
Boiling point	72.5°C
Flash point (Tag Open Cup)	-8°C
Explosive limits	2.6-13.4% by volume
Refractive index (20°C)	1.395
Absolute viscosity (20°C)	0.432 cP
Specific heat (25°C)	1.95 kJ/kg °C
Latent heat of evaporation:	379 kJ/kg
Heat of polymerization:	89.3 kJ/mol
Boiling point of azeotrope with water	66°C (7.3% water)
Solubility in water (20°C)	2%
Water solubility in vinyl acetate (20°C)	1%

Vinyl acetate may be stabilized with traces of phenolic compounds. It then has very little tendency to polymerize at ambient temperatures. A common stabilizer level is 14 ppm methyl hydroquinone. It was previously usual to remove the inhibitor by distillation prior to polymerization, but at current levels this is unnecessary.

IV. INTERNAL PLASTICIZATION AND COPOLYMERIZATION

In some adhesives a second monomer is polymerized in conjunction with the vinyl acetate to improve its properties. This is usually to soften the polymer [3-6] or to improve its alkali resistance [7]. Any emulsion polymer in the wet state takes the form of a suspension of discontinuous particles of polymer. The drying process has to remove water and also fuse the particles to create a continuous film, bridging the surfaces to be bonded.

Using a temperature gradient bar [8-9], a temperature can be found above which a continuous film is formed but below which the water evaporates to leave a powdery deposit. This temperature, the minimum filming temperature (MFT), may span a few degrees.

The MFT is strongly influenced by the glass transition temperature (T_g) of the polymer. This temperature, which may also be referred to as "GTT" or the "second-order transition temperature," is the boundary between the glassy and rubbery states of an amorphous material.

Vinyl acetate homopolymer has a T_g of 28°C. Each polymer has its own characteristic T_g, which is influenced by its main chain flexibility and the bulk of substituted groups forming side chains. Thus butyl acrylate has a lower T_g than methyl acrylate because of its larger side chains, while methyl methacrylate has a higher T_g than methyl acrylate because of the inflexibility of its main chain. These structures can be shown:

$$-CH_2-CH-CH_2-CH-CH_2-CH-$$
$$COOCH_3 \quad COOCH_3 \quad COOCH_3$$
Methyl acrylate

$$-CH_2-CH-CH_2-CH-CH_2-CH-$$
$$COOC_4H_9 \quad COOC_4H_9 \quad COOC_4H_9$$
Butyl acrylate

$$CH_3 \qquad CH_3 \qquad CH_3$$
$$-CH_2-C-CH_2-C-CH_2-C-$$
$$COOCH_3 \quad COOCH_3 \quad COOCH_3$$
Methyl methacrylate

Glass transition temperatures of copolymers formed from two or more monomers in combination are roughly in line with the mole fractions and T_g's of the individual materials. More details on the exact calculation of T_g's of copolymers are given in Section XVI.

Not all monomers polymerize well together, and traces of styrene are sufficient to almost totally prevent the polymerization of vinyl acetate. This is because of the electronic behavior of the double bond in a monomer on the approach of a free radical. Styrene adds readily to a polymer chain terminated by a vinyl acetate unit. The resonance of styrene's double bond with the benzene ring to which it is attached results, however, in the dislocation of the free electron, effectively preventing it from activating a vinyl acetate bond, and so no addition of vinyl acetate to the styrene end group occurs. The kinetics of copolymerization are discussed in more detail in Section XV.

Examples of materials that can be used successfully with vinyl acetate are dibutyl maleate [3], butyl and 2-ethylhexyl acrylates [4], and VeoVa 10 (a product of Shell) [5,7]. All these copolymers are softer and have a lower glass transition temperature than vinyl acetate homopolymer. A recent introduction to the range of suitable monomers is the polypropylene oxide adduct of hydroxyethyl methacrylate; one example is Bisomer PPM6E, from BP Chemicals.

Copolymers of vinyl acetate and ethylene manufactured under pressure have become of increasing commercial significance in recent years. A glance at the structure of a vinyl acetate-ethylene copolymer indicates the efficiency of ethylene at improving the flexibility of the main polymer chain:

$$-CH_2-CH-CH_2-CH_2-CH_2-CH-$$
$$\quad\quad | \quad\quad\quad\quad\quad\quad\quad | $$
$$\quad O-C-CH_3 \quad\quad\quad O-C-CH_3$$
$$\quad\quad \| \quad\quad\quad\quad\quad\quad\quad\quad \|$$
$$\quad\quad O \quad\quad\quad\quad\quad\quad\quad\quad O$$

More details on vinyl acetate-ethylene copolymerization are given in Section XX.

Copolymerization may also be used occasionally to increase the glass transition temperature and general hardness of the polymer. Dimethyl maleate, methyl methacrylate, VeoVa 9 (from Shell), and vinyl chloride are used for this purpose. One application of this is to control the heat seal temperature of PVA adhesives and to reduce the blocking tendency of unpigmented coatings.

There are other reasons apart from GTT to use a second monomer. VeoVa 10 gives better protection against hydrolysis. Acrylic and methacrylic acids give better mechanical stability and freeze-thaw resistance. In larger quantities, they can give solubility in aqueous alkalies. Better stability is also conferred by sodium methallyl sulfonate, sodium vinyl sulfonate, and the recently introduced polyethylene oxide adducts of hydroxyethyl acrylate. There is little need for these materials, however, if polyvinyl alcohol is used as protective colloid.

Functional groups for postreaction or to confer wet adhesion [10] can be introduced through the use of acrylamide [11], glycidyl methacrylate, and 2-dimethylamino ethyl methacrylamide [10]. Recent interest has been shown in the multifunctional monomers such as ethylene glycol dimethacrylate and trimethylol propane triacrylate [12]. Health hazards have been reduced by the introduction of short polyethylene oxide chains between the parent glycol and the unsaturated units.

From a practical viewpoint some monomers may pose handling or toxicity problems. Vinyl acetate is highly flammable and can form an explosive mixture with air in some concentrations. Ethylene and vinyl chloride are gases and have to be handled in pressure vessels, and the latter is carcinogenic. On the other hand, N-methylol acrylamide is solid, but is soluble in water and is often supplied as a solution of 50% concentration.

V. WATER

Water is the second largest component in the PVA adhesive and the least complex. Normally water of any reasonable quality free from contamination by organic or inorganic matter can be used, although for fine particle size products, softening or deionization may be necessary. Ultraviolet sterilization should be used if bacterial contamination is suspected.

VI. INITIATORS

To start the reaction, a free radical generator must be used. These can be divided into three classes: oil soluble, water soluble, and redox.

The most commonly used initiators are the water-soluble persulfates (or more correctly, the peroxydisulfates), and ammonium, sodium, and potassium salts are available. These have the structure:

$$
\begin{array}{cc}
O & O \\
\parallel & \parallel \\
O{=}S{-}O{-}O{-}S{=}O \\
| & | \\
O^- & O^-
\end{array}
$$

On heating, the oxygen-oxygen link breaks to give two sulfate radical ions:

$$
\begin{array}{c}
O \\
\parallel \\
O{=}S{-}O\cdot \\
| \\
O^-
\end{array}
$$

This can add to the double bond of vinyl acetate to initiate a chain reaction:

$$
\begin{array}{l}
O \qquad\qquad\qquad\qquad\qquad O \\
\parallel \qquad\qquad\qquad\qquad\qquad \parallel \\
O{=}S{-}O\cdot \;+\; CH_2{=}CH \longrightarrow O{=}S{-}O{-}CH_2{-}CH\cdot \\
\;| \qquad\qquad\qquad | \qquad\qquad\quad | \qquad\qquad\qquad | \\
O^- \qquad\qquad O\cdot CO\cdot CH_3 \quad O^- \qquad\qquad O\cdot CO\cdot CH_3
\end{array}
$$

$$^-SO_4-CH_2-\underset{\underset{O \cdot CO \cdot CH_3}{|}}{CH} \cdot \quad + \quad \underset{\underset{O \cdot CO \cdot CH_3}{|}}{CH_2=CH} \quad \longrightarrow \quad ^-SO_4-CH_2-\underset{\underset{O \cdot CO \cdot CH_3}{|}}{CH}-CH_2-\underset{\underset{O \cdot CO \cdot CH_3}{|}}{CH} \cdot$$

It should be noted that the ionized sulfate end group may remain on the surface of the particle and help to stabilize it.

Hydrogen peroxide and *t*-butyl hydroperoxide are further examples of water-soluble initiators, but the choice is very limited. Peroxydiphosphates have not been commercialized, and the perborates, chlorates, and bromates are rarely used.

Most organic peroxides are oil soluble and inapplicable, although benzoyl peroxide is used in small quantities to give oil-phase-initiated bead polymer of fine particle size, thus widening greatly the particle size distribution.

$$\underset{\text{Benzoyl peroxide}}{\bigcirc\!\!-\!\!\overset{\overset{O}{\|}}{C}\!\!-\!\!O\!\!-\!\!O\!\!-\!\!\underset{\underset{O}{\|}}{C}\!\!-\!\!\bigcirc} \qquad \underset{\text{t-Butyl hydroperoxide}}{\overset{CH_3}{\underset{CH_3}{\diagdown\!\!\!\diagup}}C-O-OH}$$

Benzoyl peroxide   *t*-Butyl hydroperoxide

Azo initiators have been suggested [13-15] from time to time.

Most initiators are stable under recommended storage conditions at ambient temperatures. A few—for example, the peroxydicarbonates—break down rapidly at 40-50°C and must be stored under refrigeration. They are used in the production of PVC by the emulsion route, but rarely in PVAs. Because they are strong oxidizing agents, it is usual to store peroxides away from other combustibles and away from the production area. They break down in solution progressively as the temperature rises, and the "half-life," which is the time for 50% of the material to be destroyed, can be quoted for various temperatures. For benzoyl peroxide [16] the half-life is 10 hours at 72°C, but only 1 minute at 136°C. Persulfates in aqueous solution [17] break down at a significant rate at 70°C, and very rapidly at 95°C. Persulfates must be stored dry, as the presence of water or contamination with metal salts such as iron or cobalt will cause deterioration and a significant loss of activity, even at ambient temperatures. This can be a problem in the tropics, where temperature and humidity are high.

Redox initiators are active at lower temperatures and are favored because they produce higher molecular weights, leading to stronger bond strengths. Low temperatures are difficult to use with highly viscous systems, however, because heat transfer is reduced as viscosity rises, and cooling is proportional to the difference in temperature between the reactor jacket and its contents. As most

redox systems operate at below 70°C, the cooling effect of refluxing
is also eliminated. At higher temperatures the half-life of the initiator
may be so short that it cannot provide the continuous supply of
initiating fragments required. This may stop the reaction, with
potentially serious consequences.

Redox initiators are restricted to water-soluble components
and consist of an oxidizing agent and a reducing agent. Oxidizing
agents are generally the familiar materials used as thermal initiators
(e.g., hydrogen peroxide, the persulfates, and *t*-butyl hydroperoxide). Commonly used reductants are sodium metabisulfite, sodium
formaldehyde sulfoxylate, and ascorbic acid. A transition metal,
and particularly iron, may act as a true catalyst. The simplest form
of the reaction may be illustrated with a hydrogen peroxide/sulfur
dioxide couple:

$$Fe^{2+} + H_2O_2 \rightarrow Fe^{3+} + OH^- + OH\cdot$$

$$2Fe^{3+} + SO_2 + 2H_2O \rightarrow 2Fe^{2+} + SO_4^{2-} + 4H^+$$

The energy of activation of a redox reaction is about 10 kcal/mol
against 35 kcal/mol for a thermal decomposition of the type $S_2O_8^{2-} \rightarrow$
$2SO_4^-$. This leads to a faster evolution of radicals. At 70°C, 0.1M
persulfate solutions yield 3×10^{-4} mol/liter of SO_4^- per minute,
while 0.001M persulfate/0.001M Fe^{2+} solutions, even at 10°C, produce
radicals at six times this rate [18,19].

Molecular oxygen inhibits polymerization reactions but is rapidly
desorbed from agitated solutions at thermal initiation temperatures
above 70°C, especially if there is refluxing vinyl acetate to help
remove dissolved gases. Nitrogen purging must be employed for
lower temperature redox reactions, however.

VII. SURFACE-ACTIVE AGENTS

The materials known variously as surfactants, detergents, soaps,
and wetting agents are used to disperse and stabilize monomer
droplets in water. These surface-active agents also influence the
size and stability of the finished emulsion polymer particles [20-21].

Surfactants are molecules that exhibit the opposite properties
of oil solubility and water solubility in different parts of their
structure. This behavior is described as "amphipathic." Three
classes exist: nonionic, anionic, and cationic. Some surfactants
(e.g., those with acid and amine groups) may change their nature
with pH. These are called amphoterics.

Nonionic surfactants are usually totally organic and consist of
alkyl, aryl or alkylaryl hydrophobic groups reacted with ethylene

oxide to produce a water-soluble chain of polyethylene oxide. A typical reaction is as follows:

$$C_{12}H_{25}OH + 8CH_2-CH_2 \xrightarrow{\text{NaOCH}_3} C_{12}H_{25}O(CH_2CH_2O)_7 \ CH_2CH_2OH$$
$$\diagdown O \diagup$$

The $C_{12}H_{25}$—portion of the molecule is insoluble in water, while the polyethylene oxide chain is soluble. This has led to the hydrophilic-lipophilic balance (HLB) concept, defined [22,23] as the ratio of the molecular weight of the hydrophilic portion divided by the molecular weight of the total surfactant, multiplied by 20.

HLB was used in an attempt to quantify the selection of sur- factants. This idea has had some success with simple emulsification of oils but has been of limited value in the emulsion polymer field.

In water, surfactants exhibit only a small degree of true solu- bility. Additional material concentrates first in the surface, oriented with the water-soluble ends of the molecules in the water. Finally, after the surface becomes saturated, fresh material is forced beneath the surface. Here it groups into clusters known as "micelles," with the oily, nonpolar ends turned inward. Micelles may consist of 50-100 molecules of simple surfactants such as sodium dodecyl sulfate, but comprise of only 8-10 molecules of bulky, complex surfactants such as Dowfax 2A1 (a product of Dow).

As surfactant is added to water, the surface tension is rapidly depressed, but once micelles are formed only a small further surface tension drop is observed; this is due to tighter packing of the surface molecules. The transition point at which micelles are first formed is known as the critical micelle concentration (CMC) and is a characteristic function of each individual surfactant. Some materials do not form micelles despite being of the correct general formulation, although they may show some surface activity. For example, in the straight-chain aliphatic soaps of the general formula $C_nH_{2n+1} COO^- Na^+$, 1-9 carbon units are too soluble to form micelles, while above C_{22} the oily part of the molecule is so predominant over other properties that it similarly prevents micelle formation. Intermediate chain lengths form micelles; strongly so in the $C_{12}-C_{18}$ range. Water-soluble colloids such as polyvinyl alcohol do not form micelles.

Lists of critical micelle concentrations have been prepared [24].

Introduction of nonpolar liquids such as solvents or monomers will result in a migration into the nonpolar environment of the centers of micelles. As the proportion of these materials rises, swollen micelles may give way to monomer or solvent droplets stabi- lized with a surface layer of surface-active agent whose polar groups are directed outward into the water. Conversion of monomer swollen micelles to polymer retains the surface layer of surfactant, which confers stability. A consideration of the shape of molecules provides the explanation for the observation that mixed surfactant systems give more stability than a single material. With a selection of molecular shapes and sizes, greater surface packing densities are possible.

Properties of nonionic surfactants vary with temperature because the solubility of ethylene oxide chains depends on hydrogen bonding. At elevated temperatures bonding may be disrupted, in which case solubility and surface activity are lost at a temperature known as the "cloud point."

Only nonionic and anionic surfactants are used to any extent in emulsion polymerization. The simple carboxylic acid soaps are hardly used now in the PVA field; rather sulfates, sulfonates, sulfosuccinates, and phosphates are favored.

The simplest of this class are sodium dodecyl sulfate and sodium dodecyl benzene sulfonate, while the sulfated nonionics combine the good properties of both types of surfactant and give good stability. These are manufactured by sulfating nonyl phenol polyethylene oxides and dodecyl polyethylene oxide with chlorosulfonic acid or sulfur trioxide. They are available in a number of ethylene oxide chain lengths and as sodium and ammonium salts. Products of nominally the same composition from different manufacturers can be greatly different in properties due to variation in the point of attachment of the sulfate, and so may not be interchangeable.

Another class of surfactants with a wide range of variations consists of the sulfosuccinates and sulfosuccinimates. They are now available from a number of manufacturers but were originally patented and championed by Cyanamid. A product of special interest is the disodium salt of the half-ester of a C_{10}-C_{12} straight-chain alcohol, sold by Cyanamid as Aerosol A-102.

Polymerizable surfactants or stabilizers also exist. 2-Sulfoethyl methacrylate [25], sodium vinyl sulfonate [26], sodium methallyl sulfonate [27], and the new "associative monomers" such as the nonyl phenol polyethylene oxide methacrylates [28,29] are all used in current commercial polymers. This last type of material is particularly interesting as a component of alkali-soluble thickeners because the polyethylene oxide units will be directed out from the main chain. The terminal nonyl phenol groups are oily (lipophilic) and form micelles. This association very strongly enhances low shear viscosity, while allowing shear thinning to make application easy. This property could be very useful to increase wet film weight for some applications.

VIII. COLLOIDS: POLYVINYL ALCOHOL

Polyvinyl alcohol is the major protective colloid used in PVA adhesives. The properties of the adhesive are strongly influenced by the properties and concentration of the polyvinyl alcohol used in the preparation, or added during compounding [30-32].

Polyvinyl alcohol can be manufactured by acid or alkaline hydrolysis or methanolysis of polyvinyl acetate. Vinyl alcohol monomer itself does not exist except as a transitory material, rearranging immediately to acetaldehyde.

Polyvinyl alcohol is marketed according to its degree of hydrolysis from the parent polyvinyl acetate, and also its solution viscosity. This is usually quoted as the viscosity of a 4% solution in water at 20°C. Solution viscosity has a strong relation to molecular weight, but it will vary with degree of hydrolysis and the randomness or otherwise of the residual acetate groups.

Most grades of polyvinyl alcohol used as polymerization stabilizers contain some acetate groups and are more properly considered as copolymers. The most popular level of hydrolysis for this purpose is 88%, and both random and nonrandom grades are marketed with this description. At the 23 cP viscosity level, Kuraray Poval PVA217, PVA217E, and PVA217EE designations indicate an increasing grouping of the residual acetates units, known as "blocks." The greater the proportion of blocks, the more surface activity is claimed, and higher surface activity is said to lead to finer particle sizes and improved stability [33-35].

From various manufacturers, at least 150 grades of polyvinyl alcohol are on the world market, ranging in hydrolysis from 75 to 100%, and from 1 cP to 66 cP in solution viscosity.

Dissolving is fairly easy with the lower viscosity, partially hydrolyzed grades, although it is advised to use brisk agitation and cold water in tanks with flush bottom valves. Higher viscosity grades require heating to complete the solution in a reasonable time. Fully hydrolyzed grades (say ≥98.5%) do not dissolve in water below 75°C, although solutions do not precipitate on cooling. It is prudent to heat all grades to 80-95°C to ensure complete solution before use. Solutions may increase in viscosity on prolonged storage, especially with greater degrees of hydrolysis, but viscosity may be reduced by reheating.

The insolubility of films of polyvinyl alcohols with high degrees of hydrolysis can be usefully employed to improve the water resistance of PVA adhesives, and often a combination of grades is chosen for use as protective colloids. Conversely, the presence of a sufficient concentration of a polyvinyl alcohol of a lower degree of

hydrolysis may result in dried films of the adhesive that are remoistenable.

Copolymerization of vinyl acetate with other monomers prior
to hydrolysis is possible, and various systems have been reported
[36,37]. Only one such internally plasticized polyvinyl alcohol-Mowiol
04/M1 (a product of Hoechst) is marketed as such, but the comonomer
is undisclosed. Gohsefimer Z, from Nippon Gohsei, is stated to have
functional groups and may be a product of this type. It is claimed
that Gohsefimer Z gives better water resistance than PVA emulsions
stabilized with conventional polyvinyl alcohols.

Polyvinyl alcohol solutions thicken or undergo reaction with
a number of inorganic and organic materials. Viscosity may be increased by the addition of phosphates, sulfates, and boric acid.
Borax must be avoided as a component of the emulsion polymerization
recipe because it precipitates polyvinyl alcohol solutions. This is
used as a qualitative test for the presence of polyvinyl alcohol.
An alternative test is the formation of a deep red color with iodine
[38].

IX. OTHER COLLOIDS

Probably the most used alternatives to polyvinyl alcohol in the PVA
adhesive field is the group of colloids that allow blending [39] with
dextrin for remoistenable envelope flap adhesives. Higher proportions
of polyvinyl alcohol confer the property of remoistenability, but
in common with hydroxyethyl cellulose stabilized emulsions, such
formulations are destabilized by dextrin.

Low molecular weight carboxymethyl celluloses (e.g., Blanose
9M3IF, from Hercules) have been recommended for this purpose,
and they appear to give good quality, stable blends with yellow
dextrin, free from grit, and resistant to syneresis on storage.
Blanose 7L, also from Hercules, has also been claimed to be a useful
alternative to polyvinyl alcohol at much reduced cost, but it suffers
from poor film clarity and poor water resistance. Advantages are
said to be improved mechanical and freeze-thaw stability, no foaming,
and superior heat stability of the films.

Hydroxyethyl cellulose (HEC) is used extensively in the emulsion
polymer industry as a protective colloid for paint emulsions. It has
not found favor in the adhesive field, except perhaps in tile adhesive formulations, because of its tendency to give emulsions with
finer and narrower particle sizes than polyvinyl alcohol. This leads
to pseudoplasticity in the wet film, which reduces wet tack. Hydroxyethyl cellulose is also more expensive than polyvinyl alcohol. Recent
attempts have been made [40] to modify HEC to give rheology close

to that of polyvinyl alcohol preparations and also to change its
properties by the introduction of hydrophobic groups, leading to
associations similar to that found in associative thickeners [41].

Other colloids that have been used in small quantities are starch
ethers [42], gum arabic, polyvinyl pyrrolidone, styrene-maleic
anhydride copolymers, alkali-soluble acrylics, and ethylene-maleic
anhydride copolymers. However, polyvinyl alcohol in various forms
dominates the market for the preparation of wood adhesives.

X. BUFFERS AND pH ADJUSTERS

Vinyl acetate dissolves in water to about 2% and is fairly stable
against hydrolysis in the pH range of 4.5-6.5. Sodium bicarbonate,
sodium acetate, borax, and diammonium hydrogen phosphate are
used to stabilize the pH in this range, and to prevent the self-
catalyzed acid hydrolysis of vinyl acetate to acetic acid and acetaldehyde.

Excess of sodium bicarbonate will cause alkaline hydrolysis
of some of the vinyl acetate to sodium acetate and acetaldehyde,
and so in some recipes sodium acetate-acetic acid buffers are used.
Alternatively, the minimum amount of alkali is added initially and
the balance is added as the reaction proceeds. Salts of fairly short-
chain carboxylic acids such as sodium octanoate may combine mild
alkalinity with a low level of surface activity.

Borax precipitates polyvinyl alcohol and should not be used
with this type of colloid, although sodium borate solution as a post-
addition may produce no more than a useful thickening. Phosphates
may reduce the efficiency of titanium complexes used as gelling
agents.

Presence of salts increases the particle size of non-colloid-
stabilized emulsions, and di- and trivalent cations drastically reduce
the efficiency of many surfactants.

These drawbacks explain the apparently unimaginative popularity
of sodium bicarbonate and sodium acetate as buffers in PVA preparations.

XI. EXTERNAL PLASTICIZERS, COALESCENTS,
TACKIFIERS, AND CROSSLINKING AGENTS

An alternative to using a second monomer is to add an external
plasticizer. This is best added to the hot emulsion at the end
of the reaction, for it is then readily absorbed into the polymer
particles. The choice of plasticizer is rather limited due to compati-

bility. Dibutyl phthalate is most popular, combining efficiency and low cost. Weight-for-weight, dibutyl phthalate is about twice as efficient as butyl acrylate at softening polyvinyl acetate [4], but it merely acts as a compatible extender, adding little to the properties of the polymer.

Dibutyl phthalate is also fairly permanent, but dibutyl glycol phthalate [4] and Texanol isobutyrate (a product of Eastman) have been claimed to have still slower evaporation rates. Triisobutyl phosphate is also sometimes used, giving reduced flame spread, low smell, and some defoaming action.

Often a coalescing agent is added to aid film formation. These are similar to external plasticizers, but having a lower molecular weight are expected to be lost from the film. Many proprietary esters are marketed: butyl carbitol acetate [4] (Union Carbide), Texanol (Eastman), Dowpad A [4] (Dow), and Lusolvan (BASF) being prominent examples.

Tackifiers, based on tall oil rosin or its esters with glycerol or pentaerythritol, can be obtained as dispersions in water and may be readily blended with PVA adhesives.

Where high water resistance is required, the polyvinyl alcohol can be crosslinked by means of glyoxal or by resins such as urea-formaldehyde or glyoxal-formaldehyde.

XII. FILLERS AND PIGMENTS

Adhesives are often compounded with inorganic fillers to increase viscosity, control rheology and reduce raw material cost.

Various china clays, including calcined grades, may be used (see Fig. 2.2, p. 58), as well as whiting (calcium carbonate) [43]. The latter also stabilizes the pH. Potassium sodium feldspar, gypsum, talc, mica, and barytes [44] are other possibilities.

There is rarely any requirement to include the more expensive pigments such as titanium dioxide, even if the glue line is visible. Colored pigments can be included for products such as combined tile adhesives and grouts.

XIII. FUNGICIDES AND PRESERVATIVES

Effective preservatives must be used in adhesives—especially where drying may be delayed by nonporous substrates. Mold growth is unsightly and attacks the wet adhesive, making it unfit for use. Enzyme attack may be more subtle. Celluloses in solution are destroyed by cellulase without any obvious outward signs except for severe thinning.

t-Butyltin oxide and formalin (40% formaldehyde solution) were once common but are now less in favor because of their toxicity. Carcinogenicity of formalin now appears less likely, but the material is still a severe irritant, and environmental pressures have opened the way for the more expensive but safer ranges of preservatives.

XIV. POLYMERIZATION REACTIONS AND KINETICS

The process of polymerization of vinyl acetate to give polyvinyl acetate (PVA) was described earlier as a free radical mechanism:
Initiation

$$R\cdot + CH_2{=}CH \longrightarrow R{-}CH_2{-}CH\cdot$$
$$\quad\quad\quad\quad\; | \quad\quad\quad\quad\quad\quad\quad\quad |$$
$$\quad\quad\quad O\cdot CO\cdot CH_3 \quad\quad\quad\quad O\cdot CO\cdot CH_3$$

Growth and termination

$$R{-}CH_2{-}CH\cdot + n(CH_2{=}CH) \longrightarrow$$
$$\quad\quad\quad | \quad\quad\quad\quad\quad\quad |$$
$$\quad\quad O\cdot CO\cdot CH_3 \quad\quad O\cdot CO\cdot CH_3$$

$$R{-}CH_2\; CH{-}CH_2{-\!-}CH{-}\; CH_2{-\!-}CH{-}R'$$
$$\quad\quad\quad | \quad\quad\quad\quad | \quad\quad\quad\quad |$$
$$\quad\quad O\cdot CO\cdot CH_3 \;\; O\cdot CO\cdot CH_3 \;\; O\cdot CO\cdot CH_3$$

The R' group, which terminates the growing chain and limits the molecular weight, can arise in several ways:

1. Radical termination. A radical can combine with the growing chain, eliminating the activity of both.
2. Combination. Two growing chains can add together, combining their molecular weight but preventing any further growth.
3. Chain transfer [45-47]. Reaction can take place with an easily extracted hydrogen such as the one attached to sulfur in thiols:

$$RCH_2{-}CH_2\cdot + R'{-}SH \rightarrow RCH_2{-}CH_3 + R'S\cdot$$

This reaction terminates the first chain but may initiate a new one through the R'S· radical. This reduces molecular weight but does not inhibit the reaction. Very little reaction between initiator fragments and thiols takes place in emulsion polymerization, as the generation of radicals is in the water phase where the chain transfer agents are barely soluble,

such that initiator efficiency is not greatly reduced. Some
additional chain transfer may take place to surfactants,
colloids, and even monomers [48].

4. Disproportionation.

$$R-CH_2-CH_2\cdot + R'-CH_2-CH_2\cdot \rightarrow$$

$$R-CH_2-CH_3 + R'-CH=CH_2$$

The residual double bond left after disproportionation may
take place in further polymerization, leading to branching
[49-51].

The kinetics of polymerization depend on three stages: (1)
initiation, (2) propagation, and (3) termination.

1a. The rate constant for initiator decomposition is k_d, that is:

$$I \xrightarrow{k_d} 2R\cdot$$

1b. The rate constant for radical addition to monomer is k_a,
that is:

$$R\cdot + M \xrightarrow{k_a} RM\cdot$$

2. The rate constant for addition of monomer to an existing
chain (i.e., rate of propagation) is k_p, that is:

$$RM\cdot + M \xrightarrow{k_p} RMM\cdot$$

3a. The rate constant for termination by combination is k_{tc},
that is:

$$RM_xM\cdot + R'M_yM\cdot \xrightarrow{k_{tc}} RM_{x+y+2}R'$$

3b. The rate constant for termination by disproportionation
is k_{td}, that is:

$$RM_x\cdot + RM_y\cdot \xrightarrow{k_{td}} RM_x + RM_y$$

Overall rate constant for termination is k_t.

If it is assumed that no chain transfer takes place and that the rate of propagation is independent of the chain length (including the addition of the first unit of monomer to the radical), the rate of initiator breakdown k_d becomes the sole rate-determining step.

Each initiator molecule breaks up into two initiator fragments.

If the fraction of radicals formed that successfully initiate a polymer chain is f, then we have two equations:

Rate of initiation

$$R_i = k_d f \, [I]$$

Rate of termination

$$R_t = k_t \, [R \cdot]^2$$

Under steady-state conditions, the generation of free radicals will be constant and the rate of termination will equal the rate of initiation, that is:

$$R_i = R_t - k_t \, [R \cdot]^2$$

therefore

$$[R \cdot] = \frac{R_i^{1/2}}{k_t^{1/2}}$$

The rate of propagation (R_p) is given by:

$$R_p = k_p [M] \, [R \cdot] = \frac{k_p}{k_t^{1/2}} \, R_i^{1/2} \, [M]$$

or

$$R_p = \frac{k_p k_d^{1/2}}{k_t^{1/2}} \, f^{1/2} \, [I]^{1/2} \, [M]$$

The overall rate of polymerization is controlled by the rate of propagation and is therefore proportional to the square roots of the initiator concentration $[I]^{1/2}$ and efficiency $f^{1/2}$. It is also proportional to the monomer concentration $[M]$ and the ratio of rate constants, that is, $k_p \cdot k_d^{1/2}/k_t^{1/2}$.

The factor f (the efficiency of the initiator as measured by the fraction of radicals that start chains) is dependent on the availability of monomer. It is also dependent on the rapid separation

of the two radical fragments to prevent mutual elimination (i.e.,
the efficiency of the stirring).

Addition polymerization is exothermic, and heats and entropies
of polymerization have been published [52]. In general terms, heats
of polymerization are lower as molecular weights of monomer rise,
that is, as the contribution of the double bond to the total molecular
weight of the monomer molecule falls. Steric and electronic factors
also play their part. Vinyl acetate has one of the lowest molecular
weights of the common commercial monomers at 86 and one of the
greatest heats of polymerization at 21 kcal/mol.

Molecular weights of polymers are dependent on the concentration
and efficiency of the initiator and on the rate constants of the
termination steps, as well as the extent of chain transfer. In comparison with solution polymers, the emulsion route leads to a high
molecular weight, unless chain transfer is deliberately introduced
by the addition of halogenated hydrocarbons or mercaptans [53].
Alternatively, chain branching agents containing more than one
double bond can be used to increase the molecular weight.

Molecular weights can be determined by a number of methods
including U-tube viscometers and laser light spectrometers, and
by gel permeation chromatography [54-56]. Most methods depend
on making a solution in a suitable solvent, and there is often a
proportion of insoluble gel that cannot be measured. A low molecular
weight "tail" is also frequently found, and a theory for the molecular
weight distribution consistent with the kinetic scheme has been
developed [57]. Properties in solution are strongly influenced by
the configuration of the chains, and many methods give molecular
size rather than weight.

Molecular weight may be defined in a number of ways including
number average (M_n) and weight average (M_w).

Number average is defined as:

$$M_n = \frac{\Sigma N_n M}{N_t}$$

and weight average as:

$$M_w = \frac{\Sigma W_n M}{W_t}$$

where N_n and W_n are the number of molecules and weight of molecules respectively, having n monomer units, and N_t and W_t are
the total number and total weight, respectively, of all sizes of polymer molecules; M is the molecular weight of a polymer molecule having
n monomer units.

XV. COPOLYMERIZATION KINETICS

Simple PVA adhesives are homopolymers of vinyl acetate, often with an external plasticizer.

It was noted in Section IV that an alternative and permanent method of softening vinyl acetate is to copolymerize it with a second monomer or "internal plasticizer" such as ethylene, 2-ethylhexyl acrylate or Veova 10. Other reasons mentioned included increasing molecular weight, improving stability, and providing sites for subsequent curing.

When two or more monomers polymerize together, they add to the growing chain at different rates. Ease of addition depends on whether the last monomer unit added to the existing growing chain was of the same kind. This is because of electronic and steric effects.

The different propagation rate constants are defined as follows:

$$-M_1 \cdot + M_1 \xrightarrow{K_{11}} -M_1 M_1 \cdot \qquad \text{Homopolymerization}$$

$$-M_1 \cdot + M_2 \xrightarrow{K_{12}} - M_1 M_2 \cdot \qquad \text{Copolymerization}$$

$$-M_2 \cdot + M_2 \xrightarrow{K_{22}} - M_2 M_2 \cdot \qquad \text{Homopolymerization}$$

$$-M_2 \cdot + M_1 \xrightarrow{K_{21}} - M_2 M_1 \qquad \text{Copolymerization}$$

Reactivity ratios r_1 and r_2 are defined as the ratios of the rates of homopolymerization to copolymerization, that is,

$$r_1 = \frac{K_{11}}{K_{12}} \quad \text{and} \quad r_2 = \frac{K_{22}}{K_{21}}$$

Penultimate and more distant groups will have an effect, but will average out in a practical determination of r_1 and r_2. Should one of the monomers be unable to polymerize with itself, $K_{22} = 0$ and $r_2 = 0$.

The relation between the quantity of monomer units going into the polymer at any instant, dm_1/dm_2, is given by the reactivity ratios and the molar composition of the monomer mixture, first published in 1944 by Mayo:

$$\frac{dm_1}{dm_2} = \frac{[M_1] \ (r_1 \ [M_1] + [M_2])}{[M_2] \ (r_2 \ [M_2] + [M_1])}$$

From the published values [58-61] of r_1 and r_2, the polymer composition for any monomer mixture can be calculated. If there is almost instantaneous reaction and the concentration of monomer within the reactor approaches zero, the copolymer composition is forced equal to the ratio of $[M_1]$ to $[M_2]$ independent of r_1 and r_2. These conditions are usually undesirable because initiator efficiency is low, as was discussed in Section XIV.

Starved conditions will not occur when the addition of monomer to the reactor is at a rate equal or greater than the reaction rate. The composition of the polymer can then be influenced by using a monomer mixture calculated from the r_1/r_2 ratios which is different from the ratio of units in the desired polymer. The composition of the polymer may be critical for the production of closely defined physical properties, as in the case of synthetic rubbers. In this case the reaction is "short-stopped" by addition of inhibitor while the monomers are still at their required ratio. Excess monomers are then stripped from the product by steam, vacuum, or inert gas.

In an interesting special case, the polymer composition is the same as the monomer mixture from which it is formed. This is known as "azeotropic copolymerization" [62,63].

Reactivity ratios are affected by physical factors such as temperature and pressure and by whether the values were determined in an emulsion process or in solvent. Thus the accurate determination of these ratios is not easy, and a statistical method has been proposed [64,65]. Mayo's equation assumes the equal availability of the monomer for polymerization. In emulsion polymerization very soluble and polar monomers such as acrylic acid are strongly partitioned in favor of the water phase, and allowance has to be made for this [66,67].

Should r_1/r_2 ratios not be available, they may be estimated from data called Q-e values [58,68-70], which refer to the electronic and steric properties of the monomers. The Q-e values are related to r_1/r_2 ratios by the following equations:

$$e_2 = e_1 \pm (- \ln r_1 r_2)^{1/2}$$

$$Q_2 = \frac{Q_1}{r_1} \exp\left[- e_1 \left(e_1 - e_2\right)\right]$$

The Q-e values are relative and are calculated from assumed values of styrene of $e = -0.80$ and $Q = 1.00$. This theoretical approach should be avoided if possible because accuracy is often suspect, although recent work [70] has improved reliability.

XVI. GLASS TRANSITION TEMPERATURE OF COPOLYMERS

Polymers formed by emulsion polymerization are usually amorphous and do not melt sharply. They have however a characteristic temperature at which they pass from a rubber to a glass, known as the glass transition temperature (T_g). This transition may be detected by differential scanning calorimetry, microindentometry, refractive index, specific volume, and several other techniques.

The T_g of a copolymer can be related, to about 4°C accuracy, to the proportions and T_g's of the homopolymers of the constituent monomers according to the formula:

$$\frac{W_1}{T_1} + \frac{W_2}{T_2} + \cdots + \frac{W_n}{T_n} = \frac{1}{T}$$

where T_1, T_2, and T_n are the T_g's of the first, second, $\cdots$, nth monomers of corresponding weight fractions W_1, W_2, $\cdots$, W_n, and T is the T_g of the final polymer.

Inaccuracies will arise if crystallinity is possible (e.g., with polyvinylidene chloride [71]), if the copolymer is not random, or if the number of flexible bonds (α) per monomer unit is not similar for all the monomers.

The last condition is almost always impossible to observe—for example, vinyl acetate has $\alpha = 4$, vinyl chloride has $\alpha = 2$, and butyl acrylate has $\alpha = 7$.

For a terpolymer of these monomers, a more complex formula [72] will give greater accuracy, that is,

$$\frac{W_1 \, \alpha_1 \, (T - T_1)}{M_1} + \frac{W_2 \, \alpha_2 \, (T - T_2)}{M_2} + \cdots + \frac{W_n \, \alpha_n \, (T - T_n)}{M_n} = 0$$

where W_1, $\cdots$, W_n are the weight fractions of monomers of molecular weights M_1, $\cdots$, M_n, each having number of flexible bonds α_1, $\cdots$, α_n.

The final T_g of the copolymer is T, while T_1, $\cdots$, T_n are the T_g's of homopolymers of the monomers 1, $\cdots$, n.

Chain branching and crosslinking make the chains more rigid, and crosslinking increases T_g by 4-5°C for each mole percent crosslink [73].

XVII. EMULSION POLYMERIZATION BY THE BATCH PROCESS IN THE PRESENCE OF SURFACTANT

Most of the production of PVA adhesives is by the batch process [74] (see Fig. 2.1). In all forms of the process, a stirred tank reactor be-

Fig. 2.1 Reactor for the production of emulsion polymers by the batch process.

comes filled with reactants, and when the process is complete the tank
is emptied. In some processes the raw materials are loaded before
the start of the reaction, in others the majority of material is added
in stages or continuously during reaction. While dynamic equilibria
may be found, overall there is a continuous shift in properties
as the reaction progresses and the nonvolatiles rise.

The kinetics of batch polymerization have been studied extensively. Yurzhenko [75] was one of the first to publish in the field
in 1946, but Harkins's paper [76] on the emulsion polymerization
of vinyl acetate in 1947 was of particular relevance.

The year 1948 saw the publication of Smith and Ewart's cele-
brated paper [77], which has provided both a foundation stone
and a touchstone for all subsequent theories. Three cases were
defined:

1. The number of growing chains per polymer particle is less
   than 1.
2. The number of growing chains per particle approximates
   to half.
3. The average number of growing chains per particle is large.

Case 2 is of special interest to explain the early stages of the
emulsion polymerization of a monomer such as styrene, where the
solubility both of the monomer and of low molecular weight oligomers
is very low. In this case initiation is considered to be in micelles
swollen with monomer, although this concept has been the subject
of much study [78] and debate.

At an early stage of a reaction it can be assumed that particles
are small and the mobility within them great. Under these conditions
any free radical entering a particle containing a growing chain
will rapidly collide with it and terminate the polymerization by combination. It was concluded that the instantaneous rate of polymerization (R_p) was proportional to the number of particles N, which
in turn was proportional to the 0.6 power of the concentration of
emulsifier [S], and to the 0.4 power of the rate of initiation R_i,
that is,

$$R_p \propto N \propto [S]^{0.6} [R_i]^{0.4}$$

Later ideas and developments have been proposed by Vanderhoff
[79,80], Patsiga and coworkers [81,82], Gardon [83,84], and Fitch
[85,86]. The important concepts of diffusion-controlled termination
and radical desorption from the particles were introduced. Gardon's
equations have been reworked by Dunn and Al-Shahib [87,88] and
an alternative to Smith and Ewart was proposed by Medvedev [89]

and developed further by Brodnyan [90]. Ugelstad and his coworkers have studied [91] the particle nucleation stage in detail
and have published a series of papers. Pemboss [92] has recently
examined free radical entry into particles and correlates theoretical
and experimental entry coefficients. Many theories deal with the
early stages of the reaction, but Nomura [93] has made calculations
at higher conversions.

Most work has used styrene because of its convenient low solubility of 3.7×10^{-3}, but there have been many papers published
on vinyl acetate [94-104] and on methyl acrylate [105-106] both
of which are more soluble. In such systems the concept of homogeneous nucleation—that is, the growth of polymer chains in solution
for as long as they remain soluble—is particularly likely. This rele-
gates the role of the micelles to recipients of the precipitating polymer
chains or even to the mundane function of surfactant reservoirs.
There have also been studies of specialized aspects of emulsion
polymerization such as the role of the surfactant [107-110] and
of polyvinyl alcohol [111,112]. Mathematical models of the process,
many using computer techniques, have now been developed [113-115].
These show considerable promise.

XVIII. EMULSION POLYMERIZATION OF VINYL ACETATE STABILIZED BY POLYVINYL ALCOHOL ALONE

Many PVA adhesives are polymerized using polyvinyl alcohol as
the sole stabilizer. This produces a comparatively large particle
size, which may be increased further by the introduction into the
recipe of an oil-soluble initiator such as benzoyl peroxide. This
technique is referred to as steric stabilization, as contrasted with
electrostatic stabilization, when surface-active agents are used.

The kinetics of the process have been examined empirically
by Moritz [116], who concluded that the maximum reaction rate
r_{max} and the particle number N_p were related to the concentration
[I] of the initiator 2,2'-azo-bis(2-amidinopropane) (AIBA), the
volume fraction of vinyl acetate [M], the concentration of the polyvinyl alcohol [C], and stirrer speed n by the following relationships:

$$r_{max} \sim \exp \frac{-90}{R_T} \; [I]^{0.6} \; [M]^0 \; [C]^{-0.1} \; n^{0.2}$$

$$N_p \sim \exp \frac{-30}{R_T} \; [I]^{0.2} \; [M]^0 \; [C]^{0.4} \; n^{0.1}$$

This is a most interesting result and again confirms the positive influence on rate of polymerization and particle number by increased stirrer speed. This may be related to the expected (and found) increase in both parameters of increasing initiator concentration. The negative influence on rate of polymerization of polyvinyl alcohol concentration was ascribed to either allowing a faster rate of radical escape by the stabilization of more small particles or to the hindrance of radical absorption caused by highly viscous surface layers of polyvinyl alcohol.

Using a true batch technique (i.e., all components in at the initial stage of the reaction) Smith-Ewart case I kinetics were applied, while at high conversions case III kinetics were used to model the reaction rate. By use of a process where vinyl acetate and half of the initiator and polyvinyl alcohol solutions were added over a period of 3-1/2 hours (i.e., a "semibatch" process), larger particle sizes were obtained and Smith-Ewart case III kinetics applied throughout. Rate of conversion was almost one order of magnitude smaller than in the true batch process at low conversions, and although an autoacceleration was observed, it never reached the true batch process rate. As expected from the lower average contact time, grafting was observed to be less with the semibatch process.

XIX. CONTINUOUS EMULSION POLYMERIZATION

Continuous processes with mixing or recirculation, apart from startup transients, are always in equilibrium. Raw materials are added continuously to balance an outflow of product. Conversion is stable, resulting in a constant nonvolatiles content.

Continuous processes are attractive because raw materials are added steadily, and product drawn off soon after polymerization and for as long as needed to fulfill a production requirement. In practice, however, reactor fouling, which results in loss of heat transfer, may limit the length of run.

Higher productivity seems a logical consequence of this arrangement, and World War II offered considerable impetus to the development of large-scale continuous production of styrene-butadiene rubber (SBR). Publication of references to this type of process started to appear in about 1947, referring to a series of up to 15 stirred tank reactors with product flowing from one to the next. Each reactor within the train is known as a continuous stirred tank reactor or CSTR.

The CSTR approach has not been used greatly in the PVA adhesive field, probably because the application favors smaller quantities of product with closely defined properties. The Shawinigan

Company in Canada did have a small production unit of this type,
but its use had been discontinued before 1968 when the company
(or its successor, Gulf Canada) applied for a patent on the "Loop"
continuous process [117] (Fig. 2.1).

Simple CSTR processes are different from batch processes in
several ways:

1. The formation of new particles is continuous and in the
   presence of existing particles. At equilibrium, formation
   of new particles must equal loss of older particles due to
   elution and coalescence.
2. The random nature of the overflow stream gives rise to a
   broad distribution of reactor residence times and particle
   sizes.
3. Unlike the batch reactor, where the polymerization rate was
   previously noted to be proportional to the 0.4 power of the
   initiation rate [118] and the 0.6 power of the surfactant concentration, the rate in a continuous reactor [150] is proportional
   to the zero power of the rate of initiation (R_i) and the 1.0
   power of the surfactant concentration [S], that is,
   $R_p \propto N \propto R_i^{0.0} [S]^{1.0}$.

Since molecular weight is inversely proportional to the rate
of initiation, this implies that in a continuous reactor molecular
weight can be changed by variations in initiator level without affecting polymerization rate.

Particle number and particle size are also more sensitive to
changes in surfactant concentration with a continuous process,
as opposed to a batch.

The physical nature of the simple CSTR process ensures the
presence of existing particles at the nucleation stage of new particles.
Incoming surfactant is absorbed onto undersaturated surfaces of
the previously formed polymer and a dynamic equilibrium is established. Unexpectedly however, this equilibrium may be the cause
of cycling of properties because, with a narrow particle size distribution, growth of surface area suppresses the formation of new
particles until elution reduces the surface area and demand for
surfactant. Sudden resumption of particle formation then takes
place. As the particle size broadens with several peaks from the inter-
mittant particle formation, a continuous formation process begins and
true equilibrium is reached. Many mathematical models of the process
have been developed [119-145]. Rawling's model [137] predicts that
instability occurs always above the critical micelle concentration
(CMC), although the value of the CMC controls the amplitude of
the oscillations. Instability is also more likely with higher levels

of initiator, lower levels of surface active agent, and lower mean residence times. Brooks [139] has recently predicted that long-term reactor behavior may be influenced by startup conditions.

Much work [123,128,136,137] has been directed toward the elimination of cycling, largely by the external generation of seed particles to be fed into the first reactor of a CSTR train, from a small plug-flow feeder or a CSTR. An additional advantage of an external seeder reactor is that surfactant and monomer concentrations can be arranged to create more particles than in the simple CSTR situation, leading to a finer particle size. Omi [141] calculated in 1969 that a simple continuous reactor produces only 57% of the number of particles produced by a batch reactor utilizing the same recipe.

The concept of "radical desorption" by chain transfer from the polymer chain to a small mobile monomer or chain transfer agent molecule has been examined by Poehlein [140] using a seed-fed CSTR, incorporating carbon tetrachloride and primary alkyl mercaptans in the post-seed-stage recipe. Molecular size determinations by gel permeation chromatography allowed assessment of the transfer of mercaptans of various chain lengths from the monomer droplets to the polymer particles.

Particle structure (morphology) has also been considered as an important indicator of reaction mechanism. Work by Fisher [146] using small-angle neutron scattering to detect perdeuterated methyl methacrylate shells grown on polymethyl methacrylate cores was quoted by Poehlein [140]. Ugelstad and his coworkers have published an extraordinary series of electron micrographs [147] showing particles with macropores and particles with incompatible seeds that either protrude from their surfaces or have been expelled completely, leaving jagged holes behind [148].

Excellent reviews on the subject of the mechanism of continuous polymerization have been published by Poehlein and his coworkers from time to time [149-152].

On the practical and commercial side, a number of patents have been granted [153-158] and a few systems [155] use a recirculating system of some kind.

The simplest continuous reactor using recirculation is the "Loop" process [117] (see Fig. 2.3), which is used currently for the production of PVA adhesives and differs from the continuous stirred tank reactor (CSTR) in a number of significant ways:

1. The reactor consists of a continuous length of piping connecting the outlet to the inlet of a circulation pump. This leads to a reactor whose volume is very low compared to its throughput. Average residence times may be as short as 4 minutes.

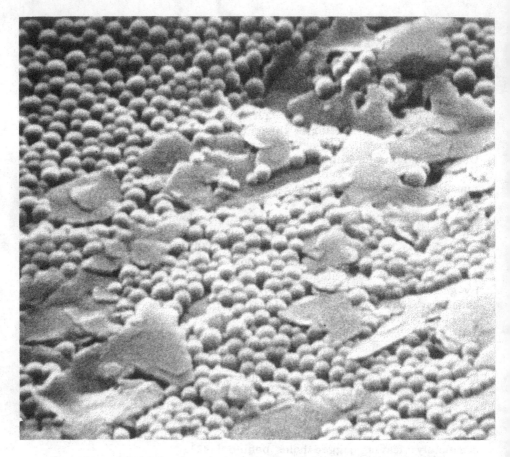

Fig. 2.2 Particles of polystyrene and clay platelets. (Magnification X 28,000.)

2. The polymerizing charge is recycled at an average rate of 10-100 times greater than the rate of input of raw materials. However, streamline flow is employed, which gives a rheology-dependent velocity profile within the pipes. With high viscosity, highly pseudoplastic emulsions, there is a tendency for much faster flow to occur in the center of the pipe compared to the walls, broadening the distribution of residence times.

3. Unlike CSTRs, where conversions in each reactor are often of the order of 10-30%, the Loop generally uses a redox initiation system that produces at least an initiating quantity

of free radicals in one circuit of the piping. Very rapid polymerization of the monomer is observed, and residual monomer at the outlet of the reactor is often less than 1%. This continues to react in the cooling tank to near completion of the reaction.

4. The reactor is completely filled and has no head space and no liquid/vapor interface. Nitrogen, which is often used for purging CSTRs and batch reactors, is not required.

5. Use of a pressure valve on the outlet line of the Loop reactor eliminates cavitation in the pump and refluxing. By the use of higher pressure, gaseous monomers can be used (e.g., for vinyl acetate-ethylene adhesives). Since there is no head space, all the ethylene is dissolved or dispersed in the reaction mixture, increasing its availability. It has been shown that the rate-determining step is the solubility of ethylene in the polymer particles [159]. Gulf also developed

Fig. 2.3 Reactor for the production of emulsion polymers by the continuous Loop process. [Photograph by permission of Gulf Polymers (Dubai).]

a unique Mn^{3+} redox catalyst for use in the Loop reactor
with ethylene [160]. Detailed published discussions of the
Loop system in practice [161-163,193] include comparisons with
batch process products intended for the same application.

Production formulas on the Loop can show cycling of properties
at startup, but within 60-90 minutes of the start of a production
run, near-equilibrium is reached. Ultimately, fouling of the reactor
restricts heat transfer, and if insufficient cooling is applied, temperature may rise with a slight broadening of the particle size
distribution.

If insufficient stabilizer and initiator are used, conversions
will be poor, and severe cycling of properties may be found with
the Loop [163], similar to those observed with the CSR.

Use of the Joyce-Loebl disc centrifuge has enabled the growth
of particles in a Loop reactor to be followed [163], and it was shown
that by extrapolation an estimate could be made of the time and
frequency of the formation of new crops of particles. As the reaction
time progressed, the sharpness of the particle size peaks lessened
and the cyclic creation of particles gave way to an equilibrium
creation process. Cycles persisted for an unexpectedly long period
of 35 residence times or more. This disc centrifuge technique should
be applicable to CSTR kinetic studies. It may also be of value to
batch processes, especially where double or multiple distributions
are desired or are found through reinitiation or coalescence at some
stage during the reaction.

XX. HIGH PRESSURE POLYMERIZATION

Any emulsion polymerization reaction may be conducted in a closed
vessel at a self-generated pressure, which will depend on the temperature and the boiling points or partial vapor pressures of the
volatile ingredients.

This arrangement suppresses reflux and prevents the loss of
toxic or expensive vapors from the condenser, eliminating the need
for fume scrubbers; but also loses the very effective cooling provided
by condensing and returning monomer.

Vinylidene chloride must be used in a closed vessel because
its boiling point is about ambient temperature (32°C), and vinyl
chloride requires a pressure of about 150 psi (10 bar) to keep
it liquid at reaction temperatures.

Ethylene introduces a further factor. It has a triple point of
9.5°C and so is impossible to liquefy at any temperature above this
value. Pressure in this case increases solubility of the gas in the

reaction mixture, and in particular greatly increases its concentration in the emulsion particles [159,164].

A detailed description of the production of "pressure polymers"— copolymers of vinyl acetate with ethylene and vinyl chloride—is beyond the scope of this review. Attention will be drawn however to some of the patent literature applying to the simple binary co-polymers of vinyl acetate and ethylene that are claimed or appear to be of value as adhesives.

Redox initiators appear to be strongly favored; presumably because they can be used at low temperatures, which decreases the partial vapor pressure of ethylene and increases solubility in the water phase.

An early process [165] employed a two-stage batch technique and used as catalyst a mixture of azobisbutyronitrile and potassium persulfate (KPS), with sodium metabisulfite as reducing agent. Pressure employed was 35 kg/cm^2 (500 psig), to give 12% ethylene incorporation. A continuous addition of vinyl acetate over an 8-hour period, together with N-methylol acrylamide, was catalyzed by a KPS/sodium formaldehyde sulfoxylate (SFS) redox couple [166]. The unusual oxidizing agent potassium peroxydiphosphate was mentioned [167] as an alternative to KPS. A similar process [168] adds both vinyl acetate and surfactant in increments.

A hydrogen peroxide-ferric chloride-SFS system run at 55°C and a pressure of 50 kg/cm^2 (710 psig) was claimed [169] to give an ethylene content of 24%. Simple oxidizing agents with bisulfite or thiosulfate [170] may be used. Another patent [171] claims 20-35% ethylene. These figures seem optimistic, since ethylene will hardly homopolymerize at these pressures, and 18% appears to be a good practical maximum. But 24% is close to the stoichiometric ratio of 1:1, and usually this can be achieved only with short grafted chains of ethylene. Some patents describe formulations that include nonionic [172] or both anionic and nonionic [173] emulsifier and polyvinyl alcohol: recipes typical of similar PVA homopolymer preparations.

Another technique applied in PVA homopolymers is the use of a proportion of a previous preparation as a preformed seed or "heel." This appears also in the context of vinyl acetate-ethylene copolymers [174].

Further familiar ingredients from PVA homopolymer production are polyoxyethylene-polyoxypropylene block copolymers [175]. Polyvinyl alcohol is employed in most adhesive preparations, some-times alone. A recent patent of this type [176] employs hydrogen peroxide/ascorbic acid or erythorbic acid as its redox couple.

There are also a number of claims for continuous processes. Bayer [177] claimed a product using 2-N-acrylamido-2 methyl propane sulfonate, a polymerizable stabilizer. Wacker [178] uses the common redox couple of *t*-butyl hydroperoxide with sodium sulfite but, unusually, claims advantages by using an excess of the reducing agent. Since sulfites add fairly readily and reversibly across double bonds, it is possible that this technique increases the solubility of the ethylene in the reaction mixture. Air Products also has a continuous process [179].

The "Loop" continuous process [117] has also been claimed to be of value in the production of vinyl acetate-ethylene copolymers for adhesives. The reactor is filled at all times. As the ethylene is injected into the vinyl acetate feed and becomes dissolved in it, the two monomers have exceptionally high availability for copolymerization, without the step of absorption from the gas to the liquid phases across the surface of the contents, as in a conventional batch reactor. Pipework can withstand high pressures readily, so the cost of a "Loop" reactor may be as little as one-tenth the cost of pressure batch reactors.

There is one important advantage claimed for all continuous processes compared with the batch approach to pressure polymers. Residual unreacted ethylene is dissolved in the water phase of the emulsion and expands to 5-10 times its volume on reducing the reactor pressure to atmospheric. Adhesives are of high viscosity and are extremely difficult to defoam without the use of excessive quantities of antifoam. Continuous production involves the handling of only small volumes of product at any one time and avoids the necessity of receiver tanks of up to 10 times the reactor volume, and the corresponding volume of explosive air-ethylene mixtures. The alternative of small scale degassing units between batch reactor and receiver is very uneconomical of reactor utilization.

XXI.   PRACTICAL ASPECTS OF PRODUCTION OF PVA ADHESIVES

Because of the high viscosity of many grades of PVA adhesives, it is advisable to have a custom-built plant for their preparation. In batch reactor plants kettle sizes vary between 1 and 60 tons capacity depending on demand for the product. Larger plants are economical on labor cost but create problems of heat transfer. To solve this, pumping through external heat exchangers is often employed. Large production volumes are also undesirable; for consistency, it is better to blend several batches. The disposal of any off-specification material will be a severe problem with large reactors. Agitators of the simple impeller, marine propeller, or Pfaudler kind are suitable for low viscosity emulsions but cannot transmit their power to more viscous preparations because shear thinning (pseudoplasticity) allows slippage. Anchor agitators with extra impellers at intervals up the shaft are therefore used, and some manufacturers add vertical bars to produce a gate stirrer. Stirrer geometry will have a profound effect on heat transfer [180]. Similarly, efficient stirrers must be provided in tanks used for making polyvinyl alcohol solutions and also in cooling tanks, particularly when additional polyvinyl alcohol is postadded or when external plasticizers are blended in.

It is essential in production-scale manufacture of polyvinyl acetate emulsions that the rate of stirring be increased to compensate for viscosity increases. If the monomer is allowed to pool on the surface, it is not available for polymerization. The buildup of monomer may be suddenly pulled beneath the surface, resulting in immmediate boiling, which may expel part of the reactor contents with great violence. Even if all monomer is incorporated successfully, an uncontrollable exotherm may subsequently occur, taking the temperature to above the cloud point of some nonionic surfactants and colloids. This can result in destabilization and coagulation of the emulsion. It is crucial, therefore, that stirring be adjusted as the reaction proceeds and be monitored on every production batch.

In batch processes with open condensers, the boiling point of the vinyl acetate-water azeotrope at 66°C ensures steady reflux. This provides useful cooling from the latent heat of evaporation, and the condensed monomer is chilled further before return. CSTR or cascade trains of reactors pose similar problems of agitation and cooling. The "Loop" has some additional problems because of the absence of reflux, but surface area is greater than in the batch process per liter of reaction mixture. Pipe diameter may be increased or a multipipe system may be employed to increase cooling area and aid efficiency. It is also found that the rate of reactor fouling is minimal with many PVA homopolymer systems, and so heat transfer requirements remain constant.

Control may be very simple. It is possible to run a batch reactor with a contents temperature gauge and a throttle valve on the cooling water flow alone, but agitator speed control is nearly essential. The next level of instrumentation will record agitator speed, agitator torque, reaction temperature, cooling jacket flow and temperature rise, monomer and initiator addition rates, and reflux return rate. Control of these functions is manual, but the data from a typical run can be used to automatically control future batches via cam controllers or, more recently, through digital data recording and microprocessor control. It must be emphasized, however, that many recipes require alert process operators, and the appearance of the surface of the emulsion (with regard to monomer pooling, foam, or graininess due to entrained pockets of liquid or vaporized vinyl acetate beneath the surface) can indicate the need for prompt action. Hence sight glasses with wiper blades and good illumination within the kettle are of great value. Attempts to force the reaction by increasing the rate of monomer addition can have a disastrous effect insofar as the rate of polymerization may fall off at higher concentrations of monomer within the reactor, giving the false impression that the cooling system is coping well with the additional heat output. Moritz [116] showed that reaction rate fell off quite sharply at volume fractions of vinyl acetate above 0.3, a situation that could easily arise locally with inadequate stirring.

With the availability of computers, automatic monitoring and control is developing at a fast pace, and a recent paper [181] reviews progress and reports experiments on a solution polymer system.

Liquid storage is straightforward, since the typical adhesive plant requires only one or two monomers in bulk. Specialty monomers are used from drums. All monomers must be kept from direct sun, and in tropical countries water sprays or internal cooling coils should be provided. The storage and use of ethylene represent a very much greater problem, requiring the highest quality pressure engineering.

Safety factors even in simple plants must be carefully considered. Vinyl acetate is highly flammable, and its vapor is explosive in most concentrations. The UK Occupational Exposure Limit of vinyl acetate is 10 ppm, and to achieve this, reactor and pipeline joints have to be leak-free and lids on cooling tanks should be well sealed. Adequate ventilation, with fume extraction at appropriate places, must be provided. Gaseous monomers, especially where toxicity problems arise as with vinyl chloride, need even greater engineering integrity coupled with constant atmospheric monitoring. Ventilation should be greater with these materials, and special extraction should be provided near pumps, cooling tanks, blenders, and vibrating strainers.

Some solids are also toxic or corrosive and may represent more subtle hazards. Acrylamide and crotonic acid are examples of crystalline materials that require careful handling. Even the dust from colloids may represent a hazard (as slippery when wetted) and should be carefully swept up if spilt.

Quality control monitoring is essential. Basic tests are for nonvolatiles, viscosity, pH, average particle size, and residual vinyl acetate monomer. Viscosity is preferably measured at at least two shear rates, so that a simple indication of rheology is gained. Other routine tests that may be carried out include film quality, particle size distribution, surface tension, and storage stability. More application-oriented checks, such as wet tack and bond strength, as well as minimum filming temperature and freeze-thaw stability [182] should be made occasionally.

XXII. FILM FORMATION AND FILM PROPERTIES

To give a good adhesive bond, it is essential that the water in the adhesive evaporates to give a well-integrated film. Provided the minimum filming temperature has been adjusted by copolymerization or the addition of a plasticizer, and the relative humidity is not too high, ready film formation will occur, but the mechanism is surprisingly complex.

First theories [183] were advanced in the late 1950s, particularly by Vanderhoff and his coworkers [184-187]. He recognized that drying takes place in a series of steps involving capillary forces and surface tension. After the film has dried, further consolidation may take place over many days, with the expulsion of incompatible surfactants onto the surface of the film. This process explains the gradual gain in bond strength and water resistance that occurs after initial drying appears to be complete. Slow evaporation of coalescing solvents will also lead to increased hardness and bond strength.

Film hardness can be assessed in a number of ways, usually with the adhesive in the form of a coating on glass or some other suitable substrate. These are mainly borrowed from coating applications and consist of such tests as pencil hardness, Sward rocker, Koenig pendulum and microindentation.

A setting time test was developed by the Paper Industry Research Association of Great Britain (PIRA) [188] in which adhesive is applied between strips of Kraft lining paper, which are then compressed and torn apart at set time intervals. Fiber tear rather than separation of the glue line occurs after the setting time. This test determines the rate at which the adhesive dried by wicking of the water into standard Kraft paper. Factors such as particle size, polyvinyl alcohol level, and the rheology change of the adhesive as the water is removed are crucial.

Wood bond strength is checked by gluing together standard wood slices [189-191]. After drying, the slices are pulled apart in a tensiometer. Acceptable bond strength gives failure of the wood, and this implies that the cohesion of the adhesive, and its adhesion to the wood surface, should be greater than the cohesion of the wood itself. Shallow penetration of the adhesive into the substrate is an obvious advantage. This can however deplete the

surface, especially on end-grain, where the porosity is very great.
A wide particle size distribution and a high viscosity, especially
at low shear rates, which will limit the speed of penetration, could
be useful. In some manufacturing operations application of adhesive
is largely viscosity controlled. Open time must be sufficient to allow
the workpieces to be brought together while the adhesive is still wet.
The setting point should then be quickly reached to minimize the time
during which surfaces must be physically held together.

The properties of the film produced on drying, and how these
properties affect the performance of adhesives and other applications,
have been the subject of many hundreds of patents and papers.
A review book [192] on emulsion polymers and their properties
and applications ran into more than 1000 pages.

The last word may be given to Alexander and Napper [78]
who made the following classic understatement: "The process of
emulsion polymerization is more complicated and rather more subtle
than was first envisaged."

REFERENCES

1. Warson, H. (1985). *Makromol. Chem., Suppl.*, 10/11, 265.
2. Anon. (1986). *Chem. Mark. Rep.*, 230, 13th October, 36.
3. Cass, R. A., and Raether, L. O. (1964). *Off. Digest*, September, 947.
4. Llewellyn, I., and Pearce, M. F. (1966). *J. Oil Colour Chem. Assoc.*, 49, 1032.
5. Oosterhoff, H. A. (1963). *J. Oil Colour Chem. Assoc. Reprint*, 48.
6. Hoy, K. L. (1973). *J. Paint Technol.*, 45, 51.
7. Reader, G. E. L. (1968). *Paint Technol.*, 32, 14.
8. Protzman, T. F., and Brown, G. L. (1960). *J. Appl. Polym. Sci.*, 4, 81.
9. Leaflet on Sheen/ICI Minimum Film Forming Temperature Apparatus, Sheen Instruments.
10. Nyquist, E. B., and Yocum, R. H. (1970). *J. Paint Technol.*, 42, 308.
11. U.K. Patent 1,092,030 to Shawinigan Chemical Corp. (1967).
12. Patella, R. F. (1978). *Modern Paint and Coatings*, July, 47.
13. Gulbekian, E. V. (1969). *Br. Polym. J.*, 1, 96.
14. Al-Shahib, W. A.-G. R., and Dunn, A. S. (1980). *Polymer*, 21, 429.
15. U.K. Patent Appl. 2,105,354 to Kuraray (1983).
16. Leaflet P15.3, Interox Chemicals Ltd.
17. Leaflet LC38, Interox Chemicals Ltd.

18. Kolthoff, I. M., and Miller, I. K. (1951). *J. Am. Chem. Soc.*, 73, 3055.
19. Fordham, J. W. L., and Williams, H. L. (1951). *J. Am. Chem. Soc.*, 73, 4855.
20. McCoy, C. E. Jr. (1963). *Off. Digest Reprint.*
21. Vegter, G. E., and Grommers, E. P., *J. Oil Colour Chem. Assoc.*, 50, 72.
22. Griffin, W. J. (1949). *J. Soc. Cosmet. Chem.*, 1, 311.
23. Griffin, W. J. (1954). *J. Soc. Cosmet. Chem.*, 5, 249.
24. Gerrens, H. (1965). In *Polymer Handbook*, 2nd ed., J. Brandup and E. Immergut, Eds. Wiley-Interscience, p. 399.
25. Mills, J. A., and Yocum, R. H. (1967). *J. Paint Technol.*, 39, 532.
26. German Patent 1,060,600 to Hoechst (1959).
27. U.K. Patent 1,350,282 to Reed International (1974).
28. U.S. Patent 4,421,902 to Rohm and Haas (1983).
29. European Patent 0109820 to Rohm and Haas (1984).
30. Nippon Gohsei (1982). *Gohsenol Polyvinyl Alcohol for Emulsions*, ref. 3051005.
31. Hoechst (1983). Technical Data Sheet on Mowiol.
32. Japanese Patent 85/23,465 to Daicel Chem. Ind. (1985).
33. Kuraray, *Polyvinyl Alcohol Stabilisers for PVAc Emulsions.*
34. Ahmed, I., and Pritchard, J. G. (1979). *Polymer*, 20, 1492.
35. Toyoshima, K. (1968). In *Properties and Applications of Polyvinyl Alcohol*, C. Finch, Ed. SCI Monograph 30, p. 154.
36. Japanese Patent 55/093972 to Takeda Chemical (1955).
37. USSR Patent 604,850 (1979).
38. Hirai, T., Okazaki, A., and Hayashi, S. (1986). *J. Appl. Polym. Sci.*, 32, 3919.
39. U.S. Patent 4,575,525 to National Starch Chemical Corporation (1986).
40. Van Arkel, J. S. (1985). In *Cellulose and Its Derivatives*, J. F. Kennedy, Ed. (Ellis Horwood), 301.
41. Craig, D. H. (1986). *Am. Chem. Soc. Div. PMSE Papers*, 54, 354.
42. Belgian Patent 877,424 to GRA N.V. (1980).
43. Romanian Patent 84,725 to Intre Prinderea Chimica (1984).
44. Cornwell, D. W., and Harrison, D. L. (1986). *Eur. Adhes. Sealant*, March, 22.
45. Flory, P. J. (1937). *J. Am. Chem. Soc.*, 59, 241.
46. Mayo, E. R. (1943). *J. Am. Chem. Soc.*, 65, 2324.
47. Okieimen, E. F. (1986). *Eur. Polym. J.*, 22, 175.
48. Devon, M. J., and Rudin, A. (1986). *J. Polym. Sci., Polym. Chem.*, 24, 2191.
49. Wheeler, O. L. et al. (1952). *J. Polym. Sci.*, 9, 157.

50. Dietz, R., and Francis, M. A. (1979). *Polymer*, 20, 450.
51. Taganov, N. G. (1985). *Polym. Sci. USSR*, 27, 828.
52. Ivin, K. J. (1985). In *Polymer Handbook*, 2nd ed., J. Brandup and E. Immergut, Eds. New York, Wiley-Interscience, p. 363.
53. Lindemann, M. K. (1966). *Acetate Abstr. Bull.*, 4, 1.
54. Ellis, R. A. (1979). *Pigment Resin Technol.*, September, 10.
55. Ellis, R. A. (1979). *Pigment Resin Technol.*, October, 4.
56. Ellis, R. A. (1979). *Pigment Resin Technol.*, November, 17.
57. Lichti, G., Gilbert, R. G., and Napper, D. H. (1978). Presented at International Polymer Latex Conference, London.
58. Ham, G. E. (1964). *High Polymers*, Vol. VIII, *Copolymerization*, New York, Wiley-Interscience.
59. Mark, H., et al. (1975). In *Polymer Handbook*, 2nd ed., J. Brandup and E. Immergut, Eds. New York, Wiley-Interscience, p. 142.
60. Young, L. J. (1975). In *Polymer Handbook*, 2nd ed., J. Brandup and E. Immergut, Eds. New York, Wiley-Interscience, p. 291.
61. Kahn, D. J., and Horowitz, H. H. (1961). *J. Polym. Sci.*, 54, 363.
62. O'Driscoll, K. (1964). *Polym. Lett.*, 2, 869.
63. Djekhaba, S., et al. (1986). *Eur. Polym. J.*, 22, 729.
64. Behnken, D. W. (1964). *J. Polym. Sci.*, A, 2, 645.
65. Khanna, R., and Sutton, R. C. (1982). Presented at 28th Macromolecular Symposium.
66. Schuller, H. (1986). In *Polymer Reaction Engineering*, K.-H. Reichert and W. Geiseler, Eds. Basel, Huthig and Wepf, p. 137.
67. Guillot, J. (1986). In *Polymer Reaction Engineering*, K.-H. Reichert and W. Geiseler, Eds. Basel, Huthig and Wepf, p. 147.
68. Alfrey, T., and Price, C. C. (1947). *J. Polym. Sci.*, 2, 101.
69. Young, L. J. (1975). In *Polymer Handbook*, 2nd ed., J. Brandup and E. Immergut, Eds. New York, Wiley-Interscience, p. 341.
70. Laurier, G. C., O'Driscoll, K. F., and Riley, P. M. (1985). *J. Polym. Sci., Polym. Symp.* No. 72, 17.
71. Elgood, B. G. (1980). *J. Oil Colour Chem. Assoc.*, 63, 103.
72. Dimarzio, E. A., and Gibbs, J. H. (1959). *J. Polym. Sci.*, 40, 121.
73. Fox, T. G., and Loshaek, S. (1955). *J. Polym. Sci.*, 15, 371.
74. U.S. Patent 3,497,521 to Gulf Oil Canada (1970).
75. Yurzhenko, A. T. (1946). *J. Gen. Chem. USSR*, 16, 1171.
76. Harkins, W. D. (1947). *J. Am. Chem. Soc.*, 69, 1428.
77. Smith, W. V., and Ewart, R. H. (1948). *J. Chem. Phys.*, 16, 592.
78. Alexander, A. E., and Napper, D. H. (1970). In *Progress in Polymer Science*, ed. Jenkins, Vol. III, Chapter 3.
79. Vanderhoff, J. W. (1962). Advances in Chemistry Series, No. 34, p. 6.

80. Vanderhoff, J. W. (1965). *J. Polym. Sci.*, 33, 487.
81. Patsiga, R. (1962). Ph.D. thesis, Syracuse University.
82. Patsiga, R., Litt, M., and Stannett, V. (1960). *J. Phys. Chem.*, 64, 801.
83. Gardon, J. L. (1968). *J. Polym. Sci.*, A-1, 6, 623, 643, 687, 2853.
84. Gardon, J. L. (1970). *Br. Polym. J.*, 2, 1.
85. Fitch, R. M., and Shih, Lih-bin. (1975). *Prog. Colloid Polym. Sci.*, 56, 1.
86. Fitch, R. M., et al. (1985). *J. Polym. Sci.*, *Polym. Symp.*, 72, 221.
87. Dunn, A. S., and Al-Shahib, W. A.-G. R. (1978). *Br. Polym. J.*, 10, 137.
88. Dunn, A. S., and Al-Shahib, W. A.-G. R. (1978). *Proc. Plast. Rubber Inst. Int. Conf. Polymer Latex*, London.
89. Medvedev, S. S. (1959). In *International Symposium on Macro-molecular Chemistry*. Elmsford, NY, Pergamon Press, p. 174.
90. Brodnyan, J. G., et al. (1963). *J. Colloid Sci.*, 18, 73.
91. Hansen, F. K., and Ugelstad, J. (1978). *J. Polym. Sci.*, 16, 1953.
92. Penboss, I. A., Gilbert, R. G., and Napper, D. H. (1986). *J. Chem. Soc., Faraday Trans. I*, 82, 2247.
93. Nomura, M., Kubo, M., and Fujita, K. (1984). *J. Appl. Polym. Sci.*, 28, 2767.
94. Priest, W. J. (1952). *J. Phys. Chem.*, 56, 1077.
95. Priest, W. J. (1956). *J. Phys. Chem.*, 60, 1250.
96. French, D. M. (1958). *J. Polym. Sci.*, 32, 395.
97. Litt, M., Patsiga, R., and Stannett, V. (1970). *J. Polym. Sci.*, A-1, 8, 3607.
98. Stannett, V., Klein, A., and Litt, M. (1975). *Br. Polym. J.*, 7, 139.
99. French, D. M. (1958). *J. Polym. Sci.*, A-1, 6, 2265.
100. Napper, D. H., and Parts, A. G. (1962). *J. Polym. Sci.*, 61, 113.
101. Elgood, B. G., et al. (1964). *Polym. Lett.*, 2, 257.
102. Fitch, R. M., et al. (1969). *J. Polym. Sci.*, C, 27, 95.
103. Zollars, R. L. (1979). *J. Appl. Polym. Sci.*, 24, 1353.
104. El-Aasser, M. S., and Vanderhoff, J. W., Eds. (1981). *Emulsion Polymerisation of Vinyl Acetate*. Barking, Applied Science Publishing.
105. Morris, C. E. M., et al. (1966). *J. Polym. Sci.*, A-1, 4, 985.
106. Banerjee, M., and Konar, R. S. (1986). *Polymer*, 27, 147.
107. McCoy, C. E. (1963). *Off. Digest Reprint*.
108. Dunn, A. S. (1983). In *Science & Technology of Polymer Colloids*, Vol. II, NATO ASI Series E, No. 68. The Hague, Martinus Nijhoff Publishers, p. 314.

109. Carra, S., et al. (1985). *Proc. Int. School Phys. Enrico Fermi*, 90, 483.
110. Dunn, A. S., and Hassan, S. A. (1986). *Am. Chem. Soc. Div. PMSE*, 54, 439.
111. Dunn, A. S., and Taylor, P. A. (1965). *Makromol. Chem.*, 83, 207.
112. Reynolds, G. E. J., and Gulbekian, E. V. (1968). In *Properties and Applications of Polyvinyl Alcohol*, C. Finch, Ed. SCI Monograph 30, p. 131.
113. Min, K. W., and Ray, W. H. (1978). *J. Appl. Polym. Sci.*, 21, 89.
114. Penlidis, A., MacGregor, J. F., and Hamielec, A. E. (1984). *Polym. Proc. Eng.*, 2, 179.
115. Dougherty, E. P. (1986). *J. Appl. Polym. Sci.*, 32, 3051, 3095.
116. Moritz, H.-U. (1986). In *Polymer Reaction Engineering*, K.-H. Reichert and W. Geiseler, Eds. Basel, Huthig & Wepf, p. 101.
117. U.K. Patent 1,220,777 to Gulf Oil Canada (1971).
118. Gershberg, D. B., and Longfield, J. F. (1961). 45th AIChE Meeting, New York.
119. Funderburk, J. O. (1969). Ph.D. Thesis, Iowa State University.
120. Gerrens, H., et al. (1971). *Chem. Ing. Tech.*, 43, 693.
121. DeGraff, A. W., and Poehlein, G. W. (1971). *J. Polym. Sci.*, A-2, 9, 1955.
122. Brooks, B. W. (1973). *Br. Polym. J.*, 5, 199.
123. Gonzalez, R. A. (1974). M.S. Thesis, Lehigh University.
124. Greene, R. K., et al. (1976). In *Emulsion Polymerization*, I. Piirma and J. L. Gardon, Eds. ACS Symposium Series No. 24. Washington, DC, American Chemical Society, p. 341.
125. Greene, R. K. (1976). Ph.D. Thesis, Lehigh University.
126. Cauley, D. A., et al. (1978). *Chem. Eng. Sci.*, 33, 979.
127. Brooks, B. W., et al. (1978). *Polymer*, 19, 193.
128. Hamielec, A. E., and MacGregor, J. E. (1978). ACS Division of Colloid and Surface Chemistry, 176th Meeting.
129. Chiang, A. S. T., and Thompson, R. W. (1979). *J. Appl. Polym. Sci.*, 24, 1935.
130. Hamielec, A. E., et al. (1979). *J. Appl. Polym. Sci.*, 23, 401.
131. Kiparissides, C., et al. (1980). *Can. J. Chem. Eng.*, 58(1), 48, 56, 65.
132. Poehlein, G. W., Dubner, W., and Lee, Hsueh-chi. (1982). *Br. Polym. J.*, December, 143.
133. Baddar, E. E., and Brooks, B. W. (1984). *Chem. Eng. Sci.*, 39, 1499.

134. Penlidis, A., MacGregor, J. F., and Hamielec, A. E. (1985). *Am. Chem. Soc. Div. PMSE*, 52, 484.

135. Heiskanen, T. (1985). *Acta Polytech. Scand., Chem. Technol. Metall. Ser.*, 165.

136. Penlidis, A., MacGregor, J. F., and Hamielec, A. E. (1986). *J. Coatings Technol.*, 58, 49.

137. Rawlings, J. B., Prindle, J. C., and Ray, W. H. (1986). In *Polymer Reaction Engineering*, K.-H. Reichert and W. Geiseler, Eds. Basel, Huthig & Wepf, p. 1.

138. Nomura, M. (1986). In *Polymer Reaction Engineering*, K.-H. Reichert and W. Geiseler, Eds. Basel, Huthig & Wepf, p. 41.

139. Brooks, B. W., Baddar, E. E., and Raman, G. (1986). In *Polymer Reaction Engineering*, K.-H. Reichert and W. Geiseler, Eds. Basel, Huthig & Wepf, p. 51.

140. Poehlein, G. W., Lee, H.-C., and Chern, C.-S. (1986). In *Polymer Reaction Engineering*, K.-H. Reichert and W. Geiseler, Eds. Basel, Huthig & Wepf, p. 59.

141. Omi, S., Veda, T., and Kubota, H. (1969). *J. Chem. Eng. Jpn.*, 2, 193.

142. Nomura, M., et al. (1971). *J. Appl. Polym. Sci.*, 15, 675.

143. Veda, T., Omi, S., and Kubota, H. (1971). *J. Chem. Eng. Jpn.*, 4, 50.

144. Lin, C.-C., and Chiu, W.-Y. (1982). *J. Appl. Polym. Sci.*, 27, 1977.

145. Penlidis, A., Hamielec, A. E., and MacGregor, J. F. (1984). *J. Vinyl Technol.*, 6, 134.

146. Fisher, unpublished work.

147. Ugelstad, J., et al. (1986). In *Polymer Reaction Engineering*, K.-H. Reichert and W. Geiseler, Eds., Basel, Huthig & Wepf, p. 77.

148. Ugelstad, J., et al. (1983). In *Science and Technology of Polymer Colloids*, G. W. Poehlein, R. H. Ottewill, and J. W. Goodwin, Eds. The Hague, Martinus Nijhoff, p. 51.

149. Poehlein, G. W., and Dougherty, D. J. (1976). *Rubber Chem. Technol.*, 50, 601.

150. Poehlein, G. W. (1982). *Br. Polym. J.*, December, 153.

151. Poehlein, G. W., Lee, H.-C., and Stubicar, N. (1985). *J. Polym. Sci., Polym. Symp. No.* 72, 207.

152. Poehlein, G. W. (1985). ACS Symposium Series No. 285, p. 131.

153. U.K. Patent 1,124,610 to BASF (1968).

154. U.K. Patent 1,297,215 to Bayer (1972).

155. U.K. Patent 1,411,465 to Bayer (1976).

156. U.K. Patent 1,462,984 to BP Chemicals (1977).

157. U.S. Patent 4,123,403 to Dow (1977).

158. U.K. Patent 1,520,440 to Montedison (1978).
159. Lohr, G. (1978). Presented at International Polymer Latex Conference, London.
160. U.K. Patent 1,376,780 to Gulf Oil Canada (1974).
161. Geddes, K. R. (1983). *Chem. Ind.*, 21st March, 223.
162. Geddes, K. R. (1986). *Polym. Paint Col. J.*, 176, 494.
163. Geddes, K. R. (1985). *Proc. XI International Coatings Conference*, Athens. In *Advances in Organic Coatings* (1987), Vol. 9, Lancaster, Technomic Publishing, p. 30.
164. Lohr, G. (1979). *Plast. Rubber Mater. Appl.*, 4, 141.
165. Japanese Patent 73/07865 to Nippon Carbide (1973).
166. U.S. Patent 3,714,105 to Borden Inc. (1969).
167. German Patent 2,301,099 to Borden Inc. (1974).
168. U.K. Patent 2,039,920 to National Distillers and Chemical Corporation (1981).
169. Japanese Patent 097563 to Dai-Nippon Ink and Chemicals (1974).
170. U.K. Patent 1,514,105 to Sumitomo Chemical Co. (1978).
171. U.S. Patent 4,128,518 to Sumitomo Chemical Co. (1978).
172. Japanese Patent 153036 to Nippon Synthetic Chemical Industry Co. (1982).
173. Japanese Patent 85/023786 to Nippon Synthetic Chemical Industry Co. (1985).
174. U.S. Patents 3,734,819 and 3,769,151 to Union Oil Co. (1972 and 1973).
175. Japanese Patent 85/031349 to Dai-Nippon Ink and Chemicals (1985).
176. European Patent Appl. 0 191 460A to Air Products and Chemicals Inc. (1986).
177. U.K. Patent 1,411,465 to Bayer AG (1976); see also German Patent 2,250,517 (1979).
178. U.S. Patent 4,035,329 to Wacker-Chemie GmbH (1977).
179. U.S. Patent 4,164,489 to Air Products and Chemicals Inc. (1974).
180. Krofta, K., and Prokopec, L. (1986). *Chem. Prum.*, 36, 62.
181. Ponnuswamy, S., Shah, S. L., and Kiparissides, C. (1986). *J. Appl. Polym. Sci.*, 32, 3239.
182. Blackley, D. C., and Teoh, S. C. (1978). Presented at International Polymer Latex Conference, London, p. 9/1.
183. Zdanowski, R. E., and Brown, G. L. (1958). *Proc. 44th Mid-Year Conf., Chem. Spec. Manuf. Assoc.*, Cincinnati.
184. Vanderhoff, J. W., and Gurnee, E. F. (1956). *TAPPI*, 39, 71.
185. Vanderhoff, J. W., and Bradford, E. B. (1963). *TAPPI*, 46, 215.
186. Vanderhoff, J. W., et al. (1966). *J. Macromol. Chem.*, 1, 361.
187. Vanderhoff, J. W. (1960). *Br. Polym. J.*, 2, 161.

188. Hine, D. J., and Streeter, A. C. (1972). Presented at International Conference on Packing Technology, p. 21:1.
189. Nippon Gohsei (1968). *Gohsenol Polyvinyl Alcohol for Adhesive Applications*, ref. 3031002, Nippon Gohsei, p. 9.
190. Nippon Gohsei (1982). *Gohsenol Polyvinyl Alcohol for Adhesives and Binders*, ref. 3031006, pp. 5-7.
191. European Committee for Standardisation, CEN/TC 103.
192. Warson, H. (1972). *The Application of Synthetic Resin Emulsions*. Ernest Benn. (Out of print.)
193. Quixley, N. E., Pizzi, A., Smith, G. S. (1985), Continuous tubular recycle reactors for PVA and other adhesives, Proc. Forest Products Research International, Pretoria, South Africa.

3

The Chemical Bonding of Wood

WILLIAM E. JOHNS / *Washington State University*
Pullman, Washington

I. INTRODUCTION

Adhesion is a universal phenomenon, daily touching the lives of
individuals. It is virtually impossible to scan an environment and
not note any number of materials or products that develop their
usefulness as a result of bonding or coating in some fashion. Yet
in spite of this ubiquitous nature, surprisingly little is known about
the primary causes of adhesion. This chapter briefly reviews the
fundamental theories of adhesion and focuses on one of them, direct
chemical bonding, as it relates to wood substrates.

Excellent reviews [1,2] outline the currently understood theories
that help explain adhesion, namely:

Mechanical entanglement
Adsorption/specific adhesion
Diffusion/molecular entanglement
Electrostatic/donor–acceptor interactions
Direct chemical bonding

Each theory has a relative importance, depending on the nature
of the adherent and adhesive. It is difficult, for instance, to see
how mechanical entanglement plays a significant role in the bonding
of glass or carbon fibers, while this theory may play a predominant
role in the bonding of paper or leather with hotmelt adhesives.
Wood researchers are faced with the problem that all the theories
above are probably active and can contribute at some level to the
formation of a proper bond. Thus, the problem lies in sorting out
the various contributions and in striving to emphasize those that
promote desirable properties.

The most widely studied approach to understanding wood bonding is that of adsorption or specific adhesion. This is not surprising in that this theory is applied to a variety of surfaces, not just wood, and does not require expensive equipment or obtuse conceptualizations. In terms of sheer volume of pages, more has been written about contact angles and the thermodynamic work of adhesion than comparable theories for wood bonding. Published work has established the importance of intimate contact between adhesive and adherent, with contact angle analysis providing that information. Care, however, is called for. The adsorption theory provides only the thermodynamic justification for whether a process can occur; it does not provide by itself a chemical or physical mechanism for how that process is realized. Recently Kinlock [2] has worked to provide mechanistic considerations to the adsorption theory of adhesion.

Chemical bonding is here defined as the process for joining two or more pieces of wood via a direct chain of covalent bonds between adjacent surfaces. The term "chemical bonding" is used throughout, recognizing that a wide variety of labels have been used to describe the process of joining wood via a covalent bond bridge. Such terms as oxidative coupling, oxidative bonding, non-conventional bonding, auto-oxidation, autohesion and, in some cases, graft polymerization bonding, have all been offered as descriptive terms for joining wood via a covalent bonding scheme. Furthermore, the word "wood" is used in a general sense, but much work has been reported on paper substrates.

It was early in the history of wood bonding that the potential for chemical bonding was acknowledged. As early as 1939 Tischer [3] pointed out that the use of neutralized oxidizing compounds such as mixtures of bichromates and sodium carbonate or nitric acid and aluminum hydroxide was responsible for changing the surface of wood so that it more easily bonded with conventional adhesives or, alternately, was capable of adhering directly to similarly treated veneer surfaces without the use of any other "glue." This was long before the development of polymer-fiber composites for structural applications or silane coupling agents. Two key factors were identified by Tischer in this early work. First was the use of oxidizing compounds, and second, the observation that load-carrying bonds could be made *without* film-forming adhesives. Historically, these two factors appear repeatedly in the literature of chemical bonding.

Chemical bonding is a captivating concept that has drawn the attention of many researchers. To bond wood with a series or network of covalent bonds would seem like the ultimate form of "adhesive" joint. A major concern of the wood adhesive technologist is the hydrolytic stability of glue joints, and it would seem that chemical bonding would provide a joint as "waterproof" and as strong as whole wood. Thus, it would seem that chemical bonding could be considered as the ultimate process of making structural, stable wood joints. This chapter reviews the various ways in which chemical bonding has been thought to develop in the formation of wood joints, and it discusses the implications, both positive and negative, real and perceived for the chemical bonding of wood.

II. VARIETIES OF CHEMICAL BONDING

A. Chemical Bonding in Polymer-Based Systems

Before discussing the chemical bonding of wood in its more exotic forms, it would be helpful to consider the evidence for chemical bonding with traditional polymerizable adhesive systems such as the family of aldehyde resins or isocyanates. At first this thought may seem like a contradiction, considering film-forming resins from a chemical bonding perspective, yet some interesting parallels exist. The implications of what goes on with conventional systems are important when considering non-film-forming approaches of chemical bonding.

To help explain why conventional systems such as phenol-formaldehyde resins work as well as they do at developing a strong, hydrolytically stable bond, some researchers have suggested that covalent bonds form between resin and wood. Since the resins themselves are thought to be held together with a series of crosslinks or covalent bonds, it could be argued that the joint itself is held together with covalent bonds. The argument goes: If covalent bonds are not involved in the joint formation, why is the joint strong and water resistant? Troughton and Chow [4,5] show in an elegant series of experiments that aldehyde-based resins do indeed have the strong potential for forming covalent bonds with wood material. This work, based on the kinetics of hydrolysis of glue-wood systems, showed a dramatic difference in reaction rates between the acid hydrolysis of glue alone and wood-glue mixtures. It was thus inferred that a new bond between wood and glue was formed.

While the work of Troughton and Chow generated some controversy, the demonstrated reaction of aldehyde resin and wood material is not surprising. Wood is a chemically reactive system. When mixed with another chemically reactive system, the adhesive, and then placed in a high temperature environment, reactions between the two should be expected. The problem becomes one of assigning the proper importance of covalent bonding in conventional systems to the overall integrity of the glue joint. When the experiment cited

above was conducted, it involved the soaking of sections of wood
50 μm thick in a many-fold excess of adhesive, squeezing out the
excess, and then heating for 40 minutes at 120°C. These are not
the conditions of manufacturing, for instance, particleboard, where
adhesive is sprayed as a fine mist, and heating is for the minimum
time necessary to achieve an acceptable panel. The reaction environment of an adhesive in a water-poor, atomized droplet smeared
over the surface of dry wood and the above-cited experiment are
significantly different. A most important difference is the chemical
environment. Wood has a buffering capacity that varies between
species [6]. This buffering capacity has been shown to have a
significant influence on both the gel time of urea-formaldehyde
resin and the properties of flake and medium density fiberboard
[7]. In the above-cited experiments by Troughton and Chow, the
chemical environment of the reaction medium was controlled by the
excesses of the aqueous mixture, not by the wood. Thus, while
the potential for chemical reaction between wood and aldehyde-based
adhesives has been shown to exist, the conditions under which
the experiments were carried out were significantly different from
chemical environment common during board manufacture. Thus,
the actual role of covalent bonds in developing mechanically sound,
hydrolytically stable wood joints with aldehyde-based resins has
yet to be addressed.

A counter to the argument that chemical bonds between phenol-
formaldehyde resin and wood are necessary for forming water-
resistant glue joints is elegantly offered by Pizzi and Eaton [8].
To explore the interactions of phenolic resins with cellulose, these
researchers completed *ab initio* computer modeling of cellulose crystals
and the interactions of such models with oligomeric forms of methylol
phenol. The results clearly show that the interactive energy for
the phenolic resin is higher than that of water with the cellulose.
These calculations show that water cannot displace phenolic moieties
from the surface of cellulose. The source of the interactions consists
solely of van der Waals and hydrogen bonding.

A much less controversial wood-adhesive reaction is that of
isocyanate systems. Here the —NCO group of the adhesive is
remarkably reactive, particularly toward hydroxyl-containing materials. Isocyanates have been repeatedly shown to provide mechanically sound, hydrolytically stable joints at very low glue spreads.
The reason for this is often thought to be the formation of a carbamate or urethane linkage with the hydroxyls on the wood surface
(Fig. 3.1).

Evidence for the reaction of isocyanates and wood material is
based on the work of Rowell and coworkers [9-11] and the work
of Morak and coworkers [12-14]. Rowell and coworkers investigated

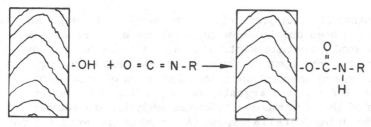

Fig. 3.1 Generalized reaction scheme showing how isocyanate resins are thought to react with wood hydroxyls.

the chemical modification of wood with isocyanates. This work was directed at better understanding the role of wood hydroxyls in the weathering and biodegradability of wood and cellulose. The technique used for this research was to treat oven-dry wood with vapor phase methyl isocyanate at 120°C, 150 psi (nitrogen pressure). Nitrogen analysis after extraction showed that a real increase in unextractable nitrogen; furthermore, the increase in nitrogen content paralleled the reaction time. Resistance to wood-destroying fungi was proportional to the increase in nitrogen content. Rowell reports that high antishrink efficiencies are realized with weight percent gains of 16-28%. It was concluded by Rowell that the dimensional stability of wood and its resistance to biological deterioration was due to the removal of active hydroxyl groups by reaction with the isocyanate group to yield a urethane linkage.

The work of Morak and Ward was directed at improving the properties of paper and liner board with isocyanate treatment. These investigators studied the use of isocyanates in solvent systems [12] and in vapor phase treatment [13], as well as the use of blocked isocyanates [14]. The success of the reaction of isocyanates with the lignocellulosic material was measured by the nitrogen content after extraction and the relative solubility of the paper products in cupriethylenediamine. These authors report that improvements in rigidity, breaking length, and moisture resistance broadly followed the uptake of isocyanate. A most interesting point was that accessibility of the isocyanate to the lignocellulosic material was quite important. The use of nonaqueous swelling agents generally improved the retention values. In almost all cases the paper products became more brittle, as evidenced in lower fold endurance.

These two bodies of research provide sound evidence to prove that low molecular weight isocyanates can chemically react with wood or cellulose and, in doing so, modify the physical and biological properties of that material. However, the parallel evidence for the reaction of isocyanate with wood as used in a typical gluing operation (i.e., the manufacture of particleboard) is curiously lacking. No one to date has been able to show the existence of a urethane bond in a typical wood-isocyanate joint. Why might this be so? For one thing, wood is a very complex chemical system. Simply looking for a single type of chemical bond amid the chaos of background noise is most difficult, but arguably not impossible. McLaughlin [15] has noted that subtractive difference analysis with Fourier transformation infrared (FTIR) spectroscopy of simple wood (as untreated waferboard flakes) and wood treated with 4-4'diphenylmethane diisocyanate (MDI) and pressed at normal moisture contents (i.e., 5%) in a conventional hot press did not show the presence of any urethane structures. On the contrary, copious amounts of polyureas (reaction scheme shown in Fig. 3.2) were found instead. Only when excess amounts of catalysts specifically designed to enhance the probability of urethane reactions were used was a trivially small amount of the urethane linkages seen in difference FTIR spectra.

In both the work of Rowell and coworkers and Morak and Ward, the need to use anhydrous wood was emphasized. Failure to ensure the oven-dry state of the wood or paper was shown to lead to "competing reactions," which resulted in failure of the isocyanate to form chemical bonds with wood. In the typical wood panel hot-pressing operation, water is present in stoichiometrically vast quantities, often sufficient to convert all the isocyanate to polyurea many times over. Furthermore, this water typically is moving through the wood mat as steam and has much greater opportunity to react with isocyanate than do hydroxyl groups of wood. In a cleverly designed experiment, Whitman [16] was able to show the presence of large amounts of carbon dioxide during the hot pressing of a wood-isocyanate mat, carbon dioxide being a by-product of the isocyanate-to-polyurea chemical reaction.

Finally, in unpublished work, Johns, Plagemann, and Noskowiak [17] were able to show that commonly used MDI has a remarkable

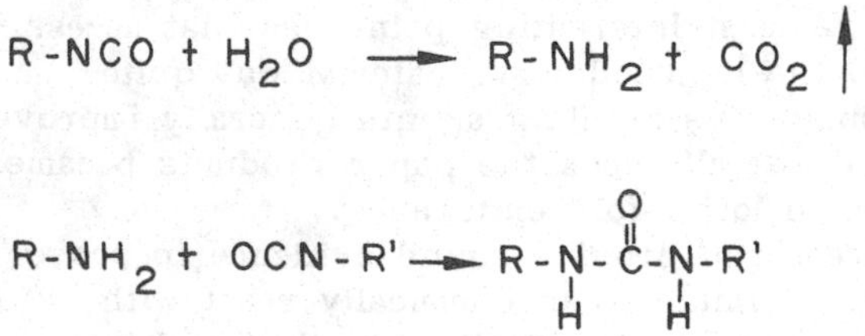

$$R\text{-}NCO + H_2O \longrightarrow R\text{-}NH_2 + CO_2 \uparrow$$

$$R\text{-}NH_2 + OCN\text{-}R' \longrightarrow R\text{-}\underset{H}{N}\text{-}\overset{\overset{\textstyle O}{\|}}{C}\text{-}\underset{H}{N}\text{-}R'$$

Fig. 3.2 Generalized reaction scheme showing how isocyanates react with water to produce polyureas.

ability to spread over the surface of wood. We noted that as little as 2 or 3% of an atomized resin is enough to completely coat the surface of wafers or strands with a continuous film of MDI; discrete droplets of isocyanate resin were not visible after hot pressing. In a comparable spray system, phenol-formaldehyde resin yielded discretely visible droplets and not continuous coverage at levels up to 6%. This mobility of MDI is due in large part to (a) the homogeneity of MDI, which is not an aqueous solution, hence its mobility will not be reduced when it loses water to wood; (b) the relatively low surface tension of MDI, approximately 50 dynes/cm, while phenolic resins are nearer to 75 dynes/cm; and (c) a low viscosity, which enhances spreading. These factors taken together, coupled with the fact that polyurea, the reaction by-product of diisocyanate-water reaction, is itself an excellent adhesive, are more than adequate to account for the excellence of MDI in bonding wood without calling on the development of chemical bonding.

B. Chemical Bonding via Heat Treatment

In work designed to generate chemical bonding of wood and paper, two general approaches have been used: either the application of heat alone or heat and chemical treatments together. First we review the history of heat treatments alone.

Heat alone is an inviting process to consider for the improvement of a material. It is direct, clean, and uncomplicated when compared with chemical processes. Because the speed of a process is so important, the application of heat is normally reserved for thin materials such as paper and thin fiberboards. There is no doubt that heat has a tremendous effect on the properties of lignocellulosic material. Stamm [18] and Hillis [19] have reviewed this subject. This discussion will be directed specifically at the use of heat to induce chemical bonding.

By far the most prolific practitioner of heat treatment of lignocellulosics is Back, who has directed his work at developing hydrolytically stable paper and fiberboard. In a series of works [20-23], Back has detailed the relationship between heat treatment of paper and fiberboards and the physical and mechanical properties of the treated material. In an attempt to understand the mechanism behind the altered properties, Back proposed a reaction scheme (Fig. 3.3) that suggests that the initial oxidation of lignocellulose to yield carbonyl groups ultimately leads to a hemiacetal linkage *between* fibers or surfaces, modifying the fundamental nature of the wood material. It should be noted that Back has never published any chemical evidence that confirms or even supports the reaction scheme offered in his several papers on the subject.

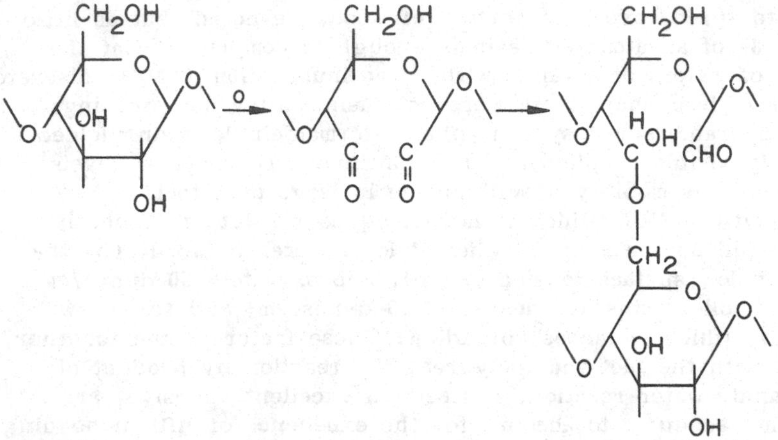

Fig. 3.3 Hypothetical reaction scheme of cellulose with oxygen
to yield hemiacetal linkage between paper fibers. (After Back [21].)

There is no question that heat treatment has a profound effect
on wood, paper, and fiberboard; however, the mechanism leading
to this modification has not been clarified. Stamm and coworkers
[24,25] noted the effect of heat on paper and concluded that the
bonds formed between surfaces were not covalent, but rather were
secondary in origin. This conclusion was based on swelling tests
conducted on heat-treated paper. A covalently bonded system should
be resistant to common solvents: the greater the crosslink density,
the more restricted the swelling. Stamm and coworkers reported
that heat-treated paper was easily swollen and readily dispersed
in several organic liquids. In recent work [26], the author noted
that heat-treated cotton linter paper that showed the expected qualities of heat-treated paper (i.e., increased wet strength, increased
stiffness, etc.) was completely soluble in both cupriammonium salts
and benzyltriethylamine solutions. These data taken together with
the conclusions of Stamm suggest that whatever the actual mechanism
for developing the modified properties of heat-treated paper is,
it is not covalent bonding.

C. Chemical Bonding via Heat plus Chemical Treatment

Far and away the most common form of chemical bonding is the
simultaneous application of some form of reactive chemistry plus
heat to create the bonding system. The reactive chemistry can
be either acidic, alkaline, or oxidative. An excellent, comprehensive

review of acidic and oxidative aspects of chemical bonding is available [27].

As noted earlier in connection with the work of Tischer [3], the whole concept of chemical bonding of wood is not particularly new, the initial patents having been issued in the late 1930s. For a period of time (late 1939-1961), work in chemical bonding was reported solely in the patent literature. Linzell [28] reported that the application of a solution of ferric salts, preferably ferric sulfate, at levels as low as 1% of the weight of the wood fiber, followed by hot pressing, produced a panel superior to panels pressed without the addition of the ferric salt. Five years later, Willey and Ruthman [29] also patented the use of ferric sulfate in the manufacture of fiber-based panels. In this case, the use of sulfuric acid, nitric acid, zinc chloride, and ammonium sulfate also reportedly yielded good results. Glab [30], in a remarkably detailed patent, outlined a method for producing a fiber or chip panel by modifying the lignocellulosic material in an autoclave, or sealed press. Reaction conditions included temperatures up to 460°F and steam pressures up to 400 psig. To actually modify the wood surface, a wide variety of materials and compounds must be mixed with the wood material. The suggested list of suitable materials is comprehensive in scope, including more than 130 compounds: mineral acids, iodine, zinc, phenol, chlorine, sulfur, numerous organic acids, styrene monomer, and dicyclopentadiene, to name a few of the preferred additives. After autoclaving, molding or pressing yields a usable product.

In discussing the use of chemical bonding techniques with various people [31], the author learned that the use of ferric salts in the manufacture of wet process fiberboard was (and may still be) a fairly common practice. Ferric sulfate was added only when, for various reasons, board properties were not up to specifications. It is commonly understood, however, that while ferric salts improve the short-term properties of panels, the long-term durability was suspect. In other work, the author [32] noted that while it is possible to make particleboard and fiberboard by simply applying strong mineral acids such as sulfuric acid or hydrochloric acid, several months of storage under common laboratory conditions will reduce the strength of the panel to levels commonly observed with charcoal. The panels also darken considerably. This seems to be related to the nonvolatile nature of by-products of the reaction.

During the early and mid-1950s, the use of hydrogen peroxide was also mentioned [31] in conjunction with the manufacture of dry process fiberboard. Here the hydrogen peroxide and its reaction by-products are volatile or innocuous, but other problems were observed. The boiling point of pure hydrogen peroxide is 150°C, while water boils at 100°C. Small spills, such as might happen in any large-scale processing situation, would tend to self-concentrate. At approximately 72% concentration, hydrogen peroxide ignites when in contact with organic matter. Thus, small fires were not uncommon when hydrogen peroxide was used without careful attention to details.

Historically the period from 1963 to 1977 was very active, if somewhat confusing, for researchers specifically interested in chemical bonding of wood veneers or chips. In 1963 Emerson [33] patented a process that became known as the Emerite process for producing a fiber or particleboard and involved the pretreatment of the wood or fiber with an acid, typically an inorganic acid such as sulfuric or nitric acid and a mixture of furfural, urea, and a lignin such as a lignosulfonate. This process was commercially developed and a full-scale production plant was built by the Kroehler Manufacturing Company [34] specifically for panel production using the Emerite process. The interesting point is that the science of this process was never subjected to review or comment in the scientific literature. One paper by Brooks [35], which included comments on the Emerite process, was primarily concerned with dielectric heating. Published in 1960, before key word computerized identification, this work was overlooked by chemical bonding researchers who followed.

In other work of this period, two centers were active in developing products based on chemical bonding. One was located at the University of California at Berkeley, and the other with the Polymer Institute at the University of Karlsruhe in West Germany.

The work at Berkeley was completed under the direction of three researchers, Zavarin, Brink, and the author, in overlapping projects aimed at developing chemical bonding. The work of Zavarin involved the use of hydrogen peroxide as an oxidative pretreatment and was completed with coworkers Stofko and Jenkins. The initial observations showed that lignin-rich material such as bark or pecky rot cedar, a cellulose-poor material, could be bonded with the application of heat and pressure and hydrogen peroxide to yield a water-resistant bond [36]. This idea was then developed by Stofko and Zavarin [37], who explored the bonding of a variety of wood-related materials with numerous oxidizing agents. The approach taken was to apply chemistry related to Fenton's reagent to induce bonding between surfaces. The process that evolved treated wood material with hydrogen peroxide in the presence of a transition metal ion ($ZrCl_4$ and $FeSO_4$) in a mildly acid environment. To duplicate natural wood, a presumption thought desirable, Stofko pressed 12.7 cm (5 in.) disks to densities approaching wood material (i.e., 1.1-1.3 g/cm^3), at temperatures ranging from 295 to 340°C and times from 5 to 20 minutes. His results showed that cellulose-to-cellulose bonding was least effective, cellulose-to-lignin was intermediate, and lignin-to-lignin was most effective.

The work of Brink and coworkers [38] was based on observations made during pulping research. Here the ". . . great difficulty incurred in rewetting and redispersion of once-dried nitric acid pulps to give suspensions of individual fibers. . . ." led to the hypothesis that ". . . functional groups or reactive moieties were formed in the lignocellulose which, on drying, formed covalent linkages with other groups in the lignocellulose." The work by Brink is remarkably similar to that of Glab in that elevated temperatures and enclosed, pressure-controlled environments are used for the treatment of the lignocellulosic material. Brink [39] reports on the specific use of nitric oxides and oxygen in gas phase reactions, as compared with Glab, who suggests the use of various liquids and salts. Brink goes on to mention the application of a wide variety of "crosslinking agents," in this case crosslinking referring to the crosslinking of one lignocellulosic surface with another, not the crosslinking that is more commonly associated with, for instance, thermoset polymerization, although several of Brink's crosslinking agents are capable of thermoset reactions.

The theory presented by Brink [38] is that by completing the activation reactions in a gas phase system, it is possible to control specific types of reaction more effectively than by simply dispersing reactants over the wood surfaces. By varying reaction conditions, the types of reaction that might happen include ". . . electrophilic substitution, electrophilic displacement of ring substituents, cleavage of alkyl-aryl ether linkages, and oxidation. . . ." The types of crosslinking agent used by Brink include hexamethylenediamine, phenylenediamine, ethylene glycol, maleic anhydride, furfuryl alcohol, formaldehyde, and resins prepared from furfuryl alcohol and formaldehyde.

In addition to gas phase reactions, Brink reports [39] on the use of acidic surface activators such as HCl, HNO_3, H_3PO_4, H_2SO_4, $FeCl_3$, and $Fe(NO_3)_3$. Often these activators are evaluated in conjunction with benzoyl or succinic peroxide.

In general, successfully completed experiments are said, by Brink, to yield boards with properties generally comparable or superior to phenol-formaldehyde bonded boards, with resin applied at the 6% level.

The final research effort completed at Berkeley was under the author's direction and was aimed at trying to develop a commercially viable flakeboard using both chemical bonding and nonpetrochemicals. The effort was initiated with the bonding of simple, flat sawn boards of white fir and pine with peroxyacids. Peroxyacids,* formed as

---

Please note: Peroxyacids are both carcinogenic and sensitive to detonation by shock. Extreme caution should be used by anyone attempting to use peroxyacids in experimentation.

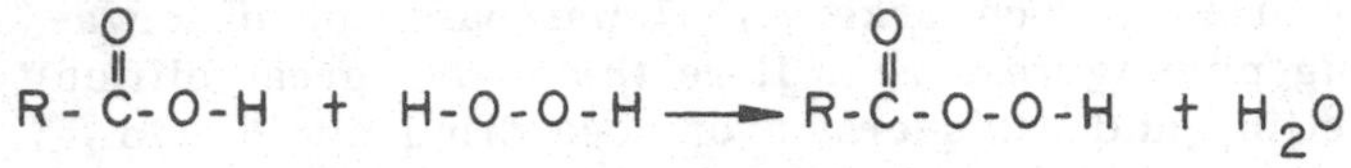

$$\underset{\parallel}{R-\overset{O}{C}-O-H} + H-O-O-H \longrightarrow \underset{\parallel}{R-\overset{O}{C}-O-O-H} + H_2O$$

Fig. 3.4 Generalized reaction scheme of organic acids with hydrogen peroxide to yield peroxyacids.

shown in Fig. 3.4, are very strong oxidizers, which offered a chance to explore the simple bonding of two pieces of wood [40]. The results were mixed in that while bonding did indeed take place, the variability was great enough to suggest that the simple oxidative bonding of two surfaces was far more complex than originally perceived. For instance, from a single parallel laminated panel, specimen joint strengths exceeding 500 psi were found next to complete delaminates. It was also seen in the course of the study that while parallel lamination of solid wood boards would lead to some level of bonding, cross-lamination of matched wood boards invariably lead to no bonding. These observations seemed to suggest that a key factor in understanding wood bonding was the proximity of bonding surfaces. In parallel constructions, it was possible for a band of early wood to be mated with a band of late wood and therefore to nest in an advantageous way. In cross-laminated constructions, late wood bands would come into close contact with only late wood bands, and meaningful level of the chemically intimate contact necessary for bonding would not be achieved.

The importance for proximity of reactants, coupled with the need to work toward a commercially viable composite panel in the 0.7 to 0.8 g/cm^3 range, suggested the need for a material that would fill or bridge gaps and hopefully also react with the surface of wood. After several tentative attempts, an aqueous mixture of ammonium lignosulfonate, furfuryl alcohol, and maleic anhydride was developed. This mixture was shown to work with both nitric acid [41,42] and hydrogen peroxide [43-45] activated wood surfaces.

This work, which involved the use of lignosulfonates, parallels the work of Nimz and coworkers [46,47]. While Nimz never claimed to have developed a chemical bonding system as such, his work is often included in discussions of the technique in that he utilized hydrogen peroxide as a part of his bonding system. In fact, Nimz's goal was to promote the polymerization of lignosulfonates to produce a thermoset, hydrolytically stable resin in situ. His work showed that the crosslink density of polymerized lignosulfonate was proportional to the amount of hydrogen peroxide in the system. To control the system in a full-scale production environment, Nimz applied the lignosulfonate and the hydrogen peroxide separately. In work we completed at the Berkeley laboratory, we saw that hydrogen peroxide, lignosulfonate, furfuryl alcohol, and maleic acid, when mixed in a beaker and heated in an oven, did not have the appearance of high molecular weight solids. Under somewhat different conditions, Nimz showed that lignosulfonates could be made to condense with peroxide. Thus the question is whether the author's earlier cited work created a lignin-based resin in situ or somehow developed a true chemically bonded system, or, alternately, whether Nimz's work, in addition to developing a resin in situ, didn't also involve a chemical bonding process step.

The most recent research effort, completed at the University of Wisconsin, attempts to chemically bond wood utilizing a novel approach. Heretofore, all attempts at chemical bonding have been completed under an acid environment. The work of Young and co-workers [48] utilizes alkaline chemistry to produce what is thought to be chemically bonded flakeboard. This work is based on chemical approaches that parallel alkaline pulping chemistry. Instead of an oxidative or acid pretreatment, wood flakes are treated first with 3N NaOH solution followed by a Kraft lignin-formaldehyde mixture, which is processed to yield a methylolated lignin in a manner similar to the manufacture of phenol-formaldehyde resin. Young reports that in initial work, the caustic treatment produced panels of acceptable dry strength but virtually no wet strength. It was only with the application of the lignin-formaldehyde mixture that adequate wet strength and excellent swelling properties were produced in a flakeboard matrix. In his published work, Young speaks of "activated" wood surfaces. No specific chemical mechanism is identified for this activation, nor is the concept of direct chemical bonding explicitly developed. It is suggested by Young that both chemical and physical processes are involved. The chemical processes include the hydrolysis of various wood components, such as cellulose and hemicellulose, and the fragmentation of lignin. The physical processes rely on the well-known capability of caustic to plasticize and swell wood, thereby enhancing the probability that all the necessary actions for adhesive mobility and proximity can be readily achieved.

III. DISCUSSION

The chemical bonding of wood is an intriguing, almost tantalizing concept, which offers seemingly limitless opportunities for creating the ultimate bonds between wood surfaces. The primary commercial division in bonded wood products is *exterior* and *interior*, with exterior products demanding much higher prices. Any thought of truly achieving a waterproof wood joint is invariably greeted with enthusiasm. Yet, as with any such potentially valuable process, scientific rigor must be maintained, perhaps even sharpened to the same degree as the enthusiasm for the possible.

This chapter started with a brief review of the theories of adhesion currently felt to be important in the development of bonded joints, without regard to the nature of the substrates. Human nature being what it is, some researchers try to explain all possible joints with a single theory, while others tend to focus more precisely on one specific type of joint and one specific theory. With this in mind, a rigorous, careful evaluation of the field of wood chemical bonding should be valuable.

A comprehensive review of the literature shows one startling fact. No researcher has yet offered unambiguous evidence for inter-surface wood bonding based on covalent (chemical) bonds with any type of chemical bonding technique; nor has the existence of such bonding been seriously questioned. This is extraordinary in view of the amount of material published in refereed journals (including numerous papers by the author) identifying chemical bonding as the major form of bonding between wood surfaces. The remainder of this chapter discusses the possible reasons for this, mentions the implications of chemical bonding as a method for bonding wood structures, and suggests possible directions for future research.

A possible source for the perceived importance and desirability of establishing a bridge of covalent bonds might very well be the nature of the wood itself. Wood swells with water contact, and at the same time, becomes weaker. Wood that has high moisture content rots and is difficult to glue and paint. A great deal of time and expense is directed to removing water as quickly and efficiently as possible. "Stabilized" wood is a premium product. Wood products such as paper are often easily dispersed in water. The shrinking and swelling of wood and the impermanence of paper bonds are attributed to "hydrogen bonds" in wood material. Hydrogen bonds are not covalent, but rather secondary bonds; thus "secondary" bonds are perceived as less than desirable, particularly when the need for permanence and strength is considered. It is not surprising that secondary bonds are held in disrepute by most traditionally trained wood scientists. With this background, it is easy to understand the acceptance of statements by Stofko [36]: "In samples found resistant to boiling water, covalent primary bonds can be assumed to have been formed because secondary bonds are not resistant to boiling water. . . ." or Brink [38]: "The unique property of nitric acid pulps compared to alkaline pulps . . . involved the great difficulty incurred in rewetting and redispersion of once-dried nitric acid pulps to give suspensions of individual fibers. . . . [Thus] it was hypothesized that functional groups or reactive moieties were formed in the lignocellulose which, on drying, formed covalent linkages with other groups in the lignocellulose."

Arguments that water resistance can be achieved only if the system is covalently bonded are entirely without merit. Table 3.1 lists the common polymers that form solids that are not soluble in boiling water and are linear (i.e., not crosslinked). The cohesive strength of these polymers is based entirely on secondary forces and chain entanglements. Since this group of polymers shows little if any swelling in water, chain entanglements really are not a consideration in the hydrolytic stability of these systems. Furthermore, several of the polymers in Table 3.1, such as nylons, delrins, and polyimides, are high strength compounds and serve as structural materials. The whole class of polyimides constitutes an interesting family of *linear* polymers that function as high strength adhesives in high temperature exposures and can bond metal substrates yielding a retention of 99% of room temperature strength after 1000 hours at 275°C [49,50]. It should also be noted that crystalline cellulose is included in Table 3.1 because cellulose is a linear polymer that associates into crystallites, which resist boiling water.

This observation raises the intriguing question about the nature of the secondary bonding that is capable of providing for the excellent hydrolytic properties of polyethylene, polytetrafluoroethylene, or delrin. The answer is that these materials are held together by secondary forces including the family of van der Waals forces (dispersive, dipole/induced dipole, and dipole/dipole) and hydrogen bonding (when possible), the general name for this type of attractive force being "coordinate bonding." Thus, secondary bonds are fully capable of providing water-durable materials that, depending on the nature of the bonding and the materials used in the backbone of the polymer, can also demonstrate very high strength properties.

Table 3.1 Common Linear Polymers That Are Resistant to Boiling Water

Polycarbonates	Polyvinyl chloride
Polystyrene	Polyethylene
Polytetrafluoroethylene	Polypropylene
Polyimides	Polyamides (nylon)
Cellulose (crystalline)	Polyesters
Polyoxymethylene	Polyvinylfluorides
Polymethyl methacrylates	Cyanoacrylates

The relative ability of a solvent to associate with the polymer materials and ultimately disrupt the bonding depends on many factors.

A possible key to the understanding of such behavior is the donor/acceptor theory as outlined by Gutmann [51]. It is not the position of this chapter to review in depth this particular approach to understanding bonding in materials. However, it must be noted that donor/acceptor interactions, also identified as Lewis acid-base neutralization or coordinate bond formation, have been identified as the fundamental mechanism underlying the bonding of inorganic materials to various polymers [52,53] and to bonding within paper [54]. A general reference to the subject of donor/acceptor bonds has been the subject of excellent reviews by Jensen [55,56].

Having shown that covalent bonds are not necessary for the development of water-resistant polymeric materials, we arrive at the most appropriate question of what holds chemically bonded wood together. The presence of covalent or chemical bonding between wood fibers, particles, or surfaces in general has not been proved, only *assumed*; thus, an important discussion relating to the implications of possible noncovalent mechanisms has never been addressed. Other explanations calling on alternate models should be offered, models that might explain, via simpler mechanisms, an alternate description of "chemical bonding."

Chemical bonds between adjacent surfaces would be extraordinarily difficult to see with any conventional chemical test. The problem is separating intrafiber bonds and interfiber bonds. Wood is a reactive material. It must be assumed that when strong acids or oxidizing materials are spread on the surface and the mixture is placed in a hot, moist environment, chemical reactions will take place. The problem is identifying, amid the complexity of wood material, the chemically altered material that is involved in new chemical bonds, and then sifting out which fraction of the new bonds form bridges between surfaces and which merely have reacted within the wood particle. A final consideration, assuming that the presence of interparticle chemical bonds is confirmed, is to assess the individual contribution of the chemical bonds to the total joint performance when grouped with all other bonds that have formed.

In a noteworthy series of papers, Theander and Popoff [57-61] detail mechanisms for the formation of furans, phenols, and other noncarbohydrates in acidic or oxidizing environments. These materials often form the primary component in traditional thermoset resins. Furans can self-polymerize or react with phenolics yielding rigid, dark, water-resistant polymers in slightly acidic environments [62]. It is possible that bonding with acidic or oxidizing materials is in part due to the formation of in situ resins similar to conventional thermosets.

Characteristic of all forms of oxidative or acidic chemical bonding schemes is the development of carbonyl groups on the wood. These groups can be easily monitored via infrared or chemical analysis. Zavarin reports on numerous methods for forming and then following the history of these groups. Jenkins, as reported in Zavarin [27], and Theander and coworkers have all shown that the effect of heat or heat plus chemicals induces carbonyl or carboxyl functionality and thus fundamentally alters the physical aspects of the surface. It is not so far-fetched an idea to suggest that chemical bonding may simply be a method for inducing chemical modification to the surface of wood, permitting the formation of stronger coordinate bonds as described by Jensen [55,56].

If chemical bonding does actually take place, it will be most difficult to prove via chemical tests. As noted earlier, the task of unambiguously separating inter- and intrafiber chemical bonds would be extremely challenging. A better approach might be to use nonchemical or physical methods. The approach outlined by Larsson and Johns [54], if applied to a variety of wood-bonded systems, might provide some new information about the nature of the mechanism by which wood is glued. Other methods, perhaps borrowed from biochemistry, biophysics, or polymer chemistry, may also be applied to this problem.

Young and coworkers [48] have shown that alkali pretreatment can lead to water-resistant wood joints when the wood surfaces are pretreated with 3N NaOH followed by the application of a gap-filling mixture of formaldehyde and Kraft lignin. It is reported that bonding is optimized when the mixture of formaldehyde and Kraft lignin is precooked under conditions similar to the preparation of a phenol-formaldehyde resin. The pretreatment of the wood surface assists the bonding, but the mechanism is not known. Young suggests that the pretreatment predisposes the wood surface to chemically react with the methylolated Kraft lignin. Alternately, the pretreated wood surface might serve the same function as flour addition to standard phenol-formaldehyde plywood resins; that is, the swollen or chemically hydrated carbohydrate retains moisture in the interphase and serves as a reservoir for the alkali catalyst for the completion of the condensation of the phenolic resin. What actually is happening is simply not yet known.

It is not necessary to chemically attach to wood to make good wood joints. The fascinating work of Goto, Saika, and Onishi [63,64] has shown that it is possible to make wood joints with properties similar to melamine-formaldehyde resin with a thermoplastic, polypropylene. In this case mechanical entanglement of synthetic with natural polymer is thought to provide the majority of the bond strength. Here neither chemical bonding nor specific adhesion is thought to be required to produce a panel with adequate mechanical strength and hydrolytic stability.

The whole point of chemical bonding is to take advantage of the perceived or anticipated qualities of having wood joined by a series of covalent bonds acting as the stress transmission device. Little consideration has been given to the thought that this may not be the most effective way to bond wood. Given any linear polymer, adding crosslinks leads to an overall reduction in volume, increased brittleness, and loss of toughness and resiliency. Wood is an excellent balance of thermoplastic lignin, linear amorphous and crystalline cellulose, and the amorphous hemicelluloses. When normalized for density, wood has a modulus of elasticity similar to that of steel. By applying the chemicals thought necessary to create active sites on the wood surface, invariably the wood materials away from the surface (i.e., intrafiber materials) are affected. This chemical treatment severs existing wood bonds and creates shorter cellulose chains. The very close proximity of reaction sites also implies the possibility/probability that intrafiber crosslinking reactions of the lignin may take place. Acidic and/or oxidizing treatments very likely condense the aromatic fraction so that the wood loses the natural mechanical qualities that makes it desirable in the first place. It may very well be that the price paid to achieve chemical bonding of wood is far too high in terms of loss of the desirable wood qualities, properties that make wood the material of choice.

IV. SUMMARY

This chapter offers a critical appraisal of the current understanding of the chemical bonding of wood. What the literature has shown is a wide range of approaches ranging from acidic to alkaline, oxidative, and free radical initiated processes. What the literature has not shown is indisputable proof of the numerous suggestions about the mechanism. With this review I hope to reopen the question of the mechanism by which wood or paper is bonded when subjected to heat or chemical treatment. Only by understanding what is happening to lignocellulosic material under the influence of heat and chemical treatment will we as researchers be in a better position to economically and efficiently exploit the concept to practical ends.

REFERENCES

1. Wake, William C. (1982). *Adhesion and the Formulation of Adhesives*, 2nd ed. New York, Applied Science Publishers.

2. Kinlock, A. J. (1987). *Adhesion and Adhesives Science and Technology.* New York, Chapman and Hall.

3. Tischer, Frank V. (1939). "Method for Treating Veneers," U.S. Patent 2,177,160, patented Oct. 24, 1939.

4. Troughton, G. E. (1967). "Kinetic Evidence for Covalent Bonding Between Wood and Formaldehyde Glues." Forest Products Laboratory, Vancouver, British Columbia, Information Report VP-X-26.

5. Troughton, G. E., and Chow, S.-Z. (1968). "Evidence for Covalent Bonding Between Melamine Formaldehyde Glue and Wood." Part I. "Bond Degradation." *J. Inst. Wood Sci.,* 21 (September), 29-34.

6. Johns, W. E., and Niazi, K. A. (1980). "Effect of pH and Buffering Capacity of Wood on the Gelation Time of Urea-Formaldehyde Resin." *Wood and Fiber,* 12(3), 255-263.

7. Johns, W. E., Myers, G. C., and Motter, W. K. (1988). "Bondability of Wood Surfaces." *Proceedings of the 1987 Particleboard Symposium,* Pullman, WA.

8. Pizzi, A., and Eaton, N.J. (1987). "A Conformational Analysis Approach to Phenol-Formaldehyde Resins Adhesion to Wood Cellulose." *J. Adhes. Sci. Technol.,* 1(3), 191-200.

9. Rowell, R. M., and Ellis, W. D. (1979). "Chemical Modification of Wood: Reaction of Methyl Isocyanate with Southern Pine." *Wood Sci.,* 12(1), 52-58.

10. Rowell, R. M., Feist, W. C., and Ellis, W. D. (1981). "Weathering of Chemically Modified Southern Pine." *Wood Sci.,* 13(4), 202-208.

11. Rowell, R. M., and Ellis, W. D. (1981). "Bonding of Isocyanates to Wood." In *Urethane Chemistry and Applications,* Kenneth N. Edwards, Ed. ACS Symposium Series No. 172. Washington, DC, American Chemical Society, pp. 263-284.

12. Morak, A. J., and Ward, K., Jr. (1970). "Cross-linking of Linerboard to Reduce Stiffness Loss." Part II. "Diisocyanates in Liquid Phase Application." *TAPPI,* 53(4), 652-656.

13. Morak, A. J., and Ward, K., Jr. (1970). "Cross-linking of Linerboard to Reduce Stiffness Loss." Part III. "Diisocyanates in Vapor-Phase Application." *TAPPI,* 53(6), 1055-1058.

14. Morak, A. J., Ward, K., Jr., and Johnson, D. C. (1970). "Cross-linking of Linerboard to Reduce Stiffness Loss." Part IV. "Application of Blocked Diisocyanates." *TAPPI,* 53(12), 2278-2283.

15. McLaughlin, A. (1981). Personal communication.

16. Whitman, O. (1976). "Wood Bonding with Isocyanate." *Holz Rohst. Werkst.,* 34, 427-431.

17. Johns, W. E., Plagemann, W., and Naskowiak, A. (1981). Unpublished data.

18. Stamm, A. J. (1964). *Wood and Cellulose Science.* New York, Ronald Press.
19. Hillis, W. E. (1984). "High Temperature and Chemical Effects on Wood Stability." Part 1. "General Considerations." *Wood Sci. Technol.*, 18, 281-293.
20. Back, E. L. (1964). "Drying Stresses in Hardboard and the Introduction of Cross-linking Stresses by Heat Treatment." *For. Prod. J.*, September, 425-429.
21. Back, E. L. (1967). "Thermal Auto-crosslinking in Cellulose Materials." *Pulp Pap. Mag. Can.*, 68, T-165.
22. Back, E. L., Thoung Htun, M., Jackson, M., and Johanson, F. (1967). "Ultrasonic Measurements of the Thermal Softening of Paper Products and the Influence of Thermal Auto-crosslinking Reactions." *TAPPI*, 50(11), 542-547.
23. Back, E. L., and Stenberg, L. E. (1976). "Web Stiffness by Heat Treatment of Running Web." Part 1. "Properties of Treated Liner and Corrugating Medium." *Pulp Pap. Can.*, 77(12), t264-70.
24. Seborg, R. M., Tarkow, H., and Stamm, A. J. (1953). "Effect of Heat upon the Dimensional Stabilization of Wood." *J. FPRS* (September), 59-67.
25. Stamm, A. J. (1959). "Dimensional Stabilization of Wood by Thermal Reactions and Formaldehyde Cross-linking." *TAPPI*, 42(1), 39-44.
26. Johns, W. E. (1986). Unpublished data.
27. Zavarin, E. (1984). "Activation of Wood Surfaces and Nonconventional Bonding." In *The Chemistry of Solid Wood*, Roger Rowell, Ed. Advances in Chemistry Series No. 207. Washington, DC, American Chemical Society, Chapter 10.
28. Linzell, H. K. (1945). "Process of Making Compressed Fiber Products." U.S. Patent 2,388,487, patented Nov. 6, 1945.
29. Willey, G. S., and Ruthman, K. S. (1950). "Laminated Product and Process of Making Same." U.S. Patent 2,495,043, patented Jan. 17, 1950.
30. Glab, W. T. (1962). "Method of Making a Lignocellulose Product and Products Resulting Therefrom." U.S. Patent 3,033,695, patented May 8, 1962.
31. Confidential communications with several mill managers.
32. Johns, W. E. Unreported laboratory work.
33. Emerson, Ralph Waldo (1963). "Lignocellulose Molding Compositions." U.S. Patent 3,097,177, patented July 9, 1963.
34. Anonymous. (1980). "Cultured Wood." *Lumberman*, May, 76-77.
35. Brooks, S. H. (1960). "Dielectric Heating and the Emerite Process." *For. Prod. J.*, 10(9), 438-440.
36. Stofko, J. (1974). "The Autohesion of Wood." Doctoral thesis submitted to the faculty of the University of California, Berkeley.

37. Stofko, J., and Zavarin, E. (1977). "Method of Bonding Solid Lignocellulosic Material, and Resulting Product. U.S. Patent 4,007,312, patented Feb. 8, 1977.
38. Brink, D. L., Collett, B. M., Pohlman, A. A., Wong, A. F., and Philippou, J. (1977). In *Wood Technology: Chemical Aspects*, Irving S. Goldstein, Ed. ACS Symposium Series No. 43. Washington, DC, American Chemical Society, Chapter 110.
39. Brink, D. L. (1975). "Lignocellulosic Molding Method and Product." U.S. Patent 3,900,334, patented Aug. 19, 1975.
40. Johns, W. E., and Nguyen, T. (1977). "Peroxyacetic Acid Bonding of Wood." *For. Prod. J.*, 27(1), 17-23.
41. Johns, W. E., Layton, H. D., Nguyen, T., and Woo, J. K. (1978). "The Nonconventional Bonding of White Fir Flakeboard Using Nitric Acid." *Holzforschung*, 32(5), 162-166.
42. Johns, W. E., and Jahan-Latibari, A. (1983). "Wood Bonding by Surface Reaction." *J. Adhes.*, 15(2), 105-115.
43. Philippou, J. L., Johns, W. E., Zavarin, E., and Nguyen, T. (1982). "Bonding of Particleboard Using Hydrogen Peroxide, Lignosulfonates, and Furfuryl Alcohol: The Effect of Process Parameters." *For. Prod. J.*, 32(3), 27-32.
44. Philippou, J. L., Johns, W. E., and Nguyen, T. (1982). "Bonding Wood by Graft Polymerization. The Effect of Hydrogen Peroxide Concentration on the Bonding and Properties of Particleboard." *Holzforschung*, 36, 37-42.
45. Philippou, J. L., Zavarin, E., Johns, W. E., and Nguyen, T. (1982). "Bonding of Particleboard Using Hydrogen Peroxide, Lignosulfonates, and Furfuryl Alcohol: Effects of Chemical Composition of Bonding Materials. *For. Prod. J.*, 32(5), 55-61.
46. Nimz, H. H., Mogharab, I., and Gurang, I. (1976). "Oxidative Coupling of Spent Sulfite Liquor." *Applied Polymer Symposium No. 28.* New York, Wiley, pp. 1225-1230.
47. Nimz, H. H., and Hitze, G. (1980). "The Application of Spent Sulfite Liquor as an Adhesive for Particleboard." *Cellul. Chem. Technol.*, 14, 371-382.
48. Young, R. A., Fujita, M., and Rivers, B. H. (1985). "New Approach to Wood Bonding. A Base-Activated Lignin Adhesive System." *Wood Sci. Technol.*, 19, 363-381.
49. Fowler, J. R. (1982). "High Temperature Adhesives for Structural Applications—A Review." *Mater. Design*, 3 (December), 602-607.
50. Serafini, T. T. (1987). *High Temperature Polymer Matrix Composites.* Noyes Data Corp., New Jersey.
51. Gutmann, V. (1978). *The Donor-Acceptor Approach to Molecular Interactions.* New York, Plenum Press.
52. Fowkes, F. M. (1980). "Donor-Acceptor Interactions at Interfaces." *Adhesion and Adsorption of Polymers*, Part A, Lieng-huang Lee, Ed. New York, Plenum Press.

53. Fowkes, F. M., and Mostafa, M. A. (1978). "Acid-Base Interactions in Polymer Adsorption." *I&EC Prod. Res. Dev.*, 17
(March), 3.

54. Larsson, A., and Johns, W. E. (1988). "Acid-Base Interactions
Between Cellulose/Lignocellulose and Organic Molecules."
J. Adhes., vol. 25, page 121-131.

55. Jensen, W. B. (1982). "Acids, Bases and Adhesion: A Synergy."
Chemtech, December, 755-764.

56. Jensen, W. B. (1980). *The Lewis Acid-Base Concepts: An Overview.* New York, Wiley.

57. Popoff, T., and Theander, O. (1976). "Formation of Aromatic
Compounds from Carbohydrates." III. "Reactions of D-Glucose
and D-Fructose in Slightly Acidic, Aqueous Solution." *Acta
Chem. Scand. B*, 30, 397-402.

58. Popoff, T., and Theander, O. (1976). "Formation of Aromatic
Compounds from Carbohydrates." IV. "Chromones from Reaction
of Hexuronic Acids in Slightly Acidic, Aqueous Solution."
Acta Chem. Scand. B, 30(8), 705-710.

59. Theander, O. (1980). "Acids and Other Oxidation Products."
In *The Carbohydrates: Chemistry/Biochemistry*, 2nd ed.,
W. Pigman and D. Horton, Eds., Vol. IB. New York, Academic
Press, pp. 1013-1099.

60. Theander, O. (1980). "Formation of Furans, Phenols and Other
Noncarbohydrate Compounds by Thermal Treatment of Sugars
in Aqueous Solution." *Proceedings, OECD Cooperative Research
Project on Food Production and Preservation.* Workshop 2: Con-
version of Lignocellulosic Materials to Simple Carbohydrates,
Amersfoort, the Netherlands, October 8-10. The Netherlands,
IBVL Wageningen, pp. 377-387.

61. Theander, O. (1981). "Novel Developments in Carmelization."
Prog. Food Nutr. Sci., 5, 471-476.

62. Dunlop, A. P., and Peters, F. N. (1953). *The Furans*. American
Chemical Society Monograph Series No. 119. New York, Reinhold.

63. Goto, T., Saika, H., and Onishi, H. (1982). "Studies on Wood
Gluing." XIII. "Gluability and Scanning Electron Microscopic
Studies of Wood-Polypropylene Bonding." *Wood Sci. Technol.*,
16, 293-303.

64. Onishi, H., Goto, T., and Saiki, H. (1983). "Studies on Wood
Gluing. XIV. "Effect of Chemical Activity of Wood Surface on
Polypropylene Bonding." *Holzforschung*, 37(1), 27-33.

4

Phenol–Formaldehyde Structures in Relation to Their Adhesion to Wood Cellulose

A. PIZZI and R. SMIT* / Council for Scientific and Industrial Research, Pretoria, South Africa

I. INTRODUCTION

Phenol-formaldehyde (PF) resins are extensively used as adhesives
and binders for lignocellulosic materials. Lignocellulosic materials,
conversely, are often used as fillers for phenol-formaldehyde resins
in many of their applications. Adhesion between an adhesive and
an adherend can be described as a physicochemical phenomenon,
and the interaction between PF resins and cellulose is no exception.
In the case of the interaction of PF resins with cellulose, the causes
of adhesion are thought to reside mostly with the attraction between
resin and adherend due to secondary forces, such as van der Waals,
H-bonding, and electrostatic, between noncovalently bonded pairs
of atoms and atom groups. Although it is well known that formaldehyde and formaldehyde-based resins to a limited extent also react
covalently with cellulose to form ether linkages and to crosslink
cellulose fibers [1], the contribution of such covalent bonds to
the phenomenon of PF adhesion in the bonding of lignocellulosic
materials does not appear, to our knowledge, to have been assessed
qualitatively or quantitatively. It has been disregarded, consequently,
also in our calculations.

Understanding of the phenomenon of adhesion between a well-
defined adhesive/adherend pair has never been attempted before
by means of calculations of all the values of the secondary interactions between the noncovalently bonded atoms of the two molecules
involved. A conformational analysis approach is ideal for this type
of calculation. It has never been attempted before because the

*Present affiliation: Rand Afrikanse Universiteit, Johannesburg,
South Africa

spatial conformations of cellulose and especially of phenol-formaldehyde condensates were not exactly known, or were not known at all. Recently, the tridimensional structure of several phenol-formaldehyde oligomers [2] and of the crystalline portion of native cellulose [3-5], such as in wood, have been determined or refined by means of *ab initio* conformational analysis methods. Although a conformational analysis study of the adhesion of simple PF condensates to crystalline cellulose I involves very extensive calculations and must, for this reason, be subjected to both physical and mathematical limitations, it was nevertheless attempted in a recent study [6].

The calculations were limited to the interaction of all the three possible phenol-formaldehyde dimers, in which the methylene bridge links two phenol nuclei ortho-ortho, para-para, and ortho-para, with an elementary cellulose I crystallite composed of five cellulose chains of four glucose residues each. Increasing the size of the PF condensates used, or of the cellulose crystallite, increases the number of degrees of freedom to such an extent that calculations, already very extensive, become far too cumbersome.

Because cellulose constitutes as much as 50% of wood, where its percentage crystallinity is as high as 70%, this study also implies partial applicability to a wood substrate.

II. THE STRUCTURE OF SOME PHENOL-FORMALDEHYDE CONDENSATES FOR WOOD ADHESIVES

Phenol-formaldehyde (PF) resins are currently produced in large quantities for many uses—for instance, the manufacture of exterior-grade wood adhesives. Many aspects of the chemistry of phenolic resins have been examined and described in the last half-century [7]. However, the structure of such systems is still unknown, with the exception of the structure of 3,3'-dichloro-4,4'-dihydroxydiphenylmethane [8] (which is related to the PF dimers to be described below).

The molecular structure of simple PF condensates like dimers and trimers is of importance to the understanding of the structure of PF resins. These molecules are too big for a formal quantum mechanical treatment or even for semiempirical molecular orbital studies, and further approximate methods are needed to allow the computation of molecular geometries. The methods of molecular mechanics focus on the internal forces in the molecules, exemplified by force constants, and should present a reasonable model.

A recent study, however, uses another approach, namely to investigate the appropriateness of an even simpler model in which the basic bond distance and interbond angle geometry of the molecule were assumed and the atom groups were allowed to rotate freely

around the methylene and OH bonds under the influence of electrostatic forces upon the atoms, van der Waals forces, and hydrogen
bond forces. This model allows calculations to be executed with
simple potential functions using a computer program [3], and the
total molecular energy can be expressed as the sum of the van
der Waals energy, the hydrogen bonding energy, and the energy
of torsion around the bonds. The systematic variation of the torsional
angles around the methylene and OH bonds allowed the determination
of the minimum total energy, hence the most stable geometrical
conformation. The model thus allows the determination of the most
probable molecular conformation as well as two-dimensional plots
of the van der Waals and total energies as functions of the
$-CH_2-$torsional angles.

The calculations showed that this very simple model predicts
the correct geometrical conformations of molecules like acetone,
which test rather stringently any geometrical investigation. The
input for the electrostatic part of the calculation consists of the
partial electrostatic charges on the various atoms.

These charges on the various atoms were obtained from semiempirical molecular orbital calculations with the MNDO/3 (QCPE
Program 485) [9]. These charges give a fair indication of the
electron distribution in the molecules and are consistent with the
bond orders found experimentally for similar compounds.

The bond lengths and interbond angles used are displayed
in Table 4.1. It is emphasized that these data differ somewhat from
those of the compound 3,3''-dichloro-4,4''-dihydroxydiphenylmethane,
the structure of which was determined by Whittaker [8] and which
shows an anomalous and inexplicable variation of the aromatic C—C
bond distances, as well as a deviation of 12° in the C—C—C interbond
angle which was taken to be 112.5° in the present study, that is, just
like the corresponding angle in diphenylmethane.

Figure 4.1 displays the structural formulas of the dimers (I-III),
trimers (IV-VI), and hexamers (VII and VIII) of these PF condensates examined in this study, as well as the relevant torsional
angles around the OH— and CH_2- groups. It is emphasized that
some of the C—O—H bonds are shown linear for the sake of clarity.

The dimers (I-III) represent all the possible PF dimers,
since the OH group on the phenyl groups can only be in a position
that is ortho or para to the methylene carbon. The three trimers
(IV-VI) studied, however, were those previously isolated from
phenol-formaldehyde condensates [10].

The hexamers (VII and VIII) represent a possible precursor
to a PF polymer, and were, therefore, examined to obtain a probable
minimum energy conformation that will allow extrapolation to the
polymeric structure.

Table 4.1 Structural Parameters for the PF Condensate Studied [2,11]

Bond	Length (Å)
$C_{phenyl\ ring}-C_{phenyl\ ring}$	1.38
$C_{phenyl\ ring}-C_{methylene}$	1.50
$C_{phenyl\ ring}-O$	1.37
C—H	0.98
O—H	0.90

Bond angle	Value
$C_{phenyl\ ring}-C_{methylene}-C_{phenyl\ ring}$	112.5°
$C_{phenyl\ ring}-C_{phenyl\ ring}-C_{phenyl\ ring}$	120°
$C_{phenyl\ ring}-O-H$	110°
$C_{phenyl\ ring}-C_{phenyl\ ring}-H$	120°

Four, seven, and two bond rotational angles were defined for
the PF dimers, trimers, and hexamers, respectively, and are indicated
on the structural formulas in Fig. 4.1. The arbitrary initial conformation (before any optimization of the structure started by the rotation
around bonds) was chosen to be the same for all PF condensates
examined, namely where all the atoms of the phenolic rings of each
PF condensate are in the same plane (only the methylene bridge
angles are not in this plane). For the initial conformation, a value
of 0° was assigned to the bond rotational angles of each PF condensate. The convention of rotation is such that the arrows on
the structural formulas in Fig. 4.1 indicate *positive rotation* from
the initial conformation. All bond rotational angles must be rotated
simultaneously to optimize the structure. This was possible for
the PF dimers and hexamers, but long computer execution times
made it impractical for the PF trimers. The bond rotational angles
of the PF trimers, therefore, had to be rotated in groups, using
procedures described in detail elsewhere [2]. All bond rotational
angles were initially rotated from 0° to 360° with a 20° increment
to identify the minima. This process was repeated successfully with
10°, 5°, 2°, and finally 1° increment, to determine the final minimum.
    There are two and four sets of optimum values for the hydroxyl
group bond rotational angles of the dimers and trimers, respectively.
For each set of optimum values for the hydroxyl group bond rotational

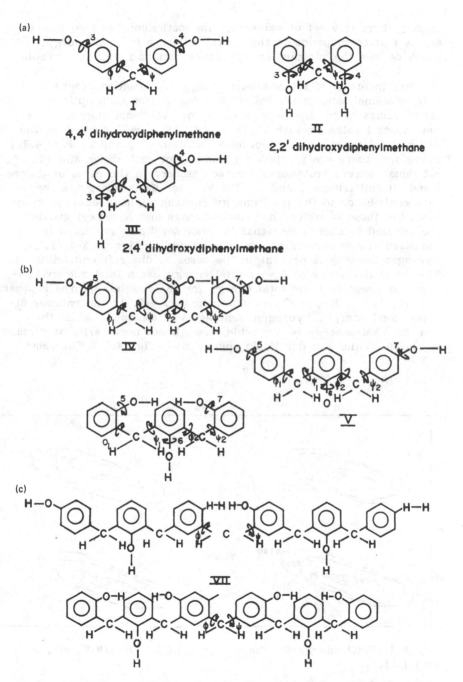

Figure 4.1 Structural formulas for the phenol-formaldehyde condensates studied: (a) dimers, (b) trimers, and (c) hexamers.

angles, there is a set of values for the methylene bond rotational
angles that corresponds to the overall minimum in total energy,
which defines the minimum energy conformations as given in Table
4.2

With inclusion of the minimum-energy conformation, there are
four rotational isomers of similar total energy for each optimum
set of values of the hydroxyl group bond rotational angles. [The
four dimer I rotamers with (3,4) = (89°, 89°) are indicated on the
plot of total energy as a function of rotation of ϕ and ψ, Fig. 4.2.]
Therefore, there are 8 rotational isomers for each dimer and 16
rotational isomers (rotamers) for each trimer. In the cases of dimers
I and III and trimers I and II, the van der Waals energy is the
main contributor to the total energy and the minimum-energy structures are those in which the phenolic rings and hydroxyl groups
are oriented in such a way that the van der Waals repulsion is
minimized (these structures are represented in Figs. 4.3-4.6).
Hydrogen bonding is present in the cases of dimer II and trimer III.
The dimer II rotamer of lowest total energy has a total energy considerably lower that the total energy for the remaining three rotamers
with (3, 4) = (-99°, -57°) as shown in Fig. 4.7. In the rotamer of
lowest total energy, hydrogen bonding is at maximum, while the
van der Waals energy is favorable, though not necessarily at minimum,
as shown on the van der Waals energy map (Fig. 4.8). The same

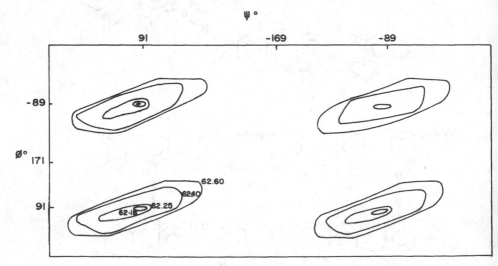

Fig. 4.2   Total energy for dimer I: (ϕ, ψ, 3, 4) = (91°, 91°, 89°,
89°) [2,11].

Table 4.2 Minimum Energy Conformation for the PF Dimers and Trimers

Dimer	Angle conformation: $(\phi, \psi, 3, 4)$ (degrees)	Total energy (kcal/mol)	Hydrogen-bonding energy (kcal/mol)
4,4'-dihydroxydiphenylmethane	I: (91, 91, 89, 89)	62.166	—
	(91, −89, 89, −89)	62.166	
2,2'-dihydroxydiphenylmethane	II: (111, 115, 80, −88)	62.159	−4.52
	(−111, 65, 81, −89)	62.157	
2,4'-dihydroxydiphenylmethane	III: (77, −107, −99, −57)	58.845	—
	(107, −77, 59, 99)	58.845	

Trimer	Angle conformation: $(\phi_1, \psi_1, \phi_2, \psi_2, 5, 6, 7)$ (degrees)	Total energy (kcal/mol)	Hydrogen-bonding energy (kcal/mol)
IV	(−84, −89, 77, 69, −91, −81, 91)	112.199	—
	(96, −89, 77, 69, 87, −81, 91)	112.201	—
	(−84, −89, 77, −111, −91, −81, −87)	112.211	—
	(96, −89, 77, −111, 87, −81, −87)	112.214	—
V	(61−, 117, −117, 61, 89, 90, 89)	111.483	—
	(−61, 117, −117, 61, −89, 90, −89)	111.485	—
	(119, 117, −117, 61, −89, 90, 89)	111.485	—
	(−61, 117, −117, −119, 89, 90, −89)	111.485	—
VI	(−73, −79, −111, −102, 57, 99, −89)	102.260	−9.44

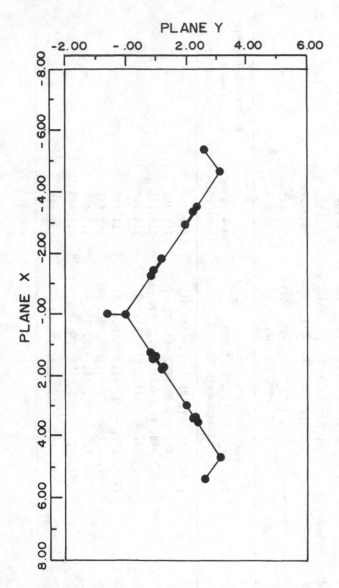

Fig. 4.3 Minimum-energy conformation for dimer I. Scale in Å
[2,11].

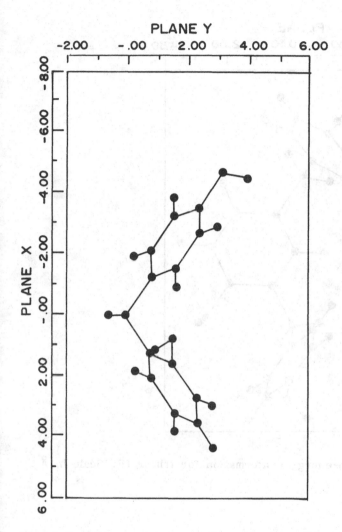

Fig. 4.4 Minimum-energy conformation for dimer III. Scale in Å [2,11].

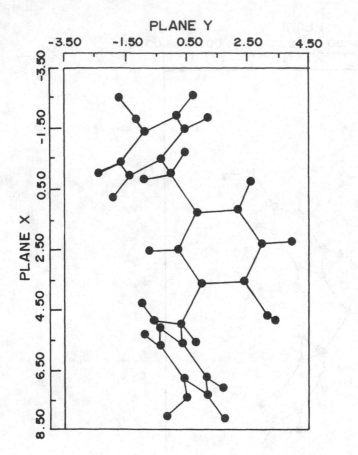

Fig. 4.5   Minimum-energy conformation for trimer IV. Scale in Å [2,11].

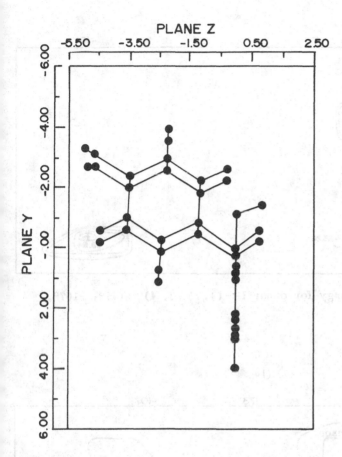

Fig. 4.6 Minimum-energy conformation for trimer V. Scale in Å [2,11].

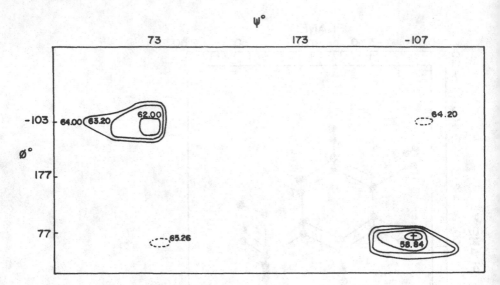

Fig. 4.7   Total energy for dimer II: (ϕ, ψ, 3, 4) = (77°, -107°, -99°, -57°) [2,11].

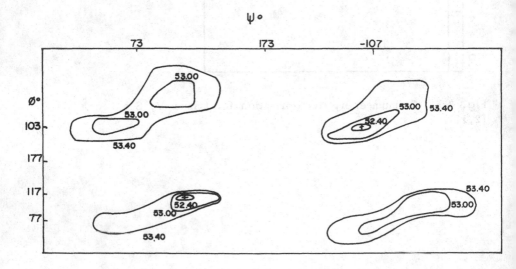

Fig. 4.8   Van der Waals energy for dimer II: (ϕ, ψ, 3, 4) = (111°, 115°, 80°, -88°) [2,11].

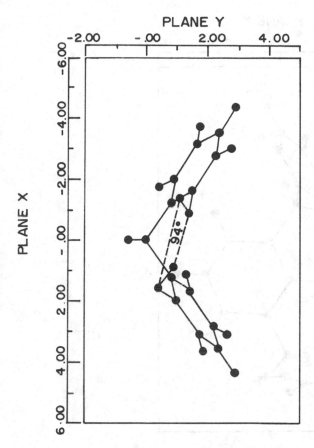

Fig. 4.9 Minimum-energy conformation for dimer II.4. Scale in Å [2,11]. H-bonds are indicated by segmented lines.

applies for trimer III. If undissolved, dimer II and trimer III probably would exist in the conformations of minimum total energy of which the structures are represented in Figs. 4.9 and 4.10. When dissolved in a solvent that masks intramolecular hydrogen bonding, the van der Waals energy would be the main contributor to the total energy, and dimer II and trimer III would change conformation to exist as the rotamers of lowest van der Waals energy.

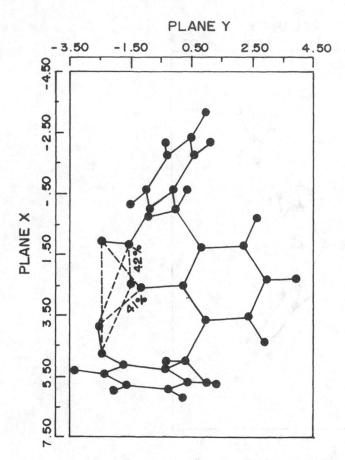

Fig. 4.10 Minimum-energy conformation for trimer VI. Scale in
Å [2,11]. H-bonds and their percentage of total H-bond value are
indicated by segmented lines.

     Since hydrogen bonding is absent in dimer I and dimer III,
trimer I and trimer II, the van der Waals energy is the main contributor to the total energy when undissolved and when dissolved
in a solvent that masks hydrogen bonding. Only the angles ϕ and
ψ involving the central methylene bridge of the PF hexamers were
rotated, and because of this restriction in rotation, no significant mean-
ing can be attached to the energy plots. Graphic representations of the
helices formed by the PF hexamers were obtained (Figs. 4.11 and 4.12).

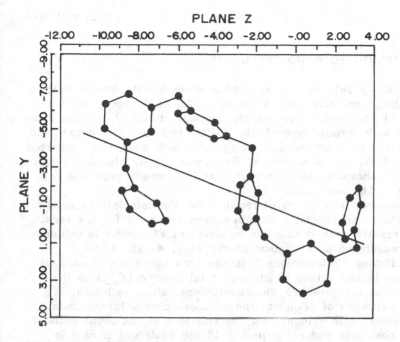

PLANE Z

Fig. 4.11 Minimum energy conformation for hexamer VII. Scale in Å [2,11]. The helix axis is indicated (continuous straight line).

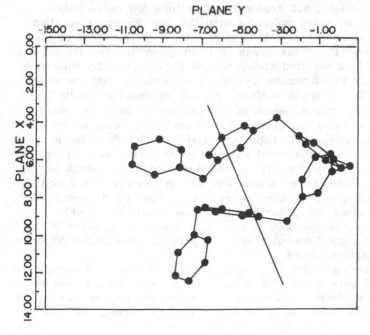

PLANE Y

Fig. 4.12 Minimum-energy conformation for hexamer VIII. Scale in Å [2,11]. The helix axis is indicated (continuous straight line).

III. ADHESION TO WOOD CELLULOSE

A recent study [6] used three phenol-formaldehyde dimers coupled
ortho-ortho, para-para, and ortho-para, hence covering all the
possible dihydroxydiphenyl methane isomers obtainable by the reaction
of phenol with formaldehyde [10], the tridimensional structure and
conformations of minimum total energy of which have been reported
[2] (Fig. 4.1). The structure of five chains, four glucoses residues
each, of a schematic elementary crystallite, already reported [5],
was used as the substrate.

The number of degrees of freedom for the calculations used is
considerable, and special techniques were employed for the task.

The results showed that the adhesion of PF resins to cellulose
is easily explained as a surface theory (Fig. 4.13). The PF
dimers/cellulose I interactions indicates that there are significant
differences in the values of minimum total energy (E_{tot}) in the
interaction of the three PF dimers with crystalline cellulose. If
the total averages of (E_{tot}) of the cellulose/phenol interactions
are compared, it is evident that the adhesion of the ortho-ortho,
(-12.329 kcal/mol) and ortho-para (-12.080 kcal/mol) dimers is
at least 20% better than for the para-para dimer (-10.203 kcal/mol).
Furthermore, in all cases the interaction of the PF dimers with
cellulose, in all positions, is more attractive than the average
for water molecules (-5.5 kcal/mol) [12]. In a few cases only,
the interaction of water molecules with the few strongest sorption
sites of the cellulose I crystal is more attractive than that of the
phenolic dimers [12]. This means that, in general, even for crystalline cellulose in a solvated state, the PF dimers, and by inference
also the higher PF oligomers, are likely to partially displace water
to adhere to the cellulose surface. This is an important factor
in wood bonding: the adhesion of the polymer resin to the wood
must be considerably better than the adhesion of water molecules
to wood. It is particularly important first for "grip" by the adhesive
of the substrate's surface and second, in the cured adhesive state,
in partly determining the level of resistance to water attack of
the interfacial bond between adhesive and adherend. The results
obtained indicate and explain, theoretically, why PF-bonded lignocellulosic materials are generally classed as weather and water
resistant: it is due not only to the imperviousness to water of the
cured PF resin itself but also to the imperviousness of the adhesive/
adherend interfacial bond.

An interesting technical point is that to minimize the energy
of interaction between the PF dimers and the cellulose, it was not
necessary to minimize *a priori* the PF dimer's internal energy in
isolation [2]. Because the dimer is left free to undergo internal

conformation adjustments during the computation of the interaction with the cellulose surface, the PF dimer's most stable conformation on the cellulose surface is always different from its most stable conformation when studied as an isolated molecule [2].

Another deduction, with some applied inferences, from the results presented here and previously [12,13] is that a PF resin impregnating wood will depress the wood cellulose water-sorption isotherm according to the number of substrate sorption sites, which, on curing, have been denied to water. The extent of the depression of the isotherm will vary according to the percentage of resin on wood. Thus, this effect is likely to be small, but still noticeable,

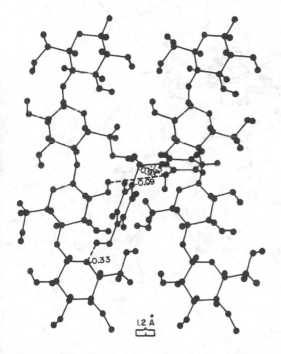

Fig. 4.13 X, Y planar projection of an ortho-para PF dimer type of interaction with the surface of the Cellulose I crystallite model. PF/cellulose H-bonds are formed. Adhesion determined by both Van der Waals forces and H-bonds. Segmented lines and negative values (kcal/mole) indicate respectively locations and values of H-bonds. Compare with similar but differently positioned interaction for the same PF dimer in Fig. 4.13b. Only cellulose surface atoms are involved as all secondary forces between PF dimer and the other three chains (not shown in figure) of the cellulose crystallite model are absent or negligible.

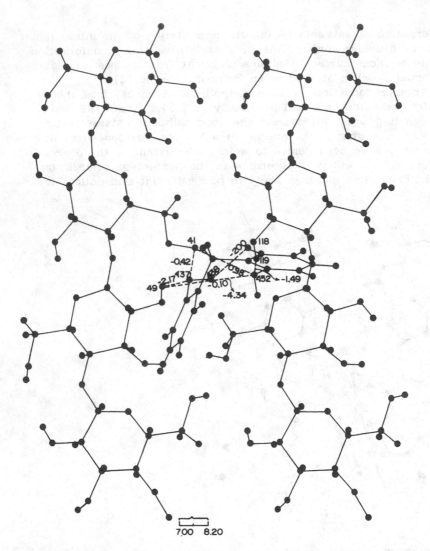

Fig. 4.13a X, Y planar projection of an ortho-ortho PF dimer type
of intersection with the surface of the Cellulose I crystallite Model.
Inter- and intramolecular H-bonds are formed (H438 to O452 in a
strong, -4.34 kcal/mole, intramolecular H-bond of the PF dimer).
All other H-bonds are PF/cellulose. Compare with case in Figs.
4.14a,b.

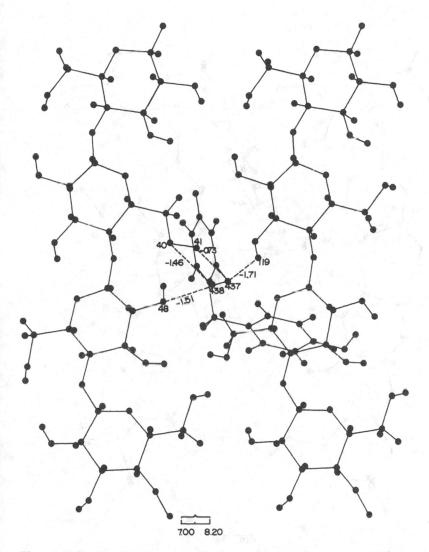

7.00 8.20

Fig. 4.13b X, Y planar projection of an ortho-para PF dimer type
of interaction with the surface of the Cellulose I crystallite model.
PF/cellulose H-bonds are formed. Adhesion is determined by both
Van der Waals forces and H-bonds. Segmented lines and negative
values (kcal/mole) indicate respectively locations and values (kcal/mole)
of H-bonds. Integer positive numbers (40, 41, 48, 437, 438) are
parts of atoms numbering system for the whole PF/cellulose crystallite
model. Van der Waals and H-bonds between the PF dimer and the
other three chains (not shown in figure) of the cellulose crystallite
model are absent or negligible. Only cellulose surface atoms are
involved. Compare with case in Fig. 4.13.

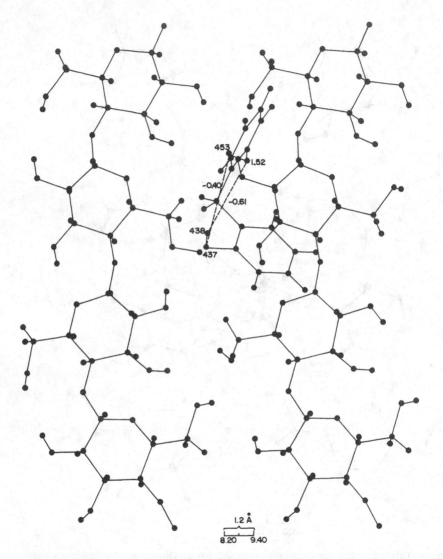

Fig. 4.14a X, Y projection of an ortho-ortho PF dimer on surface
of Cellulose I crystallite. PF/cellulose H-bonds are not formed.
Only intra-PF H-bonds are formed. Adhesion is determined in this
case by Van der Waals forces only. Segmented lines indicate positions
of H-bonds. Negative values indicate H-bonds intensity. Integer
positive numbers (437, 438, 453) are part of atom numbering system
for the whole PF/cellulose crystallite model.

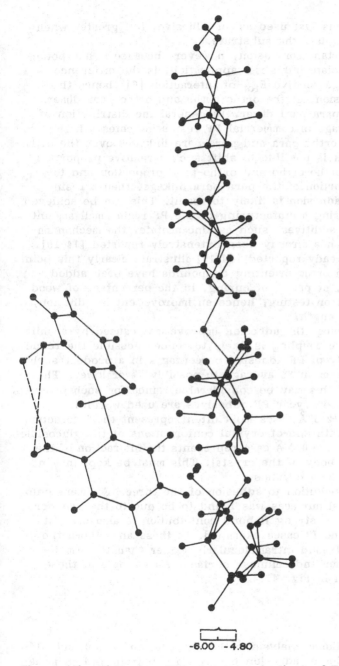

-6.00 -4.80

Fig. 4.14b Side view (YZ planar projection) of Fig. 4.14a.

when the PF resin is just used as an adhesive, but greater when it is used to impregnate the substrate.

The most important conclusion, however, because it has potentially also an immediate industrial application, is the difference between the average relative E_{tot} of interaction [6], hence the difference in adhesion, of the ortho-ortho and ortho-para dimers in relation to the para-para dimer. In general the distribution of the methylene linkage in commercial PF resins indicates a higher percentage of the ortho-para and para-para linkages over the ortho-ortho linkage. If it is possible to shift such a relative proportion to a predominant ortho-ortho and ortho-para proportion and to decrease the proportion of the para-para linkage, then a resin with much better adhesion is likely to result. This can be achieved by introducing during manufacturing of the PF resin small amounts of ortho-orienting additives, such as zinc acetate, the mechanism of action of which has already been extensively reported [14-19]. Applied results already reported [14-19] illustrate clearly this point. PF resins to which ortho-orienting compounds have been added show improvement, at parity of curing, in the percentage of wood failure of the joint on testing, hence an improvement in adhesion, and comparable strength.

As a consequence, the adhesion improvement caused by a shift in methylene bridge coupling is likely to render possible the reduction of the quantity of PF adhesive percentages in a product such as particleboard from an 8% average to possibly 7% or lower. The economic impact of this may be considerable, since for such products more than 5×10^5 tons/year PF adhesives are used worldwide.

The -2.5 and +2.5 Å cases most often represent the interaction of the PF dimers with end-of-crystal conformations of the glucosidic residues [6]. Only the 0.0 Å case represents the interaction of the PF dimer with the body of the crystal. This must be kept in mind when dealing with long crystals.

The major contribution to adhesion of the three PF dimers onto the cellulose crystal surfaces was found to be due to the van der Waals interactions. A strong H-bond contribution is also present in all but two of the 72 cases examined. In these an ortho-ortho dimer prefers to H-bond intramolecularly rather than to form H-bonds between dimer and cellulose surface. An example of these two cases is shown in Fig. 4.14.

IV. CONCLUSIONS

Ab initio conformational analysis methods appear to be well suited to theoretical studies of adhesion between a substrate and an adhesive.

In the case taken as an example, PF resins and crystalline cellulose I, the model chosen to represent a PF/cellulose system shows a stronger affinity between PF and cellulose than between water and cellulose. This implies that the resistance to weather and water of PF-bonded lignocellulosic materials is due to both the imperviousness of the cured PF resin and the imperviousness of the adhesive/adherend interfacial bond. Consequently, this implies that in adhesives where the adhesive/adherend interfacial bond is not impervious to water attack, and/or where the water resistance of the resin itself does not come into play in protecting the interfacial bond, joints of lower water resistance may well result.

The finding that configuration of the adhesive molecule has a marked influence on the energy of adsorption, hence on adhesion, is of some interest. The use of conformational analysis can indicate how the preparation of an adhesive should be changed to obtain lower adhesive consumption and higher joint strengths.

REFERENCES

1. Kottes Andrews, B. A., Reinhardt, R. M., Frick, J. G., Jr., and Bertoniere, N. R. (1986). In *Formaldehyde Release from Wood Products*, B. Meyer, B. A. Kottes Andrews, and R. M. Reinhardt, Eds. Symposium Series No. 316. Washington, DC, American Chemical Society.
2. Smit, R. (1986). "The Structure of Some Phenol-Formaldehyde Condensates for Wood Adhesives." M.Sc. thesis, University of South Africa, Pretoria.
3. Pizzi, A., and Eaton, N. J. (1984). *J. Macromol. Sci.*, *Chem. Educ.*, A21(11/12), 1443-1466.
4. Pizzi, A., and Eaton, N. J. (1985). *J. Macromol. Sci.*, *Chem. Educ.*, A22(2), 139-150.
6. Pizzi, A., and Eaton, N. J. (1987). *J. Adhes. Sci. Technol.*, 1, 3, 191-200.
7. Megson, N. J. L. (1958). *Phenolic Resin Chemistry*. London, Butterworths.
8. Whittaker, E. J. W. (1953). *Acta Crystallogr.*, 6, 714.
9. Quantum Chemistry Program Exchange, Department of Chemistry, Indiana University, Bloomington, IN 470401, Program No. 485.
10. Pizzi, A., Horak, R. M., Ferreira, D., and Roux, D. G. (1979). *Cellul. Chem. Technol.*, 13, 753.
11. Smit, R., Pizzi, A., Schutte, C. J. H., and Paul, S. O. (1987). "The Structure of Some Phenol-Formaldehyde Condensates for Wood Adhesions." *J. Macromol. Sci.*, *Chem. Educ.*, submitted.
12. Pizzi, A., Eaton, N. J., and Bariska, M. (1987). *Wood Sci. Technol.*, 21, 235-248.

13. Pizzi, A., Bariska, M., and Eaton, N. J. (1987). *Wood Sci. Technol.*, 21, 317–327.
14. Pizzi, A. (1983). In *Wood Adhesives Chemistry and Technology*, A. Pizzi, Ed. New York, Dekker.
15. Pizzi, A. (1979). *J. Appl. Polym. Sci.*, 24, 1579–1580.
16. Pizzi, A. (1979). *J. Polym. Sci.: Polym. Lett.*, 17, 489–492.
17. Pizzi, A., and van der Spuy, P. (1980). *J. Polym. Sci.*, 18, 3447–3454.
18. Pizzi, A., and Cameron, F. A. (1987). *Holz Roh Werkst.*, 38, 463–467.
19. Fraser, D. A., Hall, R. W., and Raum, A. L. J. (1957). *J. Appl. Chem.*, 7, 676.

5

The Correlation Between Preparation and Properties in Phenolic Resins

LAWRENCE GOLLOB / *Georgia-Pacific Resins, Inc.*
Decatur, Georgia

I. INTRODUCTION

The practicing resin formulation chemist uses all the scientific and observational tools available to develop products and to solve wood bonding problems. Wood bonding technology has evolved into a complex interdisciplinary science involving fields of chemistry, engineering, and materials science, as well as wood science and wood technology. The size and shapes of wood to be bonded are limited only by imagination, ranging from microscopic fibrous material in hardboard to huge structural laminated members.

The challenges and opportunities for wood bonding are derived from the nature of the interaction of the substrate with the resin. Wood is a dimensionally anisotropic, chemically heterogeneous substrate, and its properties are strongly influenced by climatic conditions such as temperature and humidity. The composite's properties are influenced by wood particle size and geometry, and by manufacturing process variables. The ability of the resin chemist to manipulate the chemistry of the resin system to fit the properties of the wood and the bonding process allows for the desired "marriage" of wood and adhesive, resulting in an effective adhesive bond and useful wood/adhesive composite. A thorough understanding of all the resin properties responsible for its performance in specific applications has not always been possible, due to insufficient resin characterization techniques. The evolution of resin technology has historically been characterized by a merger of empirical resin "art" with current tools of science. Because new chemical instrumentation and resin characterization techniques are constantly being introduced, fresh opportunities are always available for growth in our under-

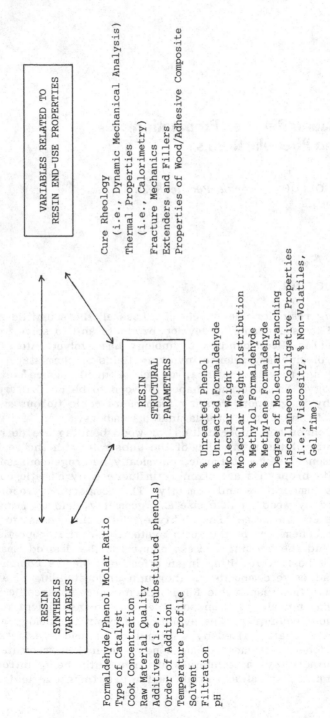

Fig. 5.1 Interaction of resin synthesis, structure, and property relationships.

standing of the chemical and physical subtleties that contribute
to superior product performance. As new or modified composite
processes appear in the marketplace, additional demands are placed
on the resins used in these products. Thus, resin chemists have
evolved into polymer engineers, using current resin characterization
techniques in conjunction with an interdisciplinary approach to
problem solving.

II. CONCEPTS OF SYNTHESIS/STRUCTURE/PROPERTY RELATIONSHIPS

The identification of variables that correlate to resin performance
is the prime objective of resin synthesis and characterization studies.
Because of the complexity and quantity of synthesis and structural
properties that may be monitored, the identification of key variables
is frequently a difficult task. One possible compilation of variables
and their interactions is shown in Fig. 5.1.

The vast majority of literature before the 1960s dealt with identification and quantitation of the initial reaction products of phenol
with formaldehyde [1-7], and reaction kinetics. During the initial
stages of reaction, it is possible to isolate pure phenol-formaldehyde
derivatives and oligomers and to study many of these as pure materials. Later in the reaction sequence, the resinous mass consists
of a mixture of molecules differing in molecular weight, structure,
and functionality. Isolation of pure compounds is difficult, and
chemical analysis yields colligative or bulk information. But because
it is the final resin mass used in the end process, these bulk chemical characteristics are precisely those that might be expected to
have a profound influence on performance. The remainder of this
chapter focuses on more recent analytical techniques to deal with
the average structure of advanced resins and the correlation of
these structural parameters both with synthesis parameters and
end-use properties.

A. The Relationship of Synthesis Parameters to Polymer Chain Growth and Polymeric Branching

Before the widespread use of high performance size exclusion
chromatography (SEC), also called gel permeation chromatography
(GPC) because many early column packings were gels, the characterization of molecular weight distribution of resins took several
days using fractional precipitation [8]. A nonsolvent was added
stepwise to a resin, resulting in the precipitation of the highest
molecular, or otherwise least soluble, species in the resin. The

stepwise fractions were dried and weighed. Modern GPC systems linked to computers can generate a molecular weight distribution in one hour or less.

GPC references in the literature generally focus on an aspect of technique development, use of the technique to characterize effects of varying resin synthesis parameters, or both. Early work by Quinn et al. [9], Aldersley et al. [10], and Armonas [11] acquainted the resin community with the power of GPC as an analytical tool. From that point in time, the technique has been applied to the study of phenolic resins with continuous refinements in technique and hardware technology. Some size exclusion chromatography systems and resin parameters examined in the literature are highlighted in Table 5.1.

Wagner and Greff [12] used Styragel columns with tetrahydrofuran (THF) as the mobile phase. Five phenol-formaldehyde (PF) resols were analyzed using polystyrene, polypropylene oxide glycols, and normal alkanes as molecular weight standards. In their resin formulations, they varied the catalyst (sodium hydroxide, barium hydroxide, and hexamethylenetetramine), as well as reaction time. They found that the use of "hexa" as a catalyst resulted in a broader initial molecular weight distribution than was achieved with the other two catalysts, but that this difference became less pronounced at longer reaction times. Their GPC column set provided good resolution of oligomeric fractions. Duval et al. [13] used Styragel columns that also emphasized resolution of oligomeric fractions. In their resin formulations, F/P molar ratio was varied between 0.92, 1.22, and 2.50, while six catalysts (hydrochloric acid, ammonia, sodium hydroxide, lithium hydroxide, barium hydroxide, and zirconium hydroxide) were examined. They concluded that a hydroxymethyl group is more reactive in the para than in the ortho position. They recognized that nonlinear conformations and associations with solvent lead to nonlinear GPC calibration curves, which they addressed by using a molecular calibration model based on the retention of nonhomologous n-alkanes, n-ethers, and rigid aromatics. Both the studies cited in this paragraph used column sets that primarily provided resolution of oligomers, and the authors concentrated their interpretations on the oligomeric fraction.

Hann [14] published a very interesting and informative paper on the effects of lithium bromide on GPC of polyester-based polyurethanes in dimethylformamide as the mobile phase. Although the samples were not phenolic, they were highly polar, as are many oligomeric phenolic resins. The samples retained polyelectrolyte characteristics because of their polarity. They exhibited artificially large hydrodynamic volumes (short retention times) at the high molecular range of the GPC chromatogram due to molecular association.

Table 5.1 Some Size Exclusion Chromatography Systems and Synthesis Parameters Examined in the Literature

Samples characterized	Chromatography system: Columns/detectors Mobile phase/standards	Resin synthesis parameters investigated	Trends identified or conclusions	Ref.
1 alkaline-catalyzed resol, 6 samples during synthesis	Bondagel/UV LiCl in DMF/polystyrene	Molecular weight increase of a single NaOH-catalyzed PF resin	MW increase of a resin can be monitored using a Bondagel column.	29
3 alkaline-catalyzed PF resols	micro-Styragel/UV THF/polystyrene and polypropylene glycol	Use of trichloroacetic acid in THF to solubilize highly alkaline resol resins	Trichloroacetic acid in THF solubilizes alkaline resols for GPC in THF.	28
19 alkaline-catalyzed PF resols, 6 samples each during synthesis	micro-Bondagel/UV + LALLS NATFAT in HFIP/LALLS	F/P ratio (1.9, 2.2, 2.5) NaOH addition level and rate, temperature profile; polymeric branching monitored with LALLS	1. MW increase of resins can be monitored using micro-Bondagel columns. 2. High initial NaOH levels yield a higher MW and less branched polymer network when cooked to the same viscosity end point in formulations with equivalent final NaOH levels but differing in rate of addition. 3. High F/P ratio (i.e., 2.5) resins yielded a higher MW when cooked to same viscosity end point. 4. The rate of the condensation reaction increases with increased NaOH levels.	20

Table 5.1 (Continued)

Samples characterized	Chromatography system: Columns/detectors Mobile phase/standards	Resin synthesis parameters investigated	Trends identified or conclusions	Ref.
Several PF resols	Styragel/RI THF/polystyrene + PFs	Resin age, acid-catalyzed cross-linking	Effects of irregular elution behavior can be compensated by use of an internal standard and relative retention.	19
18 alkaline-catalyzed PF resols, 10 samples each during synthesis	micro-Styragel/UV H_2O in THF/polystyrene + PFs	F/P ratio (1.9, 2.2, 2.5) NaOH addition level and rate, temperature profile	1. The equilibrium concentration of OMP is greater during low F/P resin cooks, but diminishes to near zero in the final resin. 2. High F/P formulations develop polymer fractions more quickly than low F/P formulations. 3. High NaOH formulations develop polymer fractions more quickly than low F/P formulations.	18
2 alkaline-catalyzed PF resols	micro-Styragel/UV + LALLS THF/polystyrene and polypropylene glycol and phenolics	Sample preparation techniques varied to include resin neutralization, freeze-drying, and acetylation	1. Light scattering detector useful at molecular weights above 500 daltons. 2. Different preparation techniques of the same resin can result in different apparent MWD's.	17

Lignin	Sepharose + Sephadex/UV LiCl in DMF/polystyrene	Kraft, synthetic, and Braun's native lignin	LiCl reduces phenolic association in DMF.	16
Southern pine bark polyphenolics	DuPont SEC/UV Methanol + aqueous salts/PEGs + polystyrenes	Solvent evaluation for association phenomena using dynamic light scattering	Aqueous sodium sulfate will disassociate alkali-soluble polyphenols.	15
Polyester-based polyurethanes	Styragel/RI LiBr in DMF/polystyrene	Association phenomena with and without LiBr	LiBr in DMF will disassociate polyester-based polyurethanes for GPC analysis.	14
5 commercial PF resols and 10(1) Lab cooks, samples during cooks	Styragel/RI THF/model based on alkanes, ethers, and aromatics	F/P ratio (0.92, 1.22, 2.50) catalyst (HCl, NH$_3$, NaOH, lithium, barium, and zirconium hydroxides)	Emphasis on oligomeric fractions; reactivity of functional groups during the polycondensation reaction.	13
5 PF resols	Styragel/RI THF/polystyrene + others	Catalyst (sodium and barium hydroxide and hexamethylenetetramine) reaction time	Hexa catalyst broadens initial MWD relative to other catalysts studied.	12

The addition of LiBr as an electrolytic salt effectively minimized artifacts in the chromatographic data by disassociating the molecules and minimizing polymer-column interactions, resulting in the disappearance of multiple peaks eluting at short retention times. The significance of Hann's work to chromatography of phenolic resins will become more apparent in later paragraphs. A key observation is that molecular association is manifested by multiple peaks at short retention times.

Wilson et al. [15] studied naturally occurring phenolic polymers, the Southern pine bark polyphenolics, using size exclusion chromatography. They used dynamic light scattering to evaluate solvent systems for association phenomena and found that aqueous sodium sulfate will disassociate alkali-soluble polyphenolics. Connors et al. [16] studied other naturally occurring phenolic polymers, lignins, using GPC. They examined 0.001-0.1 M solutions of lithium chloride in dimethylformamide and found LiCl to reduce interpolymer association in DMF.

Wellons and Gollob [17] evaluated the use of a low angle laser light scattering (LALLS) detector in series with a conventional GPC system consisting of micro-Styragel columns, THF as the mobile phase, and a UV detector. They found that polystyrene standards, polypropylene glycol standards, and dimedone monomer and dimer standards fell on three different molecular weight versus elution volume calibration curves, emphasizing that a gel permeation chromatogram is not synonymous with a molecular weight distribution if the sample's molecular weight at a given elution volume does not match that of the standard. Because the LALLS detector continuously measures molecular weight as fractions elute from the GPC columns, reliance on an external calibration curve is eliminated. We evaluated resin neutralization procedures, freeze-drying procedures, and acetylation procedures, and concluded that precipitation or drying techniques did alter the molecular weight distribution of the resin. We also observed THF to be effective in dissolving low molecular weight resin fractions, but ineffective in fully dissolving or disassociating the higher molecular weight fractions of phenol-formaldehyde resol resins.

Parker [18] examined 18 alkaline-catalyzed resols, analyzing 10 samples each, during the cooks. He used a Waters micro-Styragel column set (one 500-Å column and three 100-Å columns) which provided good resolution of oligomeric fractions. The mobile phase of 5% purified water in THF provided an improved solvent for dissolving the phenolic polymer network relative to 100% THF, but retained enough nonpolar characteristics to preserve the integrity of the micro-Styragel columns. Phenol, PF oligomers, and polystyrenes were used as molecular weight calibration standards. The F/P molar

ratio was varied between 1.9, 2.2, and 2.5. The resols were catalyzed with sodium hydroxide, which was varied both in initial
concentration and addition rate. Reaction temperature was also varied.
Among the findings were that the equilibrium concentration of ortho-
methylol phenol (OMP) was greater during low F/P ratio cooks than
high F/P cooks, but the OMP concentration diminished to near zero
in the final resin. High F/P formulations developed polymer fractions
more quickly than low F/P formulations, and formulations high in
sodium hydroxide developed polymeric fractions more quickly than
those low in sodium hydroxide.

Rudin et al. [19] examined several PF resols using Styragel
columns, 100% THF as the mobile phase, plus polystyrene and
phenolic molecular weight standards to characterize the effect of
resin age and acid-catalyzed crosslinking. They recognized that
association and other anomalies resulted in irregular elution behavior.
They introduced internal standards and the use of relative retentions,
much as commonly used in gas chromatography, to compensate for
irregular elution in GPC.

Gollob [20] used a LALLS molecular weight detector in series
with the GPC system to study the effects on weight average molecular
weight and molecular weight distribution of varying F/P ratio (1.9,
2.2, and 2.5), concentration and rate of sodium hydroxide catalyst
addition, and reaction temperature. Nineteen alkaline-catalyzed
resols, with six samples of each analyzed during synthesis, were
examined on a size exclusion chromatography (SEC) column set
consisting of micro-Bondagel E-Linear and E-125 column. Hexafluoro-
isopropanol (HFIP) was used as the mobile phase because of its
ability to solubilize polymers, such as PFs, which have intermolecular
hydrogen bonding tendencies. Sodium trifluoroacetic acid (NATFAT)
was added to the mobile phase at the rate of 0.08% to inhibit association. The SEC-LALLS system removed the dependence on external
standards for column calibration purposes and permitted a "molecular
weight versus retention" calibration curve to be generated for each
individual sample. These calibration curves were then used to estimate molecular branching based on hydrodynamic volume relationships
of linear versus branched polymers.

Because condensation is the rate-controlling step in alkaline-
catalyzed polymerization, the formation of highly methylolated phenols
is favored and numerous potential branch points are built into the
polymer structure. Not all methylols will be able to crosslink and
to form chain segments. Because size exclusion chromatography
separates molecules on the basis of their solvated size, both molecular
structure and weight are important factors in determining the retention properties of a polymer in a SEC column, as illustrated when
different classes of compounds fall on different calibration curves.

Drott and Mendelson [21] generated calibration curves of log molecular weight versus elution volume for compounds differing in known
extent of branching and observed that more highly branched polymers
were characterized by a steeper slope of the calibration curve. Such
observations are explained on the basis of the hydrodynamic volume
of the chain in solution. The hypothetical calibration curves of
Fig. 5.2 illustrate that at a given elution volume (equivalent to
a hydrodynamic molecular volume), a more highly branched molecule
is more compact and contains more mass in the same volume element
than is occupied by the less compact mass of a less highly branched
molecule. The steeper curve represents a more highly branched
polymer.

Using SEC calibrated with standards does not permit detection
of differences in hydrodynamic volume due to branching. One alternative in the literature has been to develop a universal calibration
for a column set by determining the elution volume of standards
with a narrow distribution and known intrinsic viscosity. A calibration curve formed by graphing the product [(log molecular weight) ×
(intrinsic viscosity)], also referred to as hydrodynamic volume,
versus elution volume has been shown to be relatively independent
of molecular structure [21-24]. If intrinsic viscosity of the unknown
fractions is then determined experimentally, molecular weights can
be taken from the universal calibration curve. By comparison with
a known linear polymer of similar structure, a branching coefficient
may be calculated based on a previously determined branching model
[21].

The LALLS molecular weight detector circumvents the time-
consuming need for intrinsic viscosity measurements on individual
fractions because an empirical measure of the relative branching
of an array of PF resins may be obtained from the slope of the
curve of molecular weight versus elution volume. Steeper slopes
represent a greater degree of branching.

A second estimate of molecular branching used was based on
the influence of branching in measurements of viscosity. Viscosity
is a function of molecular weight, branching, and concentration
of a polymer in solution, and may be used as a measure of molecular
weight [25]. If molecular weight is expressed per unit of viscosity,
then differences between polymers of similar monomeric constitution
in the same solvent and at the same concentration must be due
to a structural feature such as branching. Branching is directly
proportional to the value of the ratio of log of the molecular weight
to log of the viscosity. Larger ratios reflect a greater degree of
branching, and vice versa. Because the hydrodynamic volume of
a highly branched molecule is smaller than for a less highly branched
molecule, its ability to form entanglements with other molecules is

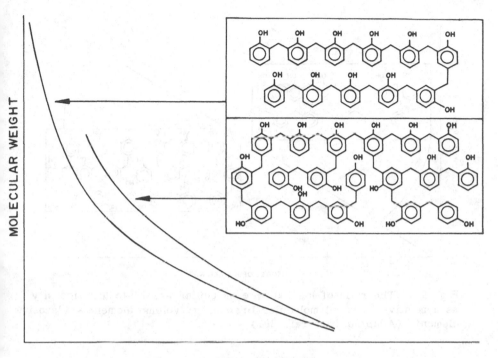

GPC ELUTION VOLUME

Fig. 5.2   Molecular weight versus elution volume: GPC calibration curves as influenced by molecular branching. A steeper slope is evident for the more highly branched polymer. (Adapted from ref. 20.)

also less. Thus, a relatively large molecular weight "per unit viscosity" is indicative of a greater degree of branching (Fig. 5.3).

Relative branching values were experimentally determined [20] directly as the log ratio of molecular weight to viscosity, and as the relative slope of the "log MW versus log retention volume" calibration curve for each sample. The trends identified in the study are summarized in Figs. 5.4-5.6. The final molecular weight reached at a constant viscosity end point was greater in formulations that had either more sodium hydroxide present during the crosslinking reaction, or a higher molar ratio of formaldehyde to phenol (Fig. 5.4). The influence of sodium hydroxide on molecular weight is explained by the ability of sodium hydroxide to confound intermolecular hydrogen bonding in phenolic polymer chains and to reduce viscosity. By reducing molecular interaction, the sodium hydroxide

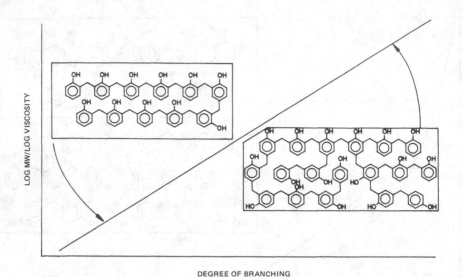

Fig. 5.3 The ratio of log average molecular weight to log viscosity as a relative index of molecular branching: volume element ~ viscosity element. (Adapted from ref. 20.)

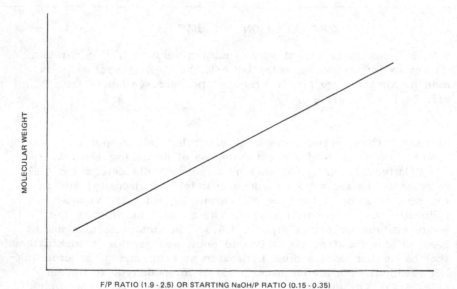

Fig. 5.4 Relationship between molecular weight ($\overline{M}_w$) and resin preparation parameters when reacted to a constant viscosity and end point. (Adapted from ref. 20.)

reduces resistance to shear and thins the resin. Because the formulations were prepared to a viscosity end point and because additional
sodium hydroxide thins the resin, the resin must be reacted to a
higher molecular weight in order to reach the viscosity end point.
Figure 5.5 summarizes the influence of sodium hydroxide in the
resin formulations. A greater solution viscosity on a "per molecular
weight unit" basis occurs when less sodium hydroxide is present
to reduce molecular interactions.

All the formulated resols were characterized to be highly
branched, but the relative degree of branching was less in formulations that had higher levels of sodium hydroxide present during
condensation (Fig. 5.6). The reduced polymeric branching pattern
was explained on the basis of chemical reactivity. Additional sodium
hydroxide catalyzes the crosslinking reaction, resulting in early
appearance of polymer fractions in the SEC during the reaction.
As the rate of crosslinking increases with additional sodium hydroxide,
internal sites become rapidly shielded from further reaction due to
steric influences [26,27]. Although still highly branched, higher
levels of sodium hydroxide result in a higher percentage of unreacted,
sterically hindered, internal methylols and a lower relative index of
molecular branching.

Bain and Wagner [28] evaluated three alkaline resols on a column
set consisting of 1000-, 500-, and 100-Å micro-Styragel columns
(Table 5.1). They used trichloroacetic acid to solubilize the resins
and concluded that trichloroacetic acid in THF is a suitable mobile
phase for the analysis of alkaline resols.

Walsh and Campbell [29] used an E-125 micro-Bondgel column
with 0.1M LiCl in DMF as the mobile phase. They also concluded
that their system was suitable for the analysis of alkaline resols.

Size exclusion chromatography continues to be the "workhorse"
in many laboratories, and it is an informative tool for understanding
phenolic adhesives.

B. The Relationship of Synthesis Parameters
   to Chemical Structure

Early work in the field of structural elucidation of phenolic resins
utilized infrared spectroscopy and gave an indication of the types
of functional groups present. The application of nuclear magnetic
resonance (NMR) spectroscopy to resin structure elucidation has
enhanced our ability to resolve isomeric structural peculiarities
and to quantitate their presence. Early NMR work was focused
on structural characterization of liquid resin samples using the
proton as the probe (i.e., ^{1}H NMR). As NMR technology became
more sophisticated, other probes could be examined, such as carbon

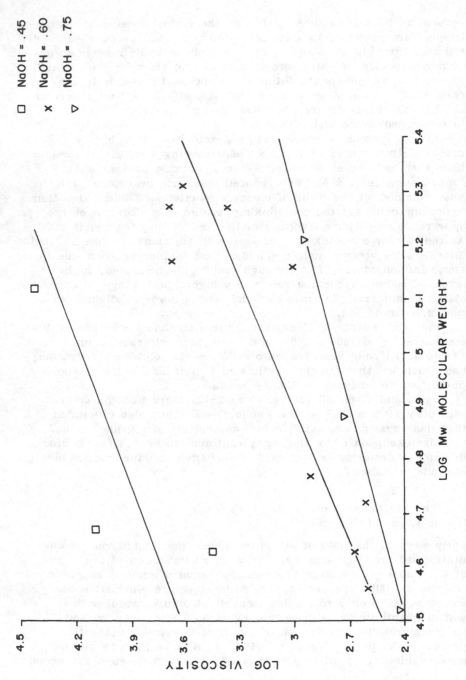

Fig. 5.5 Graph of log viscosity versus log average molecular weight as influenced by the NaOH/phenol molar ratio of the finished resin. (Adapted from ref. 20.)

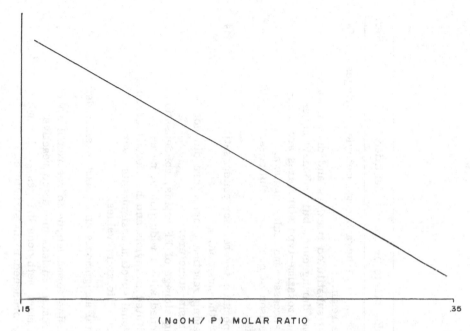

(NaOH / P) MOLAR RATIO

.15   .35

Fig. 5.6 Relationship between molecular branching and NaOH/phenol molar ratio during reaction. (Adapted from ref. 20.)

(^{13}C NMR) and nitrogen (^{15}N NMR). From liquid characterization, inferences could be made regarding the curing process. Additional refinements in instrumentation (i.e., wide bore) and technique [i.e., cross-polarization with magic angle spinning (CP-MAS)] made it possible to examine solid-state samples. The most recent applications of the NMR technique have been in structural characterization of resins during cure and in the cured solid state. Some NMR techniques applied to the study of resins are highlighted in Table 5.2.

Woodbrey et al. [30] used ^{1}H NMR to study both novolak and resol PF liquid resins that varied in molar ratio and catalyst. They provided extensive chemical shift assignments for freeze-dried and acetylated resins that have been of use to resin chemists since the paper was published in 1965. During acid catalysis, reactions at para positions were found to occur five times faster than reactions at ortho positions. Hemiformal type structures (Ph—CH_2OCH_2OH) were found to be present in many resols and were found to be stable during initial cure conditions. Hirst et al. [31] used ^{1}H NMR techniques to produce chemical shift assignments for PF dimer and trimer model compounds.

Table 5.2 Some Structural References on Phenolic Resins

Technique	Preparation variable	Major points or trends identified	Ref.
Solid-State ^{13}C- and ^{15}N-NMR	Curing of hexa-catalyzed novolak	1. The carbons of hexa become methylene bridges in the resin. 2. Benzyl-substituted triazines and diamines are present during cure, but tribenzylamine- and benzoxazime-type structures are dominant successor molecules to hexa. 3. Comparison of cured and solution-state NMR results.	46
^{13}C-NMR of liquid PF^a resins (base catalyzed)	Reaction time and aging	1. Para position favored for condensation over ortho position. 2. Increasing reaction temperature did not vary resin structure. 3. During storage of PF resins, condensation proceeded with a reduction of para-substituted methylols and hemiformal groups. 4. Additional spectral assignments over previous literature values.	44
^{13}C-NMR of liquid resorcinol-formaldehyde-^{a} resins (base catalyzed)	Reaction time; molar ratio	1. Spectral assignments and chemical shift data listed. 2. Increasing formaldehyde to resorcinol molar ratio results in less unreacted reactive sites, more methylene bridges, and a more highly branched resin structure.	45

^{13}C-NMR of liquid PF[a] and cresol oligomers	Model compounds	Extensive shift assignments based on synthe-sized moromeric and oligomeric compounds.	43
^{2}H-NMR and ^{13}C-NMR of liquid and solid resin (base catalyzed)	PF resin prepared with ^{2}H-formaldehyde or ^{2}H-phenol with ^{13}C formaldehyde	1. An increase in T_1 (spin–lattice relaxation time) with cure time parallels an increase in the methylene/methylol ratio. 2. Resin degradation occurs when cured at 120°C for 4 hours or more.	42
Photo-CIDNP ^{1}H-NMR of oligomers	Phenol–acetaldehyde (novolak) oligomers	1. —OH groups of outer rings are not accessible (to flavin dye) due to hydrogen bonding, which strengthens with increasing molecular weight. 2. —OH groups of inner rings are always accessible (to flavin dye) regardless of molecular weight.	41
Solid-state and liquid ^{13}C-NMR of PF resols and novolak	F/P ratio, catalyst level, molecular weight prior to cure	1. The phenolic hydroxy group is involved in crosslinking in the solid state. 2. There is greater involvement of phenolic hycroxy groups in low F/P ratio resins. 3. Methylene bridge carbons are further involved in crosslinking in the solid state.	40
Solid-state ^{13}C-NMR of PF novolaks	Cure time	1. Curing involves crosslinking. 2. CP-MAS experiment gives valuable informa-tion on PF novolaks.	39

(continued)

Table 5.2 (Continued)

Technique	Preparation variable	Major points or trends identified	Ref.
Solid-state ^{13}C-NMR of cured PF resols	High temperature stability	1. Thermal changes occur to cured resin at elevated (180°C) temperature. 2. Thermal degradation is characterized by a loss of methylol and methylene groups and the appearance of oxidized functionalities such as aldehydes, ketones, carboxylic acids, and anhydrides.	38
^{1}H-NMR of acetate derivatives of PF resols	F/P ratio, NaOH level	1. Higher F/P molar ratios or more NaOH in the formulation cause an increase in methylene bridge formation. 2. Ethers are present in high F/P formulations with little NaOH, but decrease if NaOH increases, if F/P decreases, or with reaction time. 3. Percent bound formaldehyde decreases with additional NaOH or high F/U ratios.	20
Solid-state ^{13}C-NMR	Cured PF resin	CP-MAS ^{13}C-NMR can be used to characterize total substitution pattern of phenol ring in cured phenolic resin.	37
^{13}C-NMR of PF[a] oligomers	PF oligomer model compounds	Chemical shift assignments of PF oligomers.	36
^{13}C-NMR of liquid	Caustic level in resin	1. Chemical shifts for 20 model compounds and PF resins were determined. 2. The influence of sample alkalinity on chemical shifts as documented.	35

Method	Resin type	Comments	Ref.
^{13}C-NMR of liquid resins	Resin type	Chemical shift assignments for PF, RF, and MF resins.	34
^1H-NMR of acetylated PF wood adhesives	F/P molar ratio (0.75–2.6), catalyst [NaOH or Ca(OH)$_2$]	1. Methylol content increased, up to F/P = 1.8, with increasing F/P ratio. 2. NaOH is a more reactive catalyst than Ca(OH)$_2$. 3. Above F/P = 1.5, formaldehyde was not recovered in PF resin.	33
^1H-NMR of novolaks	Novolak transient molecules	1. Glycol species in formalin solutions are more stable under acidic than basic conditions. 2. Spectral assignments for novolak intermediate molecules formed via novolak–hexa cure mechanisms.	32
^1H-NMR of resols	Catalyst [NaOH, Ba(OH)$_2$, or hexa], F/P molar ratio	Barium hydroxide is more ortho directing than NaOH or hexa.	12
^1H-NMR of oligomers	Linear PF polymers	Chemical shift assignments of PF dimer and trimer model compounds.	31
^1H-NMR of PF novolaks and resols	F/P molar ratio and catalyst	1. Extensive chemical shift assignments for freeze-dried and acetylated resins. 2. Acidic para-position reactions are about 5 times faster than ortho-position reactions. 3. Hemiformals are present in many resols and are stable during initial cure conditions.	30

aStabilized formaldehyde used in resin preparation.

Wagner and Greff [12] studied the effect of catalyst and F/P molar ratio on liquid resol resin structure using ^{1}H NMR spectroscopy. They concluded that barium hydroxide is more ortho-directing than either sodium hydroxide or hexamethylenetetramine. Barium hydroxide would be expected to produce a more linear polymer than the other catalysts because of its ortho-directing mechanism. Kopf and Wagner [32], studied transient molecules present during formation and cure of novolaks using ^{1}H-NMR spectroscopy, provided spectral shift assignments for intermediates formed when novolaks are cured with hexamethylenetetramine, benzoxazine, dibenzylamine, tribenzylamine, and diphenylmethane (Fig. 5.7).

Steiner [33] acetylated and studied by liquid ^{1}H-NMR spectroscopy a series of phenolic resols that had variable F/P molar ratios (0.75–2.6) and were catalyzed with either sodium hydroxide or calcium hydroxide. He found that the methylol content of the resins increased with molar ratio up to 1.8 and then leveled off. The calculated F/P molar ratio from NMR spectra indicated that above F/P of 1.5, less than 100% of the charged formaldehyde was recovered in the resin. Loss of free formaldehyde during the resin cook or during the freeze-drying stage of sample preparation were given as possible explanations for the observed formaldehyde loss. It was noted that F/P = 1.5 is also the theoretical optimum molar ratio for a fully crosslinked PF polymer network. Sodium hydroxide was found to be a more reactive catalyst that also provided better resin solubility at higher degrees of polymerization than calcium hydroxide.

De Breet et al. [34] applied ^{13}C NMR spectroscopy to analysis of formaldehyde-containing resins. Liquid samples of phenol-formaldehyde (novolak and resol), cresol-formaldehyde, resorcinol-formaldehyde, urea-formaldehyde, and melamine-formaldehyde resins were dissolved in hexadeuterated dimethyl sulfoxide (d_6-DMSO) and compared. Reference compounds and calculations were used to aid in interpretation of spectra. Chemical shift assignments were made for all the resins, and methylene carbons from bound formaldehyde were found to lie between the 20-100 ppm regions for all samples.

Kim et al. [35] applied ^{13}C-NMR spectroscopy to the characterization of chemical shifts in 20 model compounds. Subtle changes in chemical shifts were documented due to variation in sample alkalinity. The alkalinity shifts were used to characterize PF resin samples.

Sojka et al. [36] reported the chemical shifts of positional isomers of bis(hydroxybenzyl)phenols and bis(hydroxyphenyl)methanes using ^{13}C-NMR spectroscopy. The assignments for possible dimer and trimer positional isomers can be used to understand resin systems.

Fyfe et al. [37] applied the technique of solid-state ^{13}C-NMR spectroscopy to the study of cured, solid phenolic resins. They used a combination of cross-polarization, high power decoupling,

(a) Starting Materials:

Hexamethylenetetramine

+

NOVOLAK

(b) Intermediates:

*(i) 1,3,5 tribenzyl hexa hydrotriazine-type compounds

*(ii) tetra bensyldiaminomethane-type compounds

(iii) benzoxazine-type compounds

(iv) tribenzylamine-type compounds

⁺(v) dibenzylamine-type compounds

⁺(vi) benzylamine-type compounds

(c) Final Products: (vii) diphenyl methane-type compounds

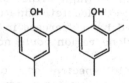

(methylene linkage carbon from hexamethylenetetramine)

Fig. 5.7 Intermediates during the cure of novolaks with hexamethylenetetramine: (a) starting materials, (b) intermediates, and (c) final products. Asterisks indicate compounds found primarily in solid-state NMR spectra; crosses indicate compounds found primarily in liquid-state NMR spectra. (Adapted from refs. 32 and 46.)

and magic angle spinning techniques to obtain high resolution ^{13}C-NMR spectra of the solid cured resin. Spectral assignments were given for phenolic carbons and methylenes assigned to free paraformaldehyde, to methylol groups, and to bridging methylenes. No methylene ethers were present in the cured resoles.

The influence of F/P molar ratio and sodium hydroxide concentration were studied by Gollob [20] on acetylated phenolic resols using ^{1}H-NMR spectra of dissolved samples. Methylene ethers were present at high F/P molar ratios, but they decreased with reaction time. The concentration of methylene ethers decreased in formulations that had a lower F/P molar ratio, or a high concentration of sodium hydroxide. An increase in methylene bridge formation was observed at higher F/P molar ratios or in formulations with a higher concentration of sodium hydroxide. As previously documented by Steiner [33], the calculated F/P molar ratio indicated that less than 100% of the charged formaldehyde was recovered in the resin structure. The percentage of bound formaldehyde, based on the initial charge, decreased at higher F/P molar ratios or in formulations with a higher concentration of sodium hydroxide. In addition to the possible loss of formaldehyde during the resin cook or during the freeze-drying and sample preparation stages, Gollob proposed that some formaldehyde may be consumed in the Cannizzaro reaction:

$$2CH_2O \xrightarrow{\text{NaOH}} CH_3OH + HCOOH$$

formalde-   methanol   formic acid
hyde

which is favored under conditions of high alkalinity and high formaldehyde concentrations.

High temperature degradation of cured phenolic resins was studied using solid state ^{13}C-NMR techniques by Fyfe et al. [38]. Thermal changes were monitored at 180°C and were characterized by a loss of methylol and methylene groups. Other thermal changes were the appearance of oxidized functionalities such as aldehydes, ketones, carboxylic acids, and anhydrides.

Bryson et al. [39] used solid state ^{13}C-NMR spectroscopy to study the effect of cure time on PF novolaks. A "high ortho" and "random" novolak were examined using hexamethylenetetramine curing catalyst at 135°C. Crosslinking was confirmed by a decrease in unsubstituted aromatic carbons, but the methylene region could not be analyzed using the normal CP-MAS spectra.

Maciel et al. [40] used solid-state ^{13}C-NMR spectroscopy to compare cured resols differing in F/P molar ratios, catalyst concentration, and molecular weight (before cure) and a novolak. In addition to the major crosslinking mechanism through methylene bridge

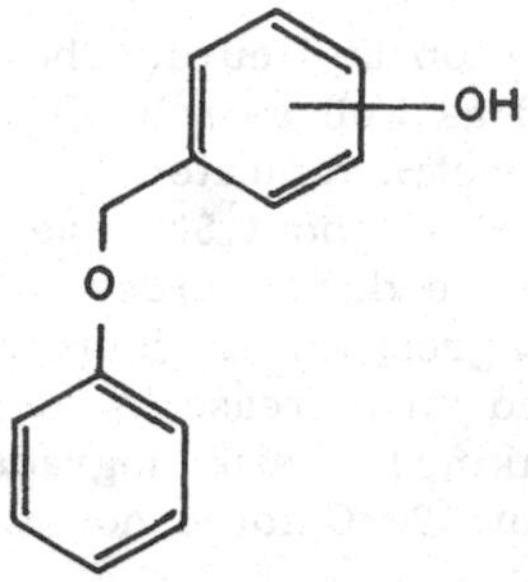

Fig. 5.8 Involvement of the phenolic hydroxy group in resin cure, as proposed by Maciel et al. [40].

formation, a minor involvement of the phenolic hydroxy group was proposed for the resols from the NMR data (Fig. 5.8). Involvement of phenolic hydroxy groups in cured resols formulated with a low F/P molar ratio (i.e., 1.9) was found to be greater than at higher F/P molar ratios (i.e., 2.2 or 2.5). Methylene bridge carbons also were found to have greater involvement in cured resols, in the solid state, than classical models (Fig. 5.9).

Conventional and photochemically induced dynamic nuclear polarization ^{1}H-NMR data were used by Zetta et al. [41] to study phenol-acetaldehyde oligomers. These investigators found that hydroxy groups of the outer rings were involved in hydrogen bonding, which made them unaccessible to a flavin dye. The outer rings became less accessible as molecular weight increased, while hydroxy groups of the inner rings were always accessible to the flavin dye regardless of molecular weight.

Kelusky et al. [42] used [2]H-labeled formaldehyde or [2]H-labeled phenol with ^{13}C-labeled formaldehyde to study the base-catalyzed

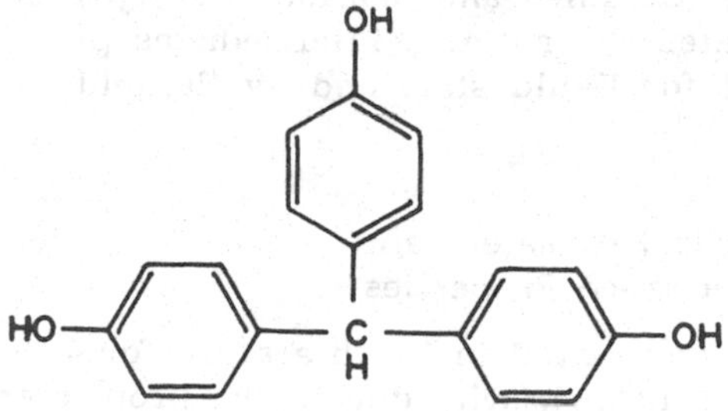

Fig. 5.9 Additional involvement of the methylene bridge carbons in resin cure, as proposed by Maciel et al. [40].

phenol-formaldehyde reactions during polymerization and cure. They used ^{2}H-NMR and ^{13}C-NMR techniques of liquid as well as solid-state samples. The line shape of the H-NMR spectra indicated that some of the prepolymer was fairly mobile, while about 50% was already rigid at a local level on the NMR time scale due to cross-linking, becoming more rigid as crosslinking progressed. An increase in the spin-lattice relaxation time (T_1) paralleled an increase in the methylene/methylol ratio (degree of crosslinking). Resin degradation products appeared if the resin was cured at 120°C for 4 hours or more.

Pethrick and Thomson [43] prepared pure compounds by acid catalysis of formaldehyde with either phenol or cresol. Extensive ^{13}C-NMR chemical shift assignments were tabulated based on the monomeric and oligomeric model compounds in solution.

Werstler [44,45] used ^{13}C-NMR spectroscopy to study the effect of reaction time on liquid samples of base-catalyzed phenol-formaldehyde and resorcinol-formaldehyde (RF) resins, respectively. He also studied the effect of varying F/R molar ratio. Extensive spectral assignments were provided for both resin systems. For PF systems, he observed as previously reported that the para position is favored for condensation relative to the ortho positions. Resin structure was unaffected by increases in reaction temperature. Aging of the resin resulted in a reduction of para-substituted methylol and hemiformal groups as condensation proceeded. In RF systems, a reduction in unreacted active sites was observed when the F/R molar ratio was increased. Higher F/R resins were more highly branched and exhibited a higher density of methylene-type linkages.

Hatfield and Maciel [46] studied the hexamethylenetetramine curing of a PF novolak using solid-state NMR spectroscopy involving the isotopes ^{13}C and ^{15}N. They found that the carbon atoms of hexa became methylene bridges in the cured resin. While tribenzylamine- and benzoxazine-type structures were the dominant successor molecules to hexa, confirming earlier ^{1}H-NMR work by Kopf and Wagner [32], benzyl-substituted triazines and diamines were present during cure. Figure 5.7 illustrates the reaction intermediates proposed by Kopf and Wagner [32] for liquid state and by Hatfield and Maciel [46] for solid state.

C. The Relationship of Synthesis Parameters and Chemical Structure to Performance Properties

Performance data are sparsely documented in the literature for varying resin synthesis formulations primarily due to the proprietary nature of such information. Product performance optimization for a given manufacturing process generally results in a unique resin for that application. It is not possible to formulate a "universal resin" for all applications. Some documented trends of structure/ property data found in the literature are listed in Table 5.3.

Pillar [47] studied the influence of molecular weight and wood moisture content on performance of resorcinol-formaldehyde resins. A "free vibrational apparatus," predecessor of the torsional braid apparatus (TBA) was used to study curing properties and to compare with gluing results on hard maple. Breaking strength generally increased with increasing moisture content, but there was a statistically significant interaction between moisture content and RF resin molecular weight. Breaking strength at low moisture contents increased with increasing molecular weight but decreased with increasing molecular weight at higher moisture contents. At 10% moisture content, the breaking strength was insensitive to molecular weight. Molecular weight was inferred from a pot time of either 20 or 50 minutes of RF resin and formaldehyde before substrate application. It was speculated that the 20-minute mix would have greater shrinkage and sensitivity to moisture content. It is also likely that branching pattern was altered as a function of pot time. If the 50-minute mix were more thermoplastic, it might have been optimized for a low moisture content gluing application due to overpenetration at higher moisture contents.

Hse [48] studied the influence of varying F/P and NaOH/P molar ratios on bond strength of experimental panels of Southern pine plywood. Resin shrinkage during cure was inversely correlated with bond quality; that is, poorer bonds were associated with greater resin shrinkage. Cure shrinkage was generally found to increase as the NaOH/P molar ratio was increased. Thicker glue lines gave superior bonds.

Takatani and Sasaki [49] applied techniques of mode I cleavage fracture mechanics to the evaluation of wood-epoxy resin bond systems. The epoxy mixture was an interesting system with which to work because the flexibility of the adhesive was easily manipulated with an added polysulfide flexibilizer. More rigid glue lines and thinner glue lines were found to have higher stress concentration. Higher stress concentration in the vicinity of crack tips results in crack propagation and a low fracture toughness. Thicker glue lines increased fracture toughness of wood-glue bond systems. Increasing glue-line flexibility effectively reduced the stress concentration at the crack tip and resulted in a glue-bond system with a higher mode I (opening mode) fracture toughness.

Torsional braid apparatus (TBA) is a method of dynamic rheological analysis useful for characterizing the transition of resins from a viscous liquid to a three-dimensional cured network. It allows the separation of elastic (rigidity) components and viscous (damping)

Table 5.3 Some Structure/Property Relationships Documented in the Literature

Resin synthesis or structural variable, or process variable		Ref.
PF's varying in F/P and NaOH/P studied by TBA	1. Initial rigidity and damping are strongly influenced by resin MW and viscosity. 2. Cured rigidity is influenced by F/P and total NaOH/P molar ratio, reaching a maximum at F/P = 2.5 and total NaOH/P = 0.75, the highest levels studied. 3. Highly branched resins were more rigid.	57
PF resins differing in MW, degree of branching, oligomer content, and process assembly time	1. MW was inversely related to bond quality at long assembly times. 2. High free phenol and orthomethylol phenol helped plasticize high MW resins. 3. Highly branched resins flow poorly at long assembly times 4. Extenders mask intrinsic resin properties. 5. Thicker glue lines perform better.	56
Cured PF resin differing in F/P, NaOH/P studied by TBA and fracture mechanics	1. Fracture toughness is inversely related to cured resin rigidity. 2. Cured resin rigidity is inversely related to NaOH/P cook molar ratio between 0.15 and 0.35, and directly related to F/P molar ratio between 2.2 and 2.5 3. Extended resins have higher fracture toughness than neat resins.	55

PF, PRF, UMF, and PVA resin systems

1. Major rigidity changes during cure were characterized using TBA.
2. TBA is a dynamic measurement that allows separation of elastic (rigidity) and viscous (damping) components of resin cure.

54

Wood-epoxy resin bond systems

1. Thinner glue lines had higher stress concentration.
2. More rigid glue lines had higher stress concentration.
3. Cleavage fracture toughness could be improved by increasing flexibility of the cured resin or by increasing glue-line thickness.

49

F/P ratio and NaOH/P ratio

1. Bond strength was generally poorer in resins with greater cure shrinkage.
2. Cure shrinkage was generally greater in resins with higher molar ratios of NaOH to phenol.

48

RF resin at low and high molecular weight, and low and high wood moisture content.

1. Bond strength increased with increasing molecular weight at low moisture content and decreased with increasing molecular weight at the high moisture content.
2. Free vibrational and rate of strength development tests are useful for in situ observations of curing properties.

47

components [50,51]. The technique can be used to obtain physical
data about resins after cure [52,53]. A fiberglass braid is impregnated with PF resin, placed in an oven, and subjected to a torsional
displacement resulting in a free oscillation of the braid-resin composite. The oscillation takes the format of a damped sin wave function,
which is recorded electronically. Because the fiberglass braid is a
support medium of low rigidity, the measured oscillation energy is
assigned to the polymer film. Because a more rigid material will
generate an oscillation with a tight sin wave function manifested
by a rapid period of oscillation, the inverse square of the period
is a measure of the relative rigidity of the PF film. Thus, a larger
value of relative rigidity reflects a more rigid cured phenolic film.

Steiner and Warren [54] used torsional braid analysis to compare
rigidity and damping responses during and after cure for four
different wood-adhesive systems that included phenol-formaldehyde,
urea-melamine-formaldehyde, polyvinyl acetate, and phenol-resorcinol-
formaldehyde resins. They obtained similar response thermograms
for the three thermosetting resins and a characteristic thermoplastic
thermogram for the polyvinyl acetate resin.

Gollob et al. [55] compared TBA thermograms obtained from
unadulterated phenol-formaldehyde resins differing only in F/P
and NaOH/P molar ratios. These were compared with mode I cleavage
fracture toughness of Douglas fir glued specimens. Fracture toughness decreased as the cured resin rigidity increased (Fig. 5.10).
Rigidity could be manipulated by varying resin synthesis parameters.
Cured resins were less rigid and had higher fracture toughness
as the NaOH/P molar ratio during condensation was increased from
0.15 to 0.35 with an equivalent NaOH/P ratio in the finished resin.
Cured resin rigidity decreased, and fracture toughness increased
as F/P molar ratio was increased from 1.9 to 2.2. The trend then
reversed as cured resin rigidity increased and fracture toughness
decreased, as F/P molar ratio increased from 2.2 to 2.5. In all
cases, resins mixed with wheat flour extenders and nutshell flour
fillers had higher fracture toughness than neat resins. Redistribution
of crack-tip stresses over a greater volume of material likely ac-
counted for the higher fracture toughness of extended/filled glue
mixes.

Gollob et al. [56] compared wood failures of parallel laminated
Douglas fir glued with PF resins differing in molecular weight,
degree of branching, and oligomer content at different assembly
times. At long assembly times, poor bonds were obtained with high
molecular weight resins because they were more prone to dry out.
High contents of free phenol and orthomethylol phenol helped to
plasticize the high molecular weight resins, enabling them to flow
at long assembly times, resulting in improved bond quality. Highly

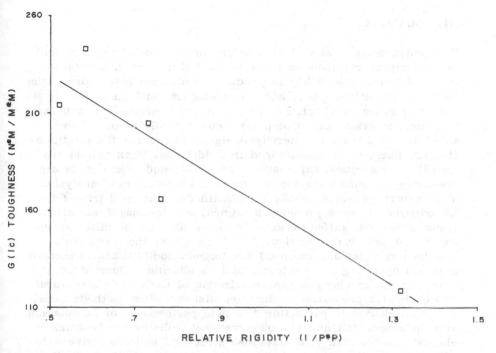

Fig. 5.10 Relationship between fracture toughness (GIc) of bonded wood composite glue line and relative rigidity (I/p^2) of cured PF resin. (From ref. 55.)

branched resins flowed poorly at long assembly times, while less highly branched resins exhibited thermoplastic behavior before cure. Thicker glue lines gave better bond quality. Extenders and fillers were able to change the performance of neat resins, effectively masking intrinsic resin properties, but improving overall glue-bond performance.

Kelley et al. [57] studied the influence of F/P and NaOH/P molar ratios and molecular weight on the development of resin rigidity during cure by torsional braid apparatus. Initial rigidity and damping were strongly influenced by resin molecular weight. Initial rigidity at 100°C increased as the resin molecular weight (weight average = Mw) increased. The highly branched resins were more rigid than the less highly branched resins. Rigidity of the cured resin was greatest at high F/P molar ratio and high NaOH/P molar ratios. Rigidity was also greatest at high NaOH/P molar ratios in the finished resin, supporting Hse's observation [48] that high NaOH/P ratio resins were subject to greater shrinkage and poorer resin bonds.

III. OUTLOOK

Thermodynamic and kinetic knowledge and synthesis/structure and structure/property theories are all included in a resin chemist's cache of skills. The ability to predict performance before production trials is the desired goal. Methods of chemical analysis, rheological characterization, mechanical analysis, fracture mechanics, multivariate statistical analysis, and pilot-plant evaluation all are useful tools. Laboratory studies are generally designed to bracket the conditions that are likely to be encountered in field trials. When trends are identified, theoretical explanations are formulated. Field trials are necessary to confirm the theory generated by the trend analysis. The same trend will generally be identified in the field trial if the lab experiments were properly designed, but the magnitude of the trend is likely to differ between the controlled lab or pilot experiments and daily mill operation. The keen eye of the formulation chemist may determine the need for formula modifications to accommodate mill operating and environmental peculiarities. Current and new methods for chemical characterization of the resin's structural and colligative properties by instrumental and other methods assist the resin chemist in predicting the field performance of formulations and their modifications. The observational skills of the formulation chemist and the ability to juxtapose structural and colligative data with mill operating and environmental peculiarities will remain the "art" in resin formulation.

ACKNOWLEDGMENTS

Thanks go to Marcia McCraney and Sandra Johnson for assistance with the preparation of the typed manuscript and figures.

REFERENCES

1. Drumm, M. F., and LeBlanc, J. R. (1972). "The Reactions of Formaldehyde with Phenols, Melamine, Aniline, and Urea." In *Step Growth Polymerizations*, Vol. 3, D. H. Solomon, Ed. New York, Dekker, pp. 157-278.
2. Ellis, C. (1935). *The Chemistry of Synthetic Resins*, Vol. I. New York, Reinhold, 829 pp.
3. Higginbottom, H. P., Culbertson, H. M., and Woodbrey, J. C. (1965). *Analy. Chem.*, 37(8):1021-1026.
4. Knop, A., and Schieb, W. (1979). *Chemistry and Application of Phenolic Resins*. New York, Springer-Verlag, 269 pp.

5. Martin, R. W. (1956). *The Chemistry of Phenolic Resins.* New York, Wiley, 298 pp.
6. Megson, N. J. L. (1958). *Phenolic Resin Chemistry.* New York, Academic Press, 323 pp.
7. Troughton, G. E., and Rozon, L. (1972). *Wood Sci.*, 4:219-224.
8. Gardikes, J. J., and Konrad, F. M. (1966). "Molecular Weight Distribution Measurements of Phenolic Resins." American Chemical Society, Division of Organic Coatings and Plastic Chemistry, *Preprints*, 26(1):131-137.
9. Quinn, E. J., Osterhoudt, H. W., Heckles, J. S., and Ziegter, D. C. (1968). *Anal. Chem.*, 40(3):547-551.
10. Aldersley, J. W., Bertram, V. M. R., Harper, G. R., and Stark, B. P. (1969). *Br. Polym. J.*, 1:101-109.
11. Armonas, J. E. (1970). *For. Prod. J.*, 23(7):22-28.
12. Wagner, E. R., and Greff, R. J. (1971). *J. Polym. Sci.*, A-1, 9:2183-2207.
13. Duval, M., Bloch, B., and Kohn, S. (1972). *J. Appl. Polym. Sci.*, 16:1585-1602.
14. Hann, N. D. (1977). *J. Polym. Sci. Polym. Chem. Ed.*, 15:1331-1339.
15. Wilson, W. W., Fang, P., and McGinnis, G. D. (1979). *J. Appl. Polym. Sci.*, 24:2195-2198.
16. Connors, W. J., Sarkanen, S., and McCarthy, L. (1980). *Holzforschung*, 34(3):80-85.
17. Wellons, J. D., and Gollob, L. (1980). *Wood Sci.*, 13(2):68-74.
18. Parker, R. J. (1982). "The Effect of Synthesis Variables on Composition and Reactivity of Phenol-Formaldehyde Resins." M.Sc. thesis, Oregon State University, Corvallis, 164 pp.
19. Rudin, A., Fyfe, C. A., and Vines, S. M. (1983). *J. Appl. Polym. Sci.*, 28:2611-2622.
20. Gollob, L. (1983). "The Interaction of Formulation Parameters with Chemical Structure and Adhesive Performance of Phenol-Formaldehyde Resins." Ph.D. Thesis, Oregon State University, Corvallis, 153 pp.
21. Drott, E. E., and Mendelson, R. A. (1970). *J. Polym. Sci.*, 8:1361-1371.
22. Harmon, D. J. (1978). "The Role of Polymer Structure in Fractionation by GPC." In *Chromatography of Synthetic and Biological Polymers*, Vol. 1, *Column Packings, GPC, FP, and Gradient Elution*, R. Epton, Ed. Ellis Harwood Ltd. Sussex, England, pp. 122-145.
23. Tung, L. H., and Moore, J. C. (1977). "Gel Permeation Chromatography." In *Fractionation of Synthetic Polymers—Principles and Practices*, L. H. Tung, Ed. New York, Dekker, pp. 545-647.
24. Hester, R. D., and Mitchell, P. H. (1980). *J. Polym. Sci., Polym. Chem. Ed.*, 18:1727-1738.

25. Flory, P. J. (1953). *Principles of Polymer Chemistry*. Ithaca, NY, Cornell University Press, 672 pp.
26. Kumar, A., Kulshreshtha, A. K., and Gupta, S. K. (1980). *Polymer*, 21:317-328.
27. Kumar, A., Phukan, U. K., Kulshreshtha, A. K., and Gupta, S. K. (1982). *Polymer*, 23:215-221.
28. Bain, D. R., and Wagner, J. D. (1984). *Polymer*, 25(3):403-404.
29. Walsh, A. R., and Campbell, A. G. (1986). *J. Appl. Polym. Sci.*, 32:4291-4293.
30. Woodbrey, J. C., Higginbottom, H. P., and Culbertson, H. M. (1965). *J. Polym. Sci.*, *A*, 3:1079-1106.
31. Hirst, R. C., Grant, D. M., Hoff, R. E., and Burke, W. J. (1965). *J. Polym. Sci.*, *A*, 3:2091-2105.
32. Kopf, P. W., and Wagner, E. R. (1973). *J. Polym. Sci.*, *Polym. Chem. Ed.*, 11:939-960.
33. Steiner, P. R. (1975). *J. Appl. Polym. Sci.*, 19:215.
34. deBreet, J. J., Dankelman, W., Huysmans, W. G. B., and deWit, J. (1977). *Angew. Makromol. Chem.*, 62:7-31.
35. Kim, M. G., Tiedeman, G. T., and Amos, L. W. (1979). "Carbon-13 NMR Study of Phenol-Formaldehyde Resins." In *Phenolic Resins: Chemistry and Application*. Proceedings of a symposium held at Tacoma, WA, June 6-8. Weyerhaeuser Science.
36. Sojka, S. A., Wolfe, R. A., Dietz, Jr., F. A., and Dannels, B. F. (1979). *Macromolecules*, 12(4):767-770.
37. Fyfe, C. A., Rudin, A., and Tchir, W. (1980). *Macromolecules*, 13:1320-1322.
38. Fyfe, C. A., McKinnon, M. S., Rudin, A., and Tchir, W. J. (1983). *Macromolecules*, 16(7):1216-1219.
39. Bryson, R. L., Hatfield, G. R., Early, T. A., Palmer, A. R., and Maciel, G. E. (1983). *Macromolecules*, 16:1669-1672.
40. Maciel, G. E., Chuang, I., and Gollob, L. (1984). *Macromolecules*, 17:1081-1087.
41. Zetta, L., DeMarco, A., Casiraghi, G., Cornia, M., and Kaptein, R. (1985). *Macromolecules*, 18:1095-1100.
42. Kelusky, E. C., Fyfe, C. A., and McKinnon, M. S. (1986). *Macromolecules*, 19:329-332.
43. Pethrick, R. A., and Thomson, B. (1986). *Br. Polym. J.*, 18(3):171-180.
44. Werstler, D. D. (1986). *Polymer*, 27:750-756.
45. Werstler, D. D. (1986). *Polymer*, 27:757-764.
46. Hatfield, G. R., and Maciel, G. E. (1987). *Macromolecules*, 20:608-615.
47. Pillar, W. O. (1966). *For. Prod. J.*, 16(6):29-37.
48. Hse, C. Y. (1971). *For. Prod. J.*, 21(1):44-52.

49. Takatani, M., and Sasaki, H. (1980). "Effect of Glueline Flexibility of Cleavage Fracture Toughness of Wood-Epoxy Resin Bond System." *Wood Res.: Bull. Wood Res. Inst. Kyoto Univ.*, 66:30-51.
50. Gillham, J. K., and Lewis, A. F. (1963). *J. Appl. Polym. Sci.*, 7:2293-2306.
51. Lewis, A. F., and Gillham, J. K. (1963). *J. Appl. Polym. Sci.*, 7:685-694.
52. Senior, A. (1966). "Dynamic Mechanical Studies on a Phenolic Resol During the Thermosetting Curing Process." American Chemical Society, Organic Coatings and Plastics Division, *Preprints*, 25(1):8-13.
53. Warfield, R. W., and Lee, G. F. (1977). *J. Appl. Polym. Sci.*, 21:123-130.
54. Steiner, P. R., and Warren, S. R. (1981). *Holzforschung*, 35(6):273-278.
55. Gollob, L., Ilcewicz, L. B., Kelley, S. S., and Maciel, G.E. (1985). "Evaluating Physical and Chemical Properties of Cured Phenolic Films." In *Wood Adhesives in 1985: Status and Needs*. Proceedings of a conference sponsored by the Forest Products Laboratory, USDA Forest Service, in cooperation with the Forest Products Research Society, May 14-16, pp. 314-327.
56. Gollob, L., Krahmer, R. L., Wellons, J. D., and Christiansen, A. W. (1985). *For. Prod. J.*, 35(3):42-48.
57. Kelley, S. S., Gollob, L., and Wellons, J. D. (1986). *Holzforschung*, 40(5):303-308.

6

Lignin Formaldehyde Wood Adhesives

GERRIT H. van der KLASHORST / Council for Scientific
and Industrial Research, Pretoria, South Africa

I. INTRODUCTION

Lignin, the second major component of wood, is produced in large
quantities as an underutilized by-product during chemical pulping.
In its natural form lignin is a three-dimensional polymer constituted
of random polymerized phenylpropane (C_9) units (Fig. 6.1) [1].
During pulping the lignin macromolecule is degraded and modified
(see below). The polymeric nature of lignin nevertheless prevails
after pulping. Consequently, the development of uses in which
the polymeric nature of the by-product lignin is exploited has
received widespread attention [2].

Chemical pulping can be grouped into two classes, namely sulfite
and alkaline pulping. Sulfite pulping is done in the presence of
sulfite under acidic or alkaline conditions [3]. The degraded lignin
fragments, called lignosulfonate, are of a rather high molecular
weight but are kept in solution by sulfonic acid groups introduced
during pulping. The polymeric character and the sulfonic acid groups
impart surface-active and binding properties to the lignosulfonate
[4]. These properties have been utilized in a number of applications
developed for lignosulfonates such as dispersants, emulsifiers, and
binders [4]. The sulfonic acid groups, however, impart a degree
of hygroscopicity to the lignosulfonate. This property and the poor
ability of lignosulfonates to co-crosslink with adhesives, such as
phenol-formaldehyde adhesives, probably lead to the poor utilization
of this material in wood adhesives. In most wood adhesive applications
that have been developed, the lignosulfonate replaces only a part
of the adhesive resin. Exceptions are the crosslinking of lignosulfonate by oxidative coupling reactions [2] or when the curing
reaction also results in the loss of the sulfonic acid groups [2].

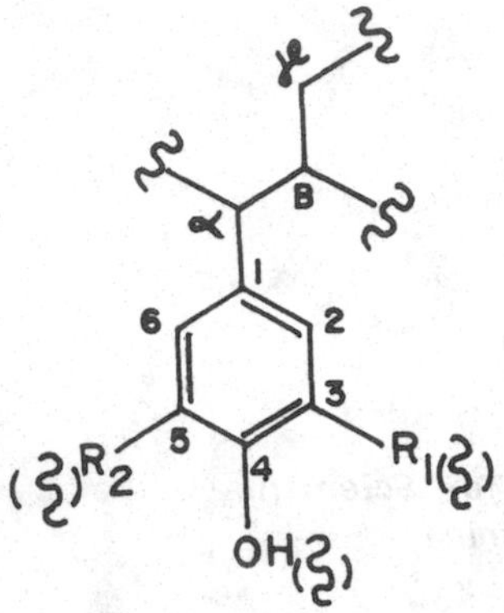

Fig. 6.1 Phenyl propanoid units of lignin: R_1, R_2 = H, OCH_3, or / and ( ) = possible linkage to other phenyl propanoid units.

Alkaline pulping is done with sodium hydroxide as the major pulping chemical [5]. Variations include Kraft pulping, in which about a third of the sodium hydroxide is replaced with sodium sulfide (Na_2S), and soda/AQ pulping, in which catalytic quantities of anthraquinone are added. During alkaline pulping the lignin is usually extensively modified. The major degradation reaction is the cleavage of alkylaryl ether (β—O—4) linkages [5]. Concomitant to the cleavage reactions, the lignin fragments also undergo condensation reactions (more about this later). The cleavage of alkylaryl ether linkages results in the formation of phenolic groups in the lignin. The pK_a value for the phenolic groups in lignin is in the region of 10. The phenolic groups are therefore dissociated under the highly alkaline conditions during pulping, rendering the lignin fragments water soluble. Alkali lignin can now be reclaimed easily from the spent pulping liquor simply by lowering of the pH of the spent liquor below the pK_a value of the lignin phenolic groups.

In practice the pH is lowered to pH 9.5 with flue gas available at the mill [6]. This results in the precipitation of the alkali lignin, which is removed from the spent liquor by filtration or decantation. The remaining liquor still contains the spent pulping chemicals and usually a large portion of organics in the form of degraded carbohydrates (mainly hemicellulose) and some lignin fragments. The organic/inorganic ratio of the remaining liquor usually allows enough energy for its combustion after concentration. Particularly since the large lignin fragments have been removed, it is possible for these liquors to be concentrated to a higher solids content than before without becoming too viscous [7]. The reclamation of alkali lignin from spent pulping liquors therefore is cheap and need not upset the recovery of the spent pulping chemicals. Alkali lignin is therefore available in large quantities as a renewable polymeric raw material.

A chapter in an earlier volume in this series [2] thoroughly discusses the utilization of lignosulfonates in wood adhesives. In this chapter we treat the use of alkali lignin in the preparation of wood adhesives with the emphasis on polymerization with formaldehyde.

II. POLYMERIZATION REACTIONS FOR ALKALI LIGNIN

To perceive the polymerization reactions that can make alkali lignin useful in wood adhesives, it is necessary to look at the chemical composition of this material. As mentioned earlier, alkali lignin contains a substantial number of phenolic hydroxy groups owing to the cleavage of arylalkyl ether linkages during pulping [5]. An important aliphatic hydroxy group introduced via this reaction is the α-hydroxy (carbinol) group. If these α-hydroxy groups are situated on phenolic C_9 units, they may form quinone methide intermediates under the highly alkaline conditions during pulping. The quinone methides react easily with reactive lignin fragments to give alkali-stable methylene linkages [8].

This last reaction is probably the most common "condensation" reaction, resulting in the repolymerization of lignin fragments during alkaline pulping (Fig. 6.2).

Clearly the chemical properties of alkali lignin are very intricate and are influenced by raw material and pulping conditions, to name two important variables. To establish a basis of discussion, we will use the tentative structure proposed in 1971 by Marton [5] for a Kraft softwood lignin (Fig. 6.3). These properties however may differ substantially from lignin to lignin.

The chemical features of alkali lignin that can be used for polymerization reactions in wood adhesive applications are as follows:

1. Phenolic hydroxy groups
2. Aliphatic hydroxy groups
3. Structures that can form quinone methide intermediates (a special case of 1 and 2)
4. Unsubstituted 3- or 5-positions on phenolic C_9 units

The phenolic hydroxy group on lignin was used for the crosslinking of model compounds with bis-acid chlorides [9] and cyanuric

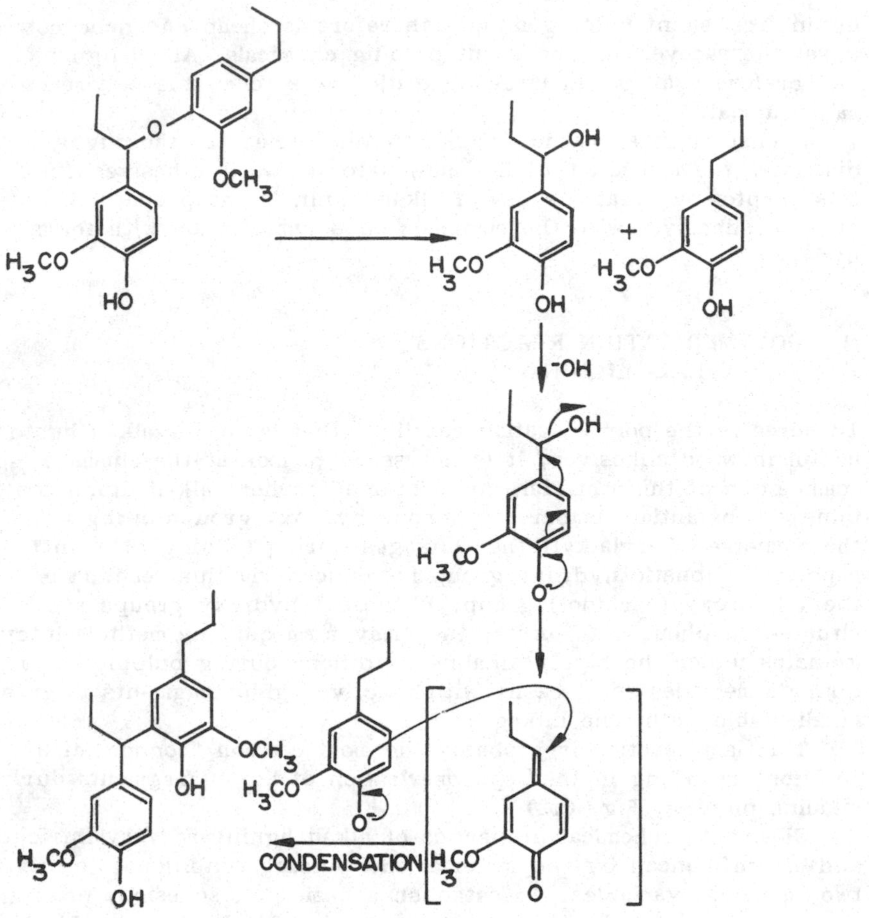

Fig. 6.2 Cleavage of the α—O—4 ethers and subsequent condensation of lignin during alkaline pulping.

chloride [10]. In both cases high crosslinking yields were obtained in aqueous solvents. The reactions however failed to give proper gelling of lignosulfonates. The phenolic hydroxy groups have also been used in attempts to graft epoxy groups onto the lignin with epichlorohydrin. The products obtained from lignosulfonate contained, however, low epoxy equivalents [11]. The epoxidation of alkali lignin with epichlorohydrin resulted in much higher epoxy yields [12]. Other reagents that can also be used for the polymerization of lignin via the phenolic hydroxy group are aziridines and iso-cyanates [13].

Di- or polyisocyanates react easily with both phenolic and aliphatic alcohols [13]. This approach to the preparation of wood adhesives is the most effective if the phenolic hydroxy groups are transformed into aliphatic hydroxy groups [14]. The final products, however, contained only a minor portion of lignin, which may obscure the interest and economic advantages of using such a cheap material.

Polymerization of lignin was also achieved by crosslinking of the electron-rich sites on the aromatic rings of lignin. Of those approaches, the use of lignin in phenol-formaldehyde adhesives has received by far the most attention. This avenue is discussed in more depth in the next section. Another approach comprises the utilization of lignin in urea-formaldehyde (UF) adhesives [2]. Model compound studies on the crosslinking of lignin with diazonium salts gave high yields of crosslinked products [15]. No durable product could however be produced from by-product lignin via this reaction [16]. Probably the most successful approach exploiting these positions consists of the free radical oxidative coupling of lignin. The exothermic reaction is brought about with hydrogen peroxide and catalytic quantities of sulfur dioxide or other catalysts. A thorough discussion is given elsewhere [2].

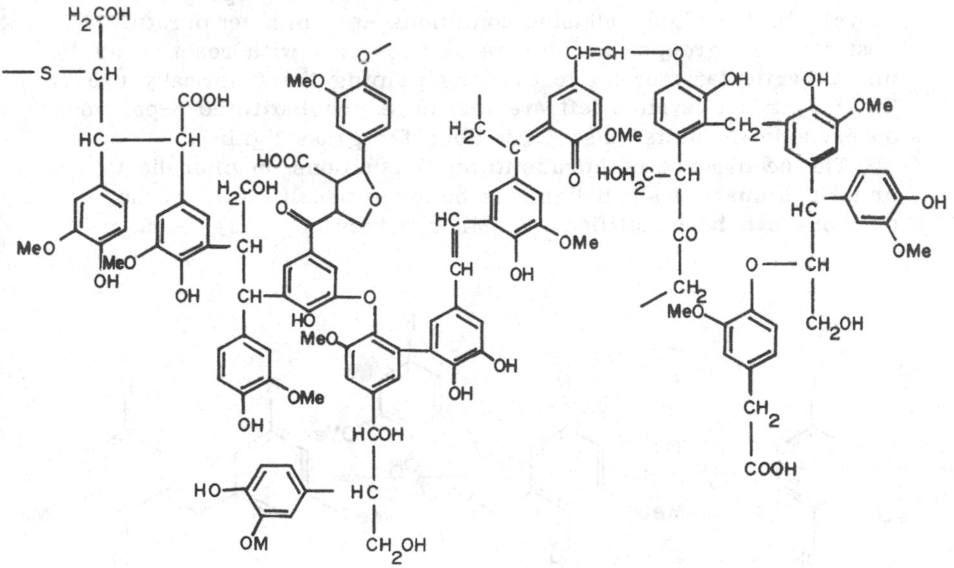

Fig. 6.3 Tentative structure of a softwood Kraft lignin by Marton [5].

III. THE UTILIZATION OF ALKALI LIGNIN IN PHENOL-FORMALDEHYDE WOOD ADHESIVES

The substitution of phenol-formaldehyde (PF) resins, and particularly phenol-formaldehyde wood adhesives, with lignin is certainly the most widely explored avenue of lignin utilization. This clearly stems from the phenolic character of lignin and therefore the expected compatibility of lignin with PF resins. The optimum situation is for the lignin to replace the largest possible portion of phenol, if not all the phenol. To attain an acceptable final product, it is therefore required that the lignin react with the PF resin in a high degree.

The reactions whereby alkali lignin react with PF resins are via unsubstituted ortho and para positions on phenolic rings and via α-carbinol groups on phenolic propanoid units. The first reaction usually occurs on the 5-position (for softwood lignin), whereby formaldehyde will react to afford a hydroxymethylated intermediate [17], which reacts (particularly at a higher temperature) with, for example, phenol, to form a methylene linkage [18]. Figure 6.4 shows this reaction. The α-carbinol groups react with phenol in a manner similar to the hydroxymethylated lignin intermediate mentioned above, again to give a methylene linkage (Fig. 6.5).

For alkali lignins, the α-carbinol groups are less important owing to their loss by this very reaction during pulping [5] (see above). Under highly alkaline conditions and high temperature, most of these groups can be expected to react with reactive lignin units, particularly owing to the long pulping times normally used. The only reactive sites left are therefore unsubstituted 5-positions on phenolic C_9 units (also 3-positions for grass lignin).

The occurrence of unsubstituted 5-positions on phenolic C_9 units in industrial alkali lignin is however usually low. These positions can be quantified by NMR spectroscopy [23]. A more

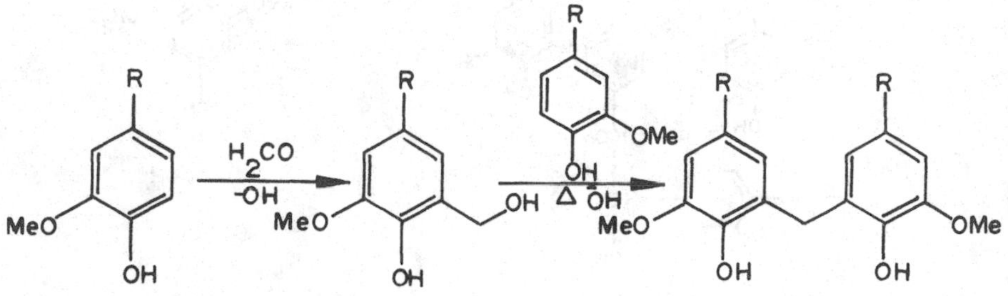

Fig. 6.4 A schematic presentation of the polymerization of lignin with formaldehyde in alkaline solution.

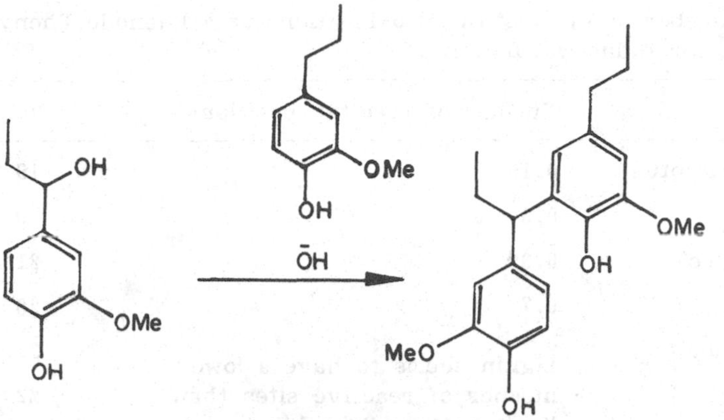

Fig. 6.5 Reaction of α-carbinol groups on phenolic C_9 units with a reactive lignin fragment to afford a methylene linkage.

robust approach is the quantified hydroxymethylation of the lignin under mild conditions [24]. The latter procedure is nevertheless very useful and gives an immediate indication of the potential degree of reactivity of the lignin.

For most industrial lignins these sites number between 0.1 and 0.3 per C_9 (Table 6.1). Exceptions are lignins obtained from the soda pulping of bagasse (which is done at 170°C for only 15 minutes).

Alkali lignin containing only 0.3 reactive site per C_9 unit clearly should have difficulty in forming an acceptable copolymerized product with phenol-formaldehyde resins. This difficulty is particularly marked for unfractionated lignins, which normally contain a substantial portion of low molecular weight fragments [21]. The low molecular weight fractions can be expected to result in the termination of the polymerization reaction of the resin owing to their large numbers and low content of reactive sites.

Not surprisingly then, most approaches to the use of industrial alkali lignin in PF resin products have been successful only when minor proportions of lignin were employed [25,26]. Lignin/PF resins usually perform better in nonwood adhesive applications such as paper laminates, where the strength of the cured resin is less critical [27].

Various avenues have been explored to circumvent this inherent inability of industrial alkali lignin to copolymerize with PF resins. For example, it has been shown that high molecular weight lignin fractions, obtained by ultrafiltration, can partially overcome this problem [28-31]. The high molecular weight lignin fragments can be expected to contain an average number of reactive sites similar

Table 6.1   Number of Unsubstituted 5-Positions of a Phenolic Phenyl Propane Unit for Industrial Lignin

Lignin	Number of reactive positions	Ref.
Soda/AQ eucalyptus	0.1	19
Kraft pine	0.3	20
Reax (Westvaco)	0.3[a]	21
Soda bagasse	0.7	20
Steam explosion ⎱	Lignin seems to have a lower number of reactive sites than Kraft softwood lignin	22
Acid hydrolysis ⎰		22

[a]Reax is a softwood (i.e., guaiacyl) lignin with about 3-4 phenolic hydroxy groups per thousand mass units [21]. Take C_9 as 180 mass units [i.e., 3.5 × (180/1000) = 0.63 phenolic hydroxy groups per C_9], of which half is condensed (i.e., phenolic C_9 units with unsubstituted 5 positions = 0.3).

to that of the whole lignin. Due to the larger number C_9 units per fragment, each fragment has a much better chance to contribute to polymerization, compared with the monomeric and dimeric fractions.

The work by Forss and coworkers resulted in high molecular weight lignin-based plywood adhesives employing 60% lignin and 40% of a commercial particleboard adhesive [30]. Plywood produced from this adhesive was within the Finnish standard when pressed for 3 minutes. Chipboards 12 mm thick complied with the German standard after pressing at 180°C for 4.5 minutes [30].

Lignin can also be chemically modified to increase its reactivity toward formaldehyde, which will enhance its incorporation into PF adhesives. This can be achieved by:

1. Substitution of the lignin fragments with phenol, thereby introducing more reaction sites
2. Activation of the aromatic nuclei of the lignin toward formaldehyde

The substitution of lignin with phenol immediately implies a reduction of the quantity of lignin in the final product. Several attempts in this direction have nevertheless been made.

Under acidic conditions phenol reacts with oxysubstituted α-carbons to afford a phenolized lignin adduct [32,33] (Fig. 6.6).

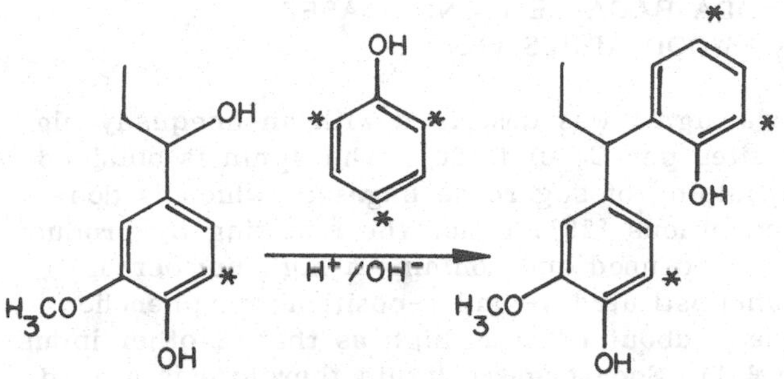

Fig. 6.6 The increase of reactive sites (*) on lignin fragments by phenolysis.

Phenolized alkali lignin can be used as a thermosetting resin for the preparation of sheets [34]. Phenolized lignosulfonate was used to prepare boil-resistant plywood adhesive when panels were pressed for 5 minutes at 140°C [33]. Various other acidic phenolysis approaches were also used on lignosulfonates [34,35]. The alkaline phenolysis of alkali lignin affords a product that can replace up to 50% of a plywood adhesive with exterior properties and a press time of 3.5 minutes for 11-mm board [35].

Activation of the aromatic nuclei of the lignin toward formaldehyde can be achieved by demethylation, resulting in an increase in the phenolic hydroxyl content of the lignin. The 2- and 6-positions of the C_9 unit are situated ortho and para to the newly formed phenol group and can now contribute in the classical reaction with formaldehyde. The demethylation of lignin hitherto studied did not prove to be viable due to the high cost of the reagents, such as hydroiodic acid [37], sodium periodate [38], and potassium dichromate [39]. Other demethylation agents that might in the future lead to a viable product include chlorine [40] and ethanolamine [41].

It is clear from the examples above that industrial by-product lignin can be used as an extender for phenol-formaldehyde resins. Addition levels are, however, usually low. When a high molecular weight lignin is used, partial substitution of 50-70% of phenol-formaldehyde resin is possible. This high compatibility is attributed to the absence of the low molecular weight components, which are, as a consequence of their small size (and therefore low number of reactive centers), unable to contribute to the polymer network. With chemical modifications of lignin aimed at producing more reactive lignins, this terminating effect was rarely taken into account. The combination of these two concepts, namely the utilization of high molecular weight lignin and an activated lignin, should have a further positive effect on the compatibility with phenol-formaldehyde resins.

IV. INDUSTRIAL SODA BAGASSE LIGNIN-BASED COLD-SETTING WOOD ADHESIVES

Recently an industrial lignin was described with an unequally high number of reactive sites per C_9 unit [20]. The lignin is obtained from the industrial pulping of sugarcane bagasse, which is done under rather mild conditions [20]. Hence the resulting by-product lignin is only mildly condensed and contains a high number of approximately 0.7-unsubstituted 3- and 5-positions on phenolic C_9 units. This value is about twice as high as that of other industrial lignins (Table 6.1). Soda bagasse lignin therefore is a good candidate to replace phenol-formaldehyde resins in high proportions in polymeric products.

Soda bagasse lignin was subsequently successfully used as the backbone polymer for the preparation of cold-cure wood adhesives [42]. The adhesives were prepared by the hydroxymethylation of the bagasse lignin followed by the reaction of resorcinol onto the hydroxymethylated lignin (Fig. 6.7). The resorcinol-lignin adduct can be crosslinked at ambient temperature by the addition of para-formaldehyde. This causes the formation of a series of methylene linkages connecting two resorcinol terminals, each grafted onto a lignin fragment.

After optimization, an adhesive was prepared that contained only 14% resorcinol on liquid adhesive and 76% hydroxymethylated

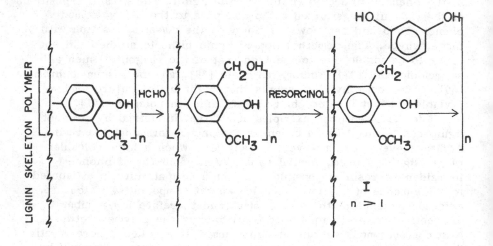

Fig. 6.7 Diagrammatic presentation of the preparation of soda bagasse lignin cold-set adhesive [42].

Table 6.2 *Pinus patula* Finger Joints: Results of Traditional-type Cold-Set Adhesive Composed of 13.6% Resorcinol Grafted onto Hydroxymethylated Soda Bagasse Lignin, and "Honeymoon" Fast-Set Adhesive Based on Hydroxymethylated Soda Bagasse Lignin Containing a Total of 13.6% Resorcinol and 19% Resorcinol (on liquid resin)

Adhesive type	Dry test		24-hr Cold soak		6-hr Boil	
	Strength (N)	Wood failure (%)	Strength (N)	Wood failure (%)	Strength (N)	Wood failure (%)
1. Lignin-based cold-set Type 1	2,560	100	2,060	90	2,010	80
2. Lignin-based "honeymoon" Type 2	1,933	82	1,383	40	2,238	97
3. Lignin-based "honeymoon" Type 3	2,533	79	2,080	95	2,117	97

Table 6.3 Beech Strip Test Results of Cold-Setting Lignin-Based Wood Adhesives

	Dry test		24-hr Cold soak		6-hr Boil	
	Strength (N)	Wood failure (%)	Strength (N)	Wood failure (%)	Strength (N)	Wood failure (%)
Lignin-based cold-set type 1[a]	3,338	33	2,473	95	1,870	80
SABS specification requirements[b]	2,500	—	2,200	75	1,500	75
BS specification requirements[c]	—	—	2,225	—	1,450	—

[a]From Table 6.2

[b]South African Bureau of Standards: SABS 0183-1981 specification for class 1 exterior-grade structural wood adhesives.

[c]British Standard: BS 1204-1965, part 2 specification for synthetic adhesives for wood, marine-grade.

lignin on total solids. The adhesive gave excellent adhesion of finger
joints (Table 6.2, type 1). The adhesion was also evaluated by the
beech strip test and gave acceptable strengths and wood failures
for exterior specification requirements (Table 6.3).

V. LIGNIN-BASED "HONEYMOON" ADHESIVES

Soda bagasse lignin was also used to prepare acceptable fast-set
"honeymoon"-type adhesives [42]. Two adhesives were prepared.
    A liquid resin identical to the cold-curing adhesive prepared above
was used for both components A and B of a "honeymoon" fast-set
adhesive (Table 6.2, type 2). Component A at pH 9.3 contained
paraformaldehyde hardener and had a pot-life exceeding 1 hour
at 25°C. Component B was the same liquid resin (same resorcinol
content) with no paraformaldehyde hardener, but at a pH of 13.
A second resin (Table 6.2, type 3) was prepared employing 0.48
part resorcinol instead of 0.32 part as before. Adhesives 2 and 3
had respectively 13.5 and 19.0% resorcinol on total liquid resin
before addition of paraformaldehyde powder as hardener. The results
obtained on *Pinus patula* finger joints having an equilibrium moisture
content of 10% (10% emc) with such a separate application system
are also shown in Table 6.2.
    The results in Tables 6.2 and 6.3 show that the cold-setting
and fast-setting wood adhesives for exterior-grade structural finger
joints and glulam based on soda bagasse lignin satisfy the relevant
international specifications [43]. In particular, the results indicate
excellent results from the lignin-based cold-set and fast-set adhesives
in which 13.6 and 19% resorcinol (on total liquid resin) has been
grafted onto soda bagasse lignin after hydroxymethylation of the
lignin. It must be noted that while a traditional-type lignin-based
cold set needs only 13.6% resorcinol to compare with synthetic
phenol-resorcinol-formaldehyde (PRF) adhesive, the fast-set type
needs as much as 19% to obtain an acceptable result.
    The use of soda bagasse lignin to replace the phenolic backbone
of phenol-resorcinol-formaldehyde cold-set adhesives clearly exempli-
fies the usefulness of this reactive industrial lignin. It was subsequently also shown to be suitable for the preparation of thermosetting
wood adhesives.

VI. SODA BAGASSE THERMOSETTING WOOD ADHESIVES

Soda bagasse lignin was initially evaluated as a thermosetting adhesive
by the beech strip test. Before being applied as adhesive, the lignin
was reacted with formaldehyde in alkali at temperatures below 60°C

to afford a hydroxymethylated lignin. The hydroxymethylation reaction was done at pH 12 and 13, and samples of the reaction mixtures were evaluated on beech strips with overlaps of 25 × 25 mm,
cured for 4 hours at 90°C and 12% equilibrium moisture content.
The results presented in Fig. 6.8 clearly indicated that the bagasse
lignin can be used to prepare a thermosetting adhesive of high
strength. The strength of the adhesive increased with increasing
hydroxymethylation times, which were as long as 15 hours.

Improvement of the performance of the lignin adhesive was subsequently attempted by the addition of several crosslinking agents.
These materials are capable of further improving the degree of
crosslinking of the lignin adhesive (Table 6.4).

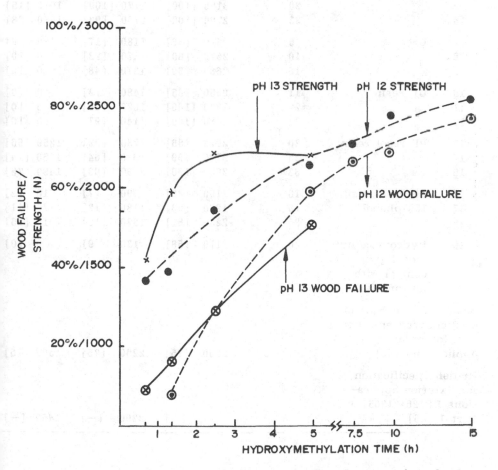

Fig. 6.8 Strength and wood failure obtained on beech strips glued
with lignin hydroxymethylated for different times at pH 13 or 12.

Table 6.4 Beech Strip Strengths of Hydroxymethylated Bagasse Lignin Adhesives Crosslinked with Various Fortifiers

Entry	Fortifier	Percentage fortifier on resin solids (%)	Adhesive strength [% wood failure]		
			Dry	Soak	Boil
1	Phenol	8	2480 [96]	1020 [5]	1160 [0]
2		15	2290 [76]	1040 [5]	0 [0]
3		25	2300 [81]	820 [5]	0 [0]
4	Melamine	11	2970 [100]	2250 [98]	670 [2]
5		20	3150 [100]	2170 [100]	1070 [15]
6		33	2700 [100]	2730 [93]	1450 [28]
7	Urea	5	2510 [56]	1180 [17]	130 [0]
8		10	2570 [60]	1270 [22]	0 [0]
9		16	2680 [90]	1370 [48]	0 [0]
10	UF resin	11	2550 [85]	1480 [2]	0 [0]
11		20	2830 [100]	1620 [35]	0 [0]
12		33	2620 [100]	140 [97]	0 [0]
13	PF resol resin	20	2970 [86]	1780 [32]	1259 [66]
14		25	2720 [99]	1920 [64]	1350 [24]
15		33	2930 [100]	2090 [83]	1490 [75]
16	PF resol resin +	10	2190 [89]	990 [7]	710 [2]
17	15% phenol	15	1960 [53]	1180 [12]	1120 [11]
18		25	2230 [61]	1320 [16]	1410 [25]
19	Hydroxymethyl-ated lignin control with no fortifier	0	2110 [58]	130 [9]	0 [0]
	South African Bureau of Standards specification for exterior applications [45]		2500 [75]	2200 [75]	1500 [75]
	British specification for exterior applications BS1204-1965, Part 1 [46]		— [−]	2200 [−]	1450 [−]

The addition of phenol resulted in only small increases in the dry strength results, as shown in the table.

The addition of a PF (resol) resin to fixed-time hydroxymethylated bagasse lignin resulted in a substantial improvement in the adhesive performance (Table 6.4, entries 13-18), while addition of phenol to the PF resin/lignin combination did not improve the dry strength of the bonded joints over the control. The strength of the adhesive obtained by the addition of 33% resol resin by mass (entry 15) practically complies with requirements of the South African and British Standard specifications for interior-type synthetic adhesives for wood, and it is not too far from the exterior-grade requirements. The addition of a resol resin to hydroxymethylated bagasse lignin therefore constitutes a versatile adhesive preparation. Employing more resol resin in the adhesive mixture, on the other hand, results in a thermosetting adhesive that marginally complies with the specification for exterior-grade resins.

The addition of melamine, urea, and UF resin resulted in each case in a substantial increase in the strength values of the hydroxymethylated lignin adhesives (Table 6.4, entries 7-12). The best results were obtained by the melamine-crosslinked adhesives, which gave an adhesive with strength values well within exterior-grade adhesive specifications.

One-layer particleboards were prepared from eucalyptus chips with 10% lignin resin on dry chips [47]. The boards were formed 12 mm thick on a laboratory press with the press platens at 170°C for press times of 15 minutes. The final density of the boards was approximately 700 kg/m^3. Different ratios of commercial PF and UF resins were added to hydroxymethylated lignin. The results listed in Table 6.5 indicate that the lignin-based adhesives give proper

Table 6.5 Internal Bond Strength of Particleboard 12 mm Thick, Prepared from Soda Bagasse Lignin Adhesives (10% on dry chips) After Pressing at 170°C for 15 Minutes

Entry		Dry internal bond strength (MPa)
1	Hydroxymethylated lignin/PF, 50:50	0.80
2	Hydroxymethylated lignin/PF, 67:33	0.63
3	Hydroxymethylated lignin/UF, 67:33	0.35
4	Hydroxymethylated lignin/UF, 80:20	0.62

From Ref. 47.

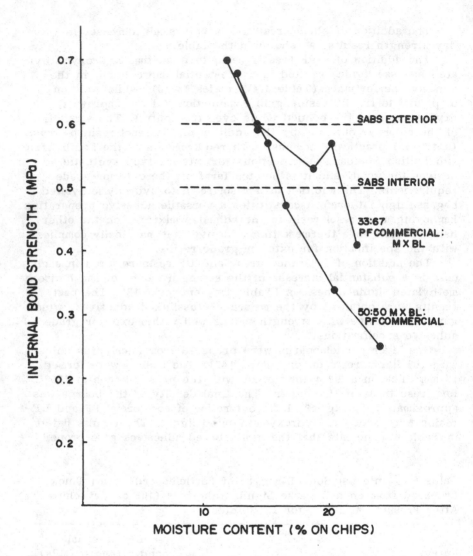

Fig. 6.9 Dependence of strength of 12-mm one-layer board glued with a mixture of hydroxymethylated lignin (MXBL) and a commercial PF (67:33) pressed at 170°C for 15 minutes [47].

bonding of particleboard. When larger proportions of lignin were
used with the PF resin, however, internal bond strength decreased
(Table 6.5, entries 1 and 2). The reverse was evident for the UF-
lignin adhesives. The poor performance of the adhesive mixture
containing the largest proportion of UF resin probably can be
attributed to degradation of the UF component due to the long
press times of 15 minutes.

The moisture dependence of the two PF extended lignin adhesives,
presented in Fig. 6.9, clearly indicate the deleterious effect of the

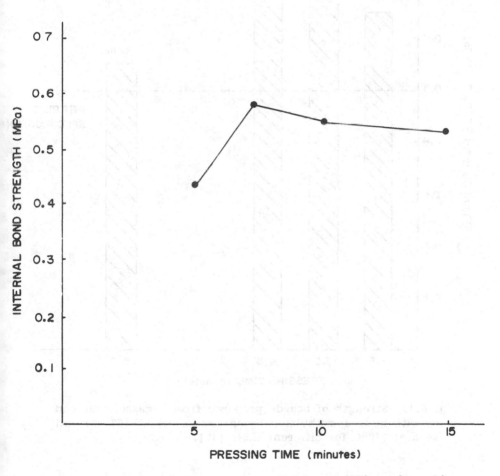

Fig. 6.10 Strength of three-layer particleboard (20% surface chips,
10% core chips) glued with hydroxymethylated bagasse lignin/commercial
PF resin (67:33) and pressed at 170-180°C for different times, 12%
chip moisture content, final density 680 kg/m³ [47].

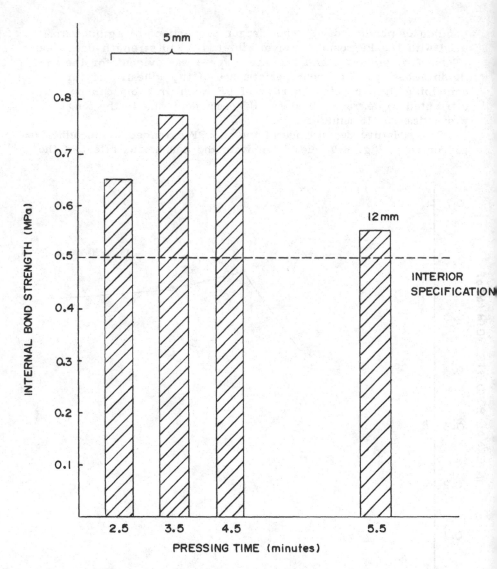

Fig. 6.11 Strength of boards prepared from bagasse fiber and glued with hydroxymethylated lignin/commercial PF (67:33) and pressed at 170°C for different times [47].

high moisture content of the resinated chips on the strength of the boards. The application of the lignin particleboard adhesive thus clearly requires strict control of the moisture content of the resinated chips. The internal bond strengths of the three-layer boards, presented in Fig. 6.10, show that times much shorter than 15 minutes can still produce acceptable boards.

The bagasse lignin/PF (67:33) adhesive was also used for the preparation of one-layer boards from bagasse fiber instead of wood chips [47]. For this material 2-1/2-minute press times were long enough to produce acceptable dry internal bond strengths for board 5 mm thick, and 5-1/2-minute pressing time for board 12 mm thick (Fig. 6.11).

The results presented indicate that industrial soda bagasse lignin can be used after hydroxymethylation as thermosetting adhesive to give acceptable adhesion of beech strips and particleboard. The lignin-based particleboard adhesive was supplemented with commercial PF resin to levels of only 33% synthetic resin (i.e., 67% lignin). For three-layer particleboard prepared from wood chips, a press time of 75 minutes was required. When bagasse fiber was used, a press time of 5-1/2 minutes was sufficient for a one-layer, 12-mm board.

The extraction of the industrial lignin is currently evaluated on a pilot-plant scale, which will enable the evaluation of the lignin particleboard adhesive on an industrial scale.

The development of the phenol-formaldehyde type of wood adhesives from reactive industrial soda bagasse lignin certainly gave new impetus to the approach. However, this industrial lignin is used mainly in Third World countries. If the current pulping operations in Europe and North America continue to be used, alkali wood lignin, with a low reactivity, will remain underutilized.

VII. META-MODIFIED LIGNIN-BASED
        WOOD ADHESIVES

It is clear from the examples presented so far that the major constraint for the utilization of lignin in phenol-formaldehyde resins is the low reactivity of most industrial by-product lignins. Industrial lignin is usually fairly condensed (see above), hence does not contain a high number of unsubstituted 5-positions on phenolic C_9 units to contribute in crosslinking reactions with formaldehyde.

The 2- and 6-positions of the lignin phenyl propane units are mainly unsubstituted in natural lignin [1]. During pulping, these positions usually remains unsubstituted [5]. Industrial by-product lignin therefore contains a high number of unsubstituted 2- and

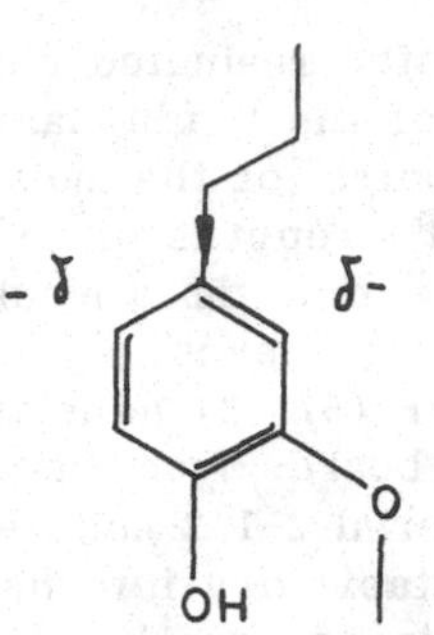

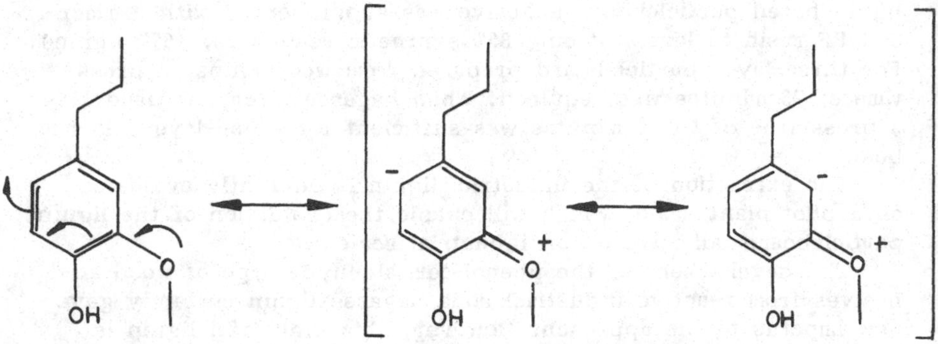

Fig. 6.12 Mechanisms responsible for the higher electron density
on positions 2 and 6 on lignin phenylpropane units in acidic medium.

6-positions. These positions are situated meta to the phenolic group
and are hence not influenced by the ortho-para directing effect
of the phenol group. Under alkaline or mild acidic conditions, the
2- and 6-positions contribute hardly at all to any electrophilic substitution reaction.

Under acidic conditions the situation is however different for
the lignin C_9 units with 1-alkyl substituents. The 1-alkyl groups
are ortho-para electron donating; that is, positions 2, 6, and 4
become more electron rich. Secondly, the electron pair on methoxy
group 5 also increases the electron density on positions 2, 6, and
4 by resonance (Fig. 6.12).

Sarkanen, Erickson, and Suziki [48] showed lignin model compounds to undergo protodedeuteration in acidic medium preferably
on the 2- and 6-positions.

Electrophiles other than protons were indeed shown to react with lignin model compounds at the 2- and 6-positions in acidic media. Kratzl and Wagner [49] reacted the benzyl alcohol 1 with 4-hydroxy-3-methoxyphenylmethane (2) under acidic conditions to yield the meta-linked product 3. Under alkaline conditions the classical product 4 was obtained (Fig. 6.13).

Under the conditions used for Klason lignin determinations, Yasuda and Terashima [50,51] showed that vanilyl alcohol and other model compounds may also give meta-crosslinked products.

The polymerization of lignin model compounds at the meta positions was subsequently studied to carry out this reaction in a controlled manner [52]. Hardwood and softwood lignin model compounds were polymerized on the 2- and 6-positions with formaldehyde in acidic aqueous dioxane to afford dimers, tetramers, and higher oligomers (Fig. 6.14). Phenolic and nonphenolic models both reacted to form methylene-bridged products. Kinetic experiments (Table 6.6) showed that the hardwood models react somewhat faster than the softwood models. The presence of the second methoxy group thus enhances the electron density of the 2- and 6-positions even more than the contribution of the single methoxy group on softwood models.

A mechanistic formulation of the reaction of formaldehyde with 3,5-dimethoxy-substituted 4-hydroxyphenyl alkanes is given in Fig. 6.15. For stoichiometric ratios of model compound and formaldehyde in 50% aqueous dioxane containing 0,44N HCl, the reaction was of second order. No benzyl alcohol (21) was isolated, and the total yield of the starting material and methylene-linked products

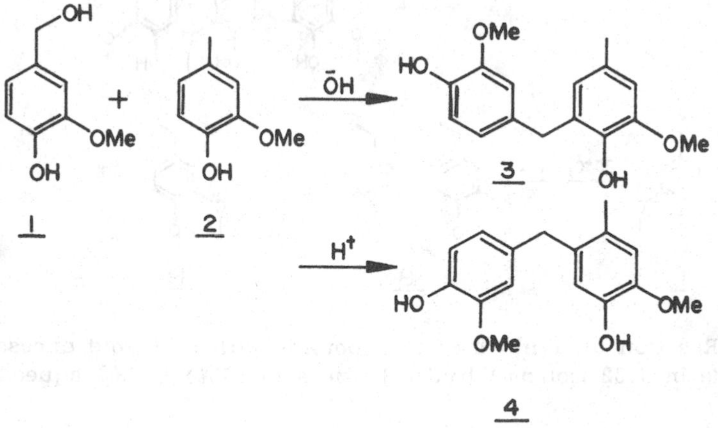

Fig. 6.13 Reaction of vanilyl alcohol with 4-hydroxy-3-methoxyphenyl methane [49].

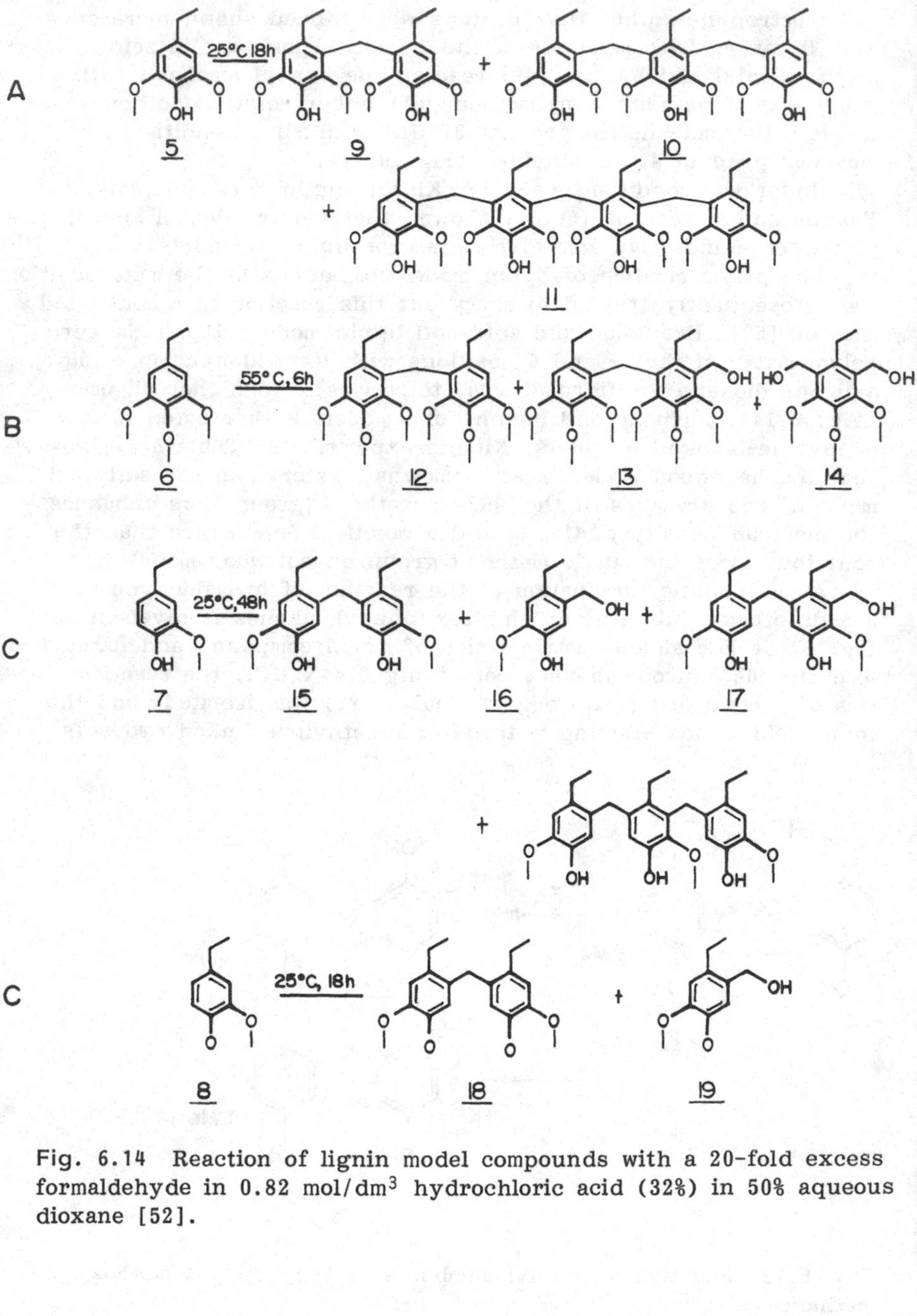

Fig. 6.14 Reaction of lignin model compounds with a 20-fold excess formaldehyde in 0.82 mol/dm^3 hydrochloric acid (32%) in 50% aqueous dioxane [52].

Table 6.6 Reactions of Formaldehyde H_2CO with Lignin Model Compounds in Acidic Aqueous Dioxane [52]

Entry	SM[a]	SM concentration (mol/liter)	SM/H_2CO ratio	Acid concentration (mol/dm^3)	Reaction temperature (°C)	Reaction rate constant k (liter/s-mol) (dimer formation)
1	5	2.005	1:1	0.44	80	2.3×10^{-3}
2	5	2.070	1:1	2.2	80	2.4×10^{-2}
3	5	1.857	1:1	2.2	30	$(4.2 \times 10^{-7}$ mol/liter-s)[b]
4	5	1.857	1:20	2.2	30	$(1.62 \times 10^{-6}$ mol/liter-s)[b]
5	6	1.777	1:1	2.2	80	4.8×10^{-2}
6	7	2.187	1:1	2.2	80	1.0×10^{-2}
7	8	2.211	1:1	2.2	80	6.1×10^{-3}

[a]SM = starting material.
[b]Zeroth-order reaction.

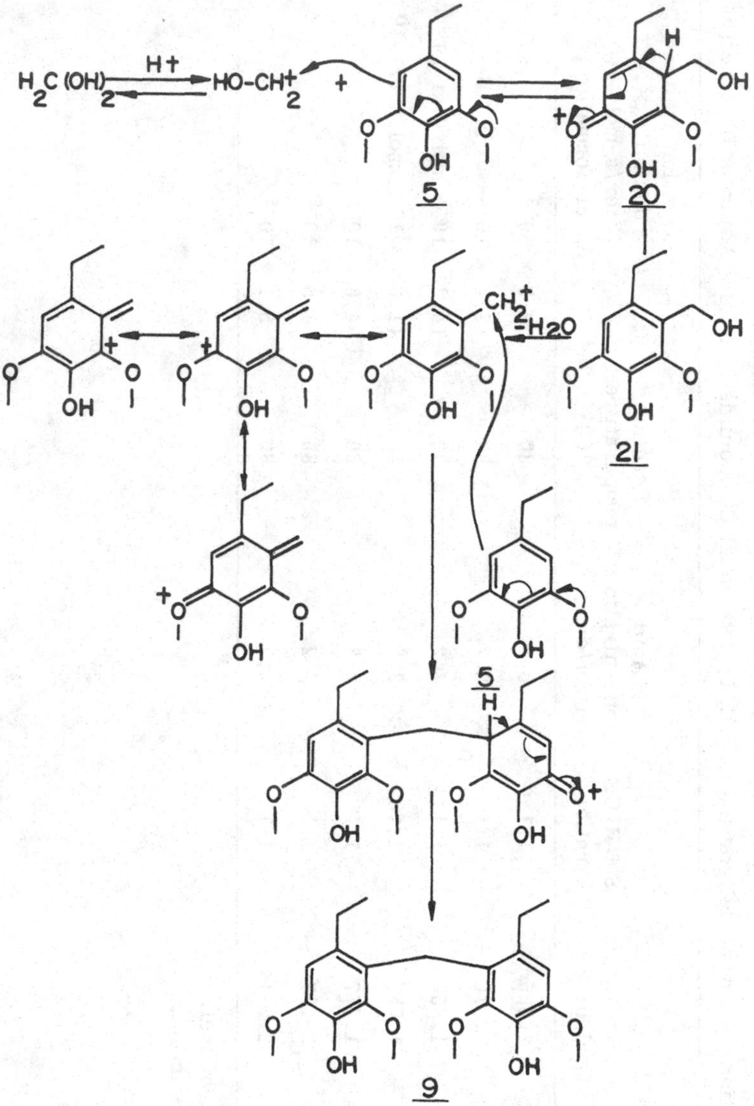

Fig. 6.15 Mechanistic formulation of the reaction of formaldehyde with 3,5-methyoxy-4-hydroxyphenyl alkanes [52].

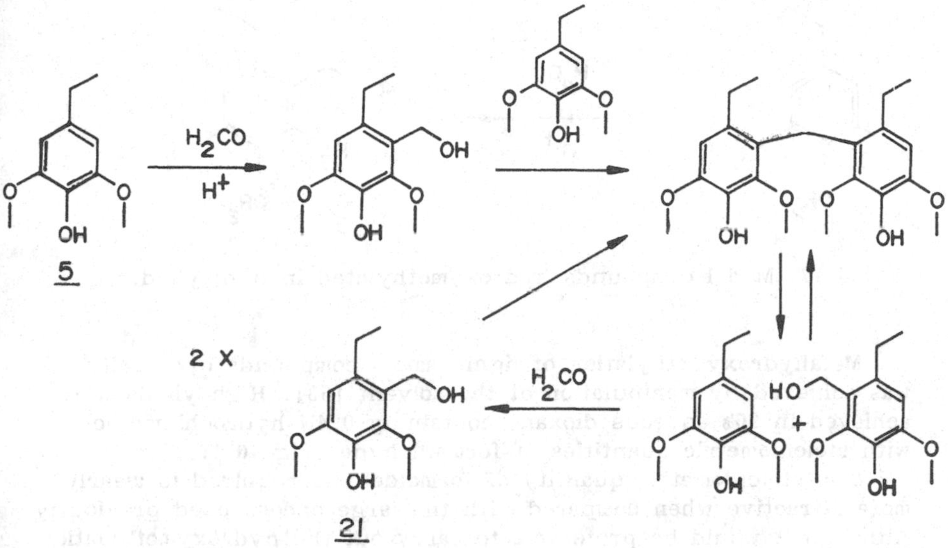

Fig. 6.16 Proposed approach for the metahydroxymethylation of the lignin model compound (5). Monosubstitution is used for simplicity.

always was close to 100%. This indicates that benzyl alcohol reacts with another phenol much faster than it is formed. The rate-determining step therefore is the hydroxymethylation of positions 2 and 6 of the model compounds.

The crosslinking of lignin at positions 2 and 6 can certainly lead to the utilization of lignin in a number of polymeric products. However, this reaction cannot be applied directly to wood adhesives owing to the low pH employed. The hydroxymethylation step was subsequently investigated, to achieve a highly hydroxymethylated product that could be expected to be reactive toward phenols under much milder acidic conditions.

The hydroxymethylation of model compounds on the meta position was first attempted by use of an excess of formaldehyde. Under these conditions the reaction is expected to proceed to the formation of the methylene linkages [53]. In highly acidic aqueous dioxane solutions, protonolysis of the methylene linkages do occur. If the formed unsubstituted 5-position is hydroxymethylated fast enough, the meta-hydroxymethylated product should be produced in high yield (Fig. 6.16). However, none of the variations around this theme gave high meta-hydroxymethylation yields—instead, methylene linkage formation was dominant.

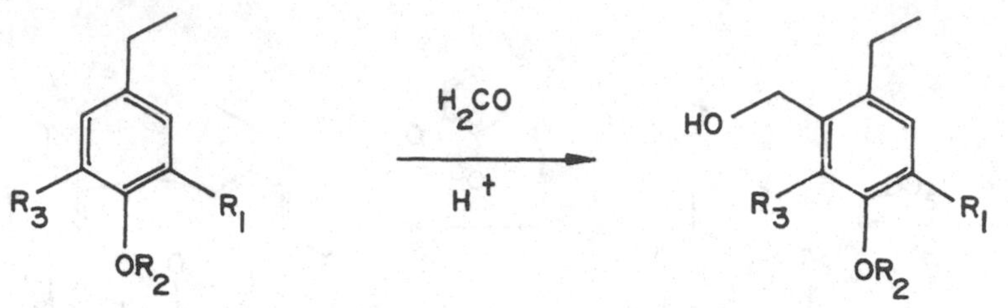

Fig. 6.17 Model compounds hydroxymethylated in high yield.

Metahydroxymethylation of lignin model compounds eventually was achieved by manipulation of the solvent [53]. High yields were achieved in 90% aqueous dioxane containing 0.4N hydrochloric acid with stoichiometric quantities of formaldehyde (Fig. 6.17).

The stoichiometric quantity of formaldehyde required is clearly more attractive when compared with the large excess used previously. Although it would be preferable to carry out the hydroxymethylation of lignin in water, to eliminate costly organic solvents, this problem is not insurmountable because organic solvent recovery by evaporation and condensation is frequently practiced in phenol-formaldehyde resin preparation procedures [43].

The results of the model compound investigations are summarized in Table 6.7. Clearly the degree of metahydroxymethylation versus methylene linkage formation of 5 can be controlled in either direction (Fig. 6.16). Phenolic and nonphenolic softwood as well as nonphenolic hardwood model compounds gave similar products [53].

Table 6.7 The Reaction of 4-Hydroxy-3,5-dimethoxyphenylethane (5) with Formaldehyde in Acidic Aqueous Dioxane at 80°C for 6 Hours [53]

Mole ratio, $H_2CO/(5)$	Solvent	$[H^+]$	Product
1:1	50% aq dioxane	2.2N HCl	Mainly methylene linkage, no hydroxymethylated compounds were isolated
1:2	23% aq dioxane	1.0N HCl	100% monohydroxymethylation + 20% methylene linkages
1:2	10% aq dioxane	0.4N HCl	100% monohydroxymethylation

VIII. REACTION OF INDUSTRIAL LIGNIN WITH FORMALDEHYDE AT THE META POSITIONS

The suitability of the different conditions found for the metahydroxy-methylation of the model compounds (Table 6.7) has been evaluated on three alkali lignins from different industrial origins (Table 6.8). The lignins selected varied widely in their properties.

The industrial lignins were reacted with an excess formaldehyde of 1.35 mol of H_2CO per mole of C_9 of each lignin [54]. When the reactions were performed in 2.2N hydrochloric acid in 50% aqueous dioxane, the lignins polymerized as expected. The polymerization occurred after only a small portion of the formaldehyde had reacted (Fig. 6.18) indicating that extensive methylene linkage formation (crosslinking) had occurred. After gelling of the lignins had occurred, formaldehyde was still consumed by the crosslinked lignins, though at a lower rate. The latter portion of formaldehyde can be expected to result in further methylene linkage formation or in hydroxymethylation of the lignin fragments if steric hindrance prevents the formation of the linkages.

Each lignin was also reacted with formaldehyde in 10% aqueous dioxane containing 0.4N hydrochloric acid (Fig. 6.19). Under these conditions, only monometahydroxymethylation is expected. The soda

Table 6.8 Properties of Three Industrial Lignins [19,20]

	Lignin		
	Kraft	Soda/AQ	Soda
Type	Softwood	Hardwood	Grass
Plant material	Pine	Eucalyptus	Bagasse
Number of phenolic groups per C_9	0.63	0.77	0.63
Number of unsubstituted 3/5 positions per C_9	0.48	0.10	1.05
Calculated number of unsubstituted 3/5 positions on aromatic rings per C_9	0.3	0.1	0.7
Average molecular mass per C_9	176	184	175
Approximate degree of polymerization (dp)	5.7	6.1	6.3

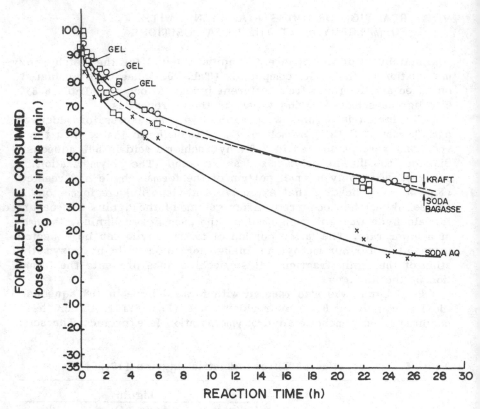

Fig. 6.18   Reaction of industrial lignin with formaldehyde in 50%
aqueous dioxane at 80°C [54].

bagasse and soda/AQ lignin reacted very fast and consumed about
half the theoretical amount of formaldehyde in 1 hour. The rate
of formaldehyde consumption leveled off after about 5 hours. The
total quantity consumed for both the soda/AQ and soda bagasse
lignin was approximately 85% of the calculated theoretical amount.
The formaldehyde consumed by the Kraft lignin leveled off after
60% of the stoichiometric calculated quantity had been consumed.
Some crosslinking of the lignin, however, still occurred, since the
soda bagasse lignin gelled after about 9 hours and the soda/AQ
lignin after 20 hours. The Kraft lignin showed no gelling after
28 hours reaction time. The results obtained nevertheless show
that lignin can be extensively metahydroxymethylated without crosslinking. Prolonged reaction times should, however, be excluded.

The high formaldehyde consumption by the bagasse lignin probably can be attributed to monohydroxymethylation of syringyl and guaiacyl phenylpropanoid units at the 2- and 6-positions. The parahydroxyphenyl units of the lignin were probably hydroxymethylated at the ortho (3- or 5-) positions. The formaldehyde consumption of the soda/AQ lignin can only be attributed to metahydroxymethylation, owing to the low availability of unsubstituted 5-positions. The extensive consumption of formaldehyde by this lignin and its

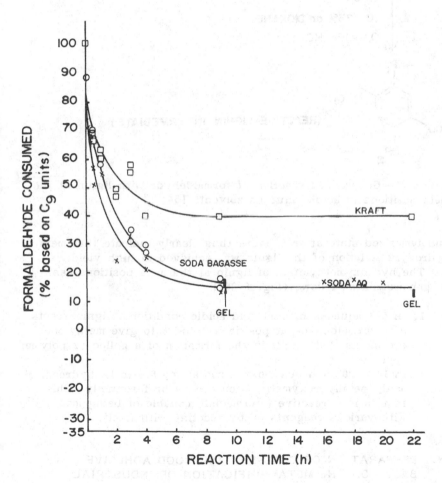

Fig. 6.19 Reaction of lignin with formaldehyde in 0.4N hydrochloric acid in 10% aqueous dioxane at 80°C [54].

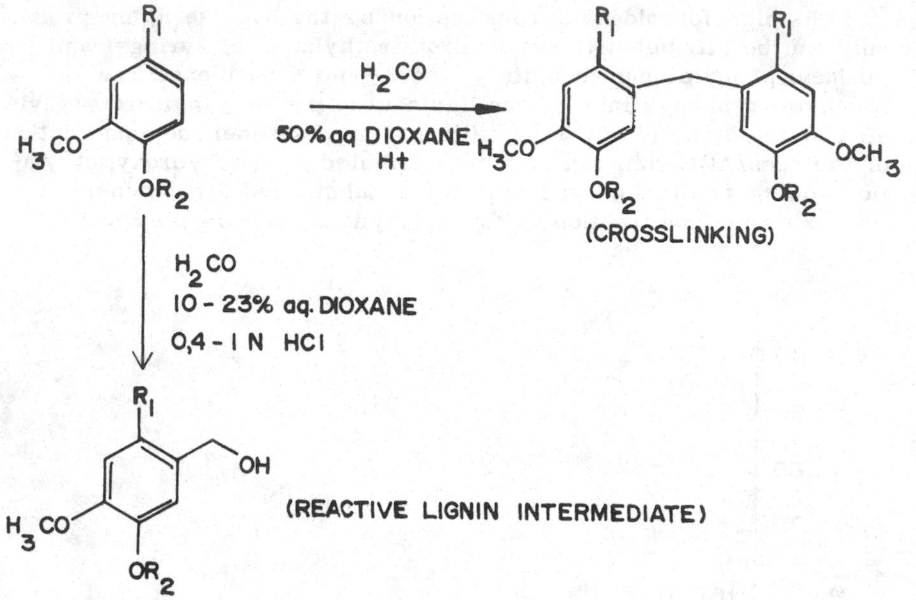

Fig. 6.20 Generalized reaction of formaldehyde with lignin at the meta positions in acidic aqueous solvents [54].

unpolymerized state after 6 hours thus clearly indicate that meta-hydroxymethylation of the lignin was achieved in high yield.

The hydroxymethylation of lignin at the meta positions can be generalized as follows (Fig. 6.20):

1. In 50% aqueous dioxane and acidic conditions, lignin reacts with formaldehyde at positions 2 and 6 to give methylene crosslinks that result in the formation of a gelled or polymerized product.
2. In 10 to 20% aqueous dioxane containing 0.4 to 1N hydrochloric acid, metahydroxymethylation occurs preferentially. This results in a reactive intermediate capable of being modified with various reagents or by reacting with itself.

IX. PREPARATION OF COLD-SETTING WOOD ADHESIVE BASED ON THE METAMODIFICATION OF INDUSTRIAL PINE KRAFT LIGNIN

To illustrate the potential of the modification of lignin at the meta positions, a cold-setting wood adhesive was prepared [55]. A purified

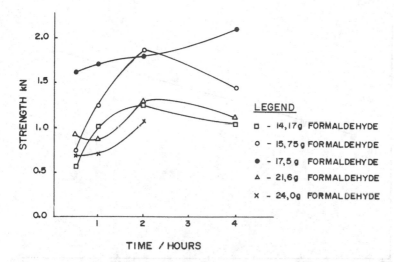

Fig. 6.21 Tensile strength of beech strips glued with resorcinol-grafted meta-hydroxymethylated Kraft lignin adhesive after curing at ambient temperature. Time = time of hydroxymethylation [54].

industrial pine Kraft lignin (described before: Table 6.1) was hydroxymethylated in 90% aqueous dioxane and 0.4N hydrochloric acid. The formaldehyde concentration was varied and the metahydroxymethylation reaction performed for different times (Fig. 6.21). After cooling to ambient temperature, resorcinol was added and a large portion of water to facilitate the addition of the resorcinol to the hydroxymethylated lignin. The reaction mixtures were neutralized with sodium hydroxide to pH 6 and the dioxane removed under reduced pressure.

The cold-curing adhesives were finally prepared by adjustment of the pH to 9 and the addition of fillers. The adhesives were cured with paraformaldehyde and evaluated by the beech strip test.

Selected beech strips 125 × 25 × 3 mm were glued with a 25 × 25 mm^2 overlap and cured at 25°C for 7 days at a relative humidity of 12% emc. The dry strengths of the beech strips are depicted in Fig. 6.22.

The amount of formaldehyde used and the reaction time clearly have a marked effect on the strength of the lignin adhesive. The optimum conditions were subsequently chosen at 17.5% formaldehyde and a reaction time of 4 hours; the resorcinol content varied from 13 to 9% on liquid adhesive [54]. The dry strength of the adhesive increases dramatically when more resorcinol is used to strengths in excess of 3 kN [54]. It is clear from the dry strengths of the meta-Kraft lignin wood adhesives prepared that this approach does have

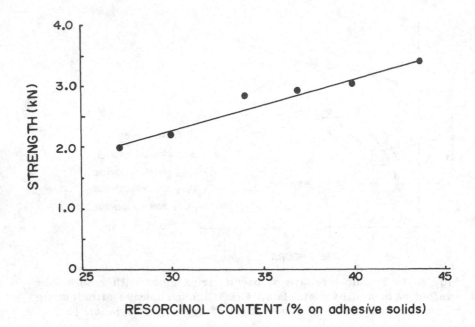

Fig. 6.22 Tensile strength of beech strips glued with resorcinol-grafted meta-hydroxymethylated Kraft lignin after curing at ambient temperature for 7 days [54].

considerable potential to be used for industrial adhesives. More research is currently under way to optimize the adhesive preparations. A preliminary patent has also been registered.

X. CONCLUSION

The utilization of alkali lignin in phenol-formaldehyde wood adhesives can be achieved via various approaches. The best results are achieved with modified or reactive industrial lignins. Of these, the modification of the lignin on the 2- and 6-positions of the aromatic nuclei is an exciting recent development. This approach could lead to a number of applications over and above the use of lignin as a wood adhesive. The preferred approach nevertheless should remain to use by-product alkali lignin, which already has the required properties. The reactive soda bagasse lignin discussed in this chapter is available only in limited quantities. Clearly future pulping processes should be designed to yield more reactive by-product lignin to ensure the better use of wood biomass.

REFERENCES

1. Sarkanen, K. V., and Hergert, H. L. (1971). In *Lignins: Occurrence, Formation, Structure, and Reactions*, K. V. Sarkanen and C. H. Ludwig, Eds. New York, Wiley-Interscience, Chapter 4, pp. 95ff.
2. Nimz, H. (1983). "Lignin-Based Wood Adhesives," in *Wood Adhesives Chemistry and Technology*, Pizzi, A., Ed. New York, Dekker, Chapter 5, pp. 248ff.
3. Glennie, D. W. (1971). In *Lignin: Occurrence, Formation, Structure, and Reactions*, K. V. Sarkanen and C. H. Ludwig, Eds., New York, Wiley-Interscience, Chapter 15, pp. 597ff.
4. Hoyt, C. H., and Goheen, D. W. (1971). In *Lignin: Occurrence, Formation, Structure, and Reactions*, K. V. Sarkanen and C. H. Ludwig, Eds., New York, Wiley-Interscience, Chapter 20, pp. 833ff.
5. Marton, J. (1971). In *Lignin: Occurrence, Formation, Structure, and Reactions*, K. V. Sarkanen and C. H. Ludwig, Eds., New York, Wiley-Interscience, Chapter 16, pp. 639ff.
6. Merewether, J. W. T. (1962). *TAPPI*, 45(2), 159.
7. Van der Klashorst, G. H. (1985). Advances in the Chemical Utilization of Alkali Lignin. Ph.D. thesis, University of Pretoria, Chapter 3, p. 41.
8. Ishizu, A., Makano, J., Oya, H., and Migata, N. (1958). *J. Jpn. Wood Res. Soc.*, 4, 176.
9. Van der Klashorst, G. H., Forbes, C. P., and Psotta, K. (1983). *Holzforschung*, 37(6), 279-286.
10. Forbes, C. P., Van der Klashorst, G. H., and Psotta, K. (1984). *Holzforschung*, 38(1), 42-46.
11. Tai, S., Nakano, J., and Migata, M. (1967). *Nippon Mokuzai Gakkashi*, 13, 257.
12. Ito, H. and Shiraishi, N. (1987). *Mokuzai Gakkaishi*, 33(5), 393-399.
13. Kratzl, K., Buchtela, K., Gratzl, J., Zauner, J., and Ettinghausen, O. (1962). *TAPPI*, 45(2), 113.
14. Glasser, W. G., and Hsu, O. H.-H. (1977). U.S. Patent 4,017,474.
15. Psotta, K., and Forbes, C. P. (1983). *Holzforschung*, 37(2), 91-99.
16. Forbes, C. P., and Psotta, K. (1983). *Holzforschung*, 37(1), 101-106.
17. Marton, J., Marton, T., Falkehag, S. I., and Adler, E. (1966). *Adv. Chem. Ser.*, 59, 125-144.
18. Knop, A., and Scheib, W. (1979). *Chemistry and Application of Phenolic Resins*. Springer-Verlag, Berlin, Chapter 3, pp. 28ff.

19. Van der Klashorst, G. H., and Strauss, H. F. (1987). *Holz-forschung*, 41(3), 185-189.
20. Van der Klashorst, G. H., and Strauss, H. F. (1986). *Holz-forschung*, 40(6), 375-382.
21. Schulerud, C. F., and Doughty, J. B. (1961). *TAPPI*, 44, 823.
22. Muller, P. C., and Glasser, W. G. (1984). *J. Adhes.*, 17, 157-174.
23. Ludwig, C. H., Nist, B. J., and McCarthy, J. L. (1964). *J. Am. Chem. Soc.*, 86, 1186-1196.
24. Goring, D. A. I. In *Lignin: Occurrence, Formation, Structure, and Reactions* (1971), K. V. Sarkanen and C. H. Ludwig, Eds. New York, Wiley-Interscience, Chapter 17, pp. 695ff.
25. Roffael, E. (1979). *Adhäsion*, 11, 334.
26. Johansson, I. (1978). U.S. Patent 4,113,542.
27. Ball, F. J., and Doughty, J. B. (1963). U.S. Patent 3,090,700.
28. Forss, K. G., and Fuhrmann, A. (1979). *For. Prod. J.*, 29(7), 39.
29. Fuhrmann, A. (1976). *Papp. Trä*, 11, 817.
30. Forss, K., and Fuhrmann, A. (1980). *Int. Congr. Scand. Chem. Eng.* (Copenhagen), 5, 526-535.
31. Metsaliiton Teollissuus Oy (1981). Finnish Patent 1,601,751.
32. Schweers, W. H. M., and Vorher, W. (1979). IAEA Symposium, Possibilities of an Economic and Non-Polluting Utilization of Lignin, 211/45, 85.
33. Sakakibara, P. (1974). Japanese Patent 7 401 642.
34. Rieche, A., Redinger, L. (1962). East German Patent 24,060.
35. Foster, N. C. (1979). Adhesive Formulation via Thermolytic Modification of Lignin, Ph.D. thesis, University of Washington.
36. Ludwig, G. H., and Stout, A. H. (1972). U.S. Patent 3,658,638.
37. Gupta, R. C., and Sehgal, V. K. (1978). *Holzforsch. Holzver-wert.*, 30, 85.
38. Marton, J., and Adler, E. (1963). U.S. Patent 3,071,570.
39. Jolly, S., Singh, S. P., and Gupta, R. C. (1982). *Cell. Chem. Technol.*, 16, 185-193.
40. Enkvist, T., Turunen, J., and Ashort, T. (1962). *TAPPI*, 45(2), 128.
41. Wallis, A. F. A. (1976). *Cell. Chem. Technol.*, 10, 345-355.
42. Van der Klashorst, G. H., Cameron, F. A., and Pizzi, A. (1985). *Holz Rohst. Werkst.*, 43, 477-481.
43. Pizzi, A. (1983). "Phenolic Wood Adhesives," in *Wood Adhesives Chemistry and Technology*, A. Pizzi, Ed. New York, Dekker, Chapter 3, pp. 105-173.
44. Walker, J. E. (1975). *Formaldehyde*, 3rd ed. New York, Krüger, Chapter 2, pp. 264.
45. South African Bureau of Standards. (1981). SABS 0183-1981.

46. British Specification, BS 1204-1965 (part 2).
47. Pizzi, A., Cameron, F. A., and Van der Klashorst, G. H. (1987). "Preliminary Results with Soda Bagasse Lignin Thermosetting Adhesives for Particleboard." Presented at the American Chemical Society Symposium, New Orleans, August-September.
48. Sarkanen, K. V., Ericksson, B., and Suziki, J. (1966). *Adv. Chem. Ser.*, 59, 38.
49. Kratzl, K., and Wagner, I. (1972). *Holzforsch. Holzverwert.*, 24(3), 56.
50. Yasuda, S., and Terashima, N. (1982). *Mokuzai Gakkaishi*, 28(6), 383.
51. Yasuda, S., and Terashima, N. (1982). *Mokuzai Gakkaishi*, 26(8), 552.
52. Van der Klashorst, G. H., and Strauss, H. F. (1986). *J. Appl. Polym. Sci. (Chem. Ed.)*, 24, 2143-2169.
53. Van der Klashorst, G. H. (1988). *J. Wood Sci. Technol.* 8(2), 209-220.
54. Van der Klashorst, G. H., and Jackson, S. A. "The Modification of Lignin at Positions 2 and 6 of the Phenyl Propanoid Nuclei." Part III. "Hydroxymethylation of Industrial Lignin." *J. Wood Sci. Technol.* In press.
55. Truter, P., Van der Klashorst, G. H., and Pizzi, A. (1988). "The Modification of Lignin at Positions 2 and 6 of the Phenyl Propanoid Nuclei." Part VIII. "Lignin-Based Coldset Adhesive." Unpublished data. CSIR Special Report C/WOOD 2, P.O. Box 395, Pretoria 0001. Republic of South Africa.

7

Low-Resorcinol PRF Cold-Set Adhesives— The Branching Principle

A. PIZZI / Council for Scientific and Industrial Research
Pretoria, South Africa

I. INTRODUCTION

Phenol-resorcinol-formaldehyde (PRF) cold-setting adhesives have
always been used in the manufacture of structural glulam, finger
joints, and other exterior timber structures. They produce bonds
not only of high strength but also of outstanding water and weather
resistance when exposed to many climatic conditions [1,2]. PRF
resins are mainly prepared by grafting resorcinol onto the active
methylol groups of the low condensation resols obtained by the
reaction of phenol with formaldehyde. Resorcinol is the chemical
species that gives to PRF adhesives their characteristic cold-setting
behavior. It gives accelerated and improved crosslinking, at ambient
temperature and on addition of a hardener, to the phenol-formaldehyde
resins onto which it has being grafted by chemical reaction during
resin manufacture. Resorcinol is an expensive chemical, produced
in only a very few locations around the world (to date only three
commercial plants are known to be operative: in the United States,
West Germany, and Japan), and as such it is the determining factor
in the cost of PRF adhesives. The resorcinol manufacturing process
is a difficult one, and the use of certain solvents during production
and purification of this chemical enhances the risks of dramatic
plant breakdowns. This, from time to time, may endanger world
supply, with unwanted critical consequences to the bigger users
of resorcinol, namely the adhesives and timber industries depending
on it. Disruptions in the world supply of resorcinol have occurred
several times in the past, and although the last major crisis was
nearly 14 years ago, there is never any guarantee that another
crisis in the supply of this chemical is not just around the corner.

Thermosetting phenol-formaldehyde resins can be used in place
of PRF adhesives during such a crisis, but the cost of equipping
a glulam or finger-jointing plant with adequate machinery, such
as high temperature, moisture-controlled ovens or high capacity
radiofrequency presses, is economically unwarranted. The performance
of the PF resins is also not as good as that of PRF resins. Since
resorcinol is such an expensive, strategic chemical, it is also the
determining factor in the cost of PRF adhesives. Significant reductions in the cost of such adhesives must, by necessity, be coupled
with a decrease in their resorcinol content.

In the past, significant reductions in resorcinol content have
been achieved: from pure resorcinol-formaldehyde resins, to PRF
resins in which phenol and resorcinol were used in equal or comparable amounts, to the modern-day commercial resins for glulam and
finger jointing in which the percentage, by mass, of resorcinol
on liquid resin is of the order of 16-18%. Although such resins
have been commercial for more than 20 years, no system has been
found to further and significantly decrease the resins' resorcinol
content without loss of performance. The only step forward has
been the recent development and commercialization of the "honeymoon"
fast-set system [3], described in Chapter 9 of this volume, coupled
with the use of tannin extracts or tannin-based adhesives. This
was a "system" improvement, not an advance on the basic formulation
of PRF resins. The preparation of cold-set phenolic adhesives containing no resorcinol is also possible, based on acid-setting PF
adhesives [4], although these have other disadvantages [5], or
on the use of modified vegetable tannin extracts as a resorcinol
substitute [6]. The application of such resins or their final development is still far away, however, even assuming that the problems
associated with them can be completely overcome. Meanwhile, during
the past 10 years, the price of resorcinol has been continuously
increasing faster than the other PRF resins components, and so
has the price of PRF resins.

Recently, however, a step forward has been taken in the formulation of PRF adhesives of lower resorcinol content. In the preparation of bagasse alkali lignin based cold-set adhesives, also described
in Chapter 6, it was noticed that the optimum resorcinol requirements
of such adhesives [7] was much lower than what was used both in
the best phenol-resorcinol-formaldehyde and in tannin-resorcinol-
formaldehyde resins of commerce. The reason for this was found
to be that the bagasse alkali lignin used for such experiments was
not a mainly linear polymer but was considerably branched. This
is unusual because natural and synthetic PF resins, from the literature [8], appear to be mostly linear. The "branching" principle
so fortuitously discovered was then applied to the preparation of

synthetic PRF adhesives [9] and resulted in a major decrease in
their resorcinol content without loss of performance. The application
of such a "branching" concept decreases resorcinol content in the
finished adhesive in two distinct manners; to better understand
these, it is necessary to illustrate schematically the chemistry of
manufacture of PRF adhesives and how this is affected by branching.

II. CHEMISTRY OF PRF RESINS AND BRANCHING

PRF adhesives are generally prepared by reaction of phenol with
formaldehyde to form a polymer that has been proved to be in the
greatest percentage, and often completely, linear [8]. This can
be represented as follows:

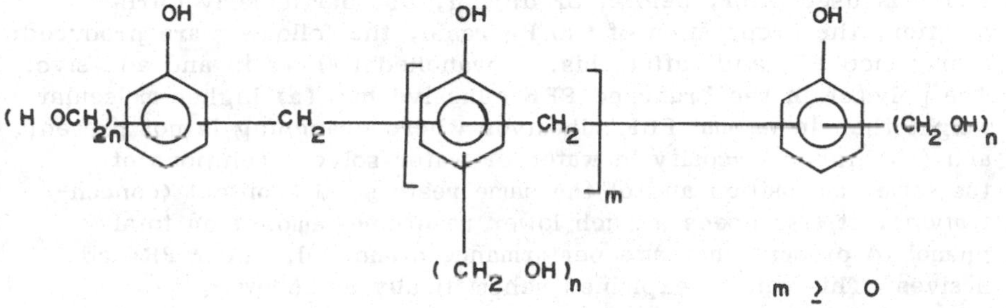

$m \geq 0$
IN INTEGER NUMBERS
$0 \leq n \leq 2$ IN
INTEGER NUMBERS

I

   In the reaction, the resorcinol chemical is added in excess,
in a suitable manner, to polymer I to react with the $-CH_2OH$ groups
to form polymers of the following type, in which the terminal resorcinol groups can be resorcinol chemical or any type of resorcinol-
formaldehyde polymer.

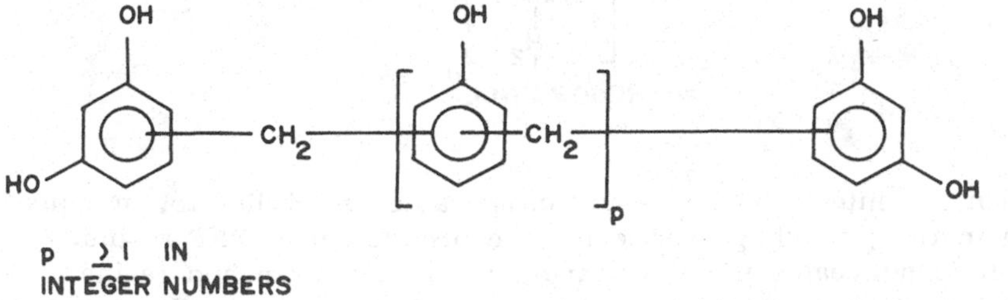

$p \geq 1$ IN
INTEGER NUMBERS

The linear phenol-formaldehyde (PF) polymers formed are terminated by resorcinol for the preparation of the PRF adhesive. Admitting that the average degree of polymerization of the PF resin before resorcinol addition is n, then the PRF resins produced are shown schematically as follows.

resorcinol $-CH_2-[-phenol-CH_2-]_n-$resorcinol

resorcinol $-CH_2-[-phenol-CH_2-]_n-$resorcinol

resorcinol $-CH_2-[-phenol-CH_2-]_n-$resorcinol

n in integer numbers

If a chemical molecule capable of extensively branching (three or more effective reaction sites with an aldehyde) the PF and PRF resins is used after, before, or during, but particularly during or after, the preparation of the PF resin, the following are produced: a branched PF, and, after this, a branched PRF resin and adhesive. The polymer in the branched PRF adhesive has (a) higher molecular weight than in normal PRF adhesives where branching is not present, and (b) higher viscosity in water or water/solvent solutions of the same composition and of the same resin solids content (concentration). It also needs a much lower resorcinol amount on total phenol to present the same performance of normal, linear PRF adhesives. This can be explained schematically as follows:

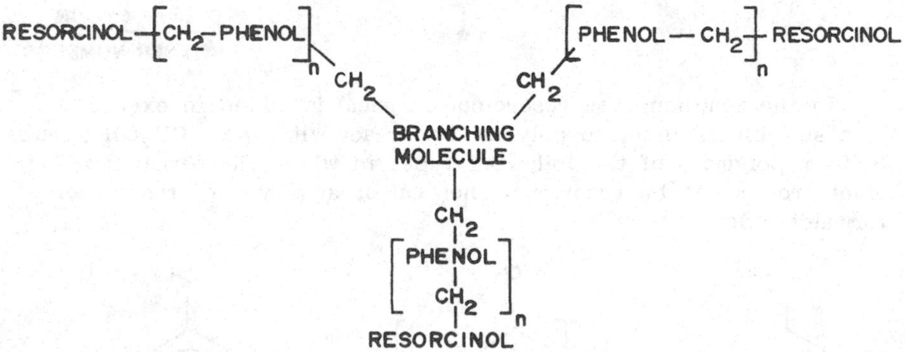

with $n \geq 1$ integer numbers and comparable to, similar to, or equal to n in the preceding scheme for the production of PRF resins.

It is noticeable when comparing linear and branched resins that for every n molecules of phenol used in the particular schematic

examples shown, 2 molecules of resorcinol are used in the case
of a normal, traditional, linear PRF adhesive, while only 1 molecule
of resorcinol for n molecules of phenol is used in the case of a
"branched" PRF adhesive. The amount of resorcinol has then been
halved or approximately halved in the case of the "branched" PRF
resin.

A second effect caused by the "branching" is the noticeable
increase in the degree of polymerization of the resin. This causes
a considerable increase in the viscosity of the liquid adhesive solution.
Because PRF adhesives must be used within fairly narrow viscosity
limits, to return the viscosity of the liquid PRF adhesive within
these certain limits, the resin solids content in the adhesive must
be considerably lowered, with a consequent further decrease on
total liquid resin of the amount of resorcinol, and of the other
materials except solvents and water. This further decreases the
cost of the resin without decreasing its performance.

Thus, to conclude, the decrease of resorcinol by branching
of the resin is based on two effects, namely:

1. The decrease of resorcinol percentage in the polymer itself,
   hence on the resin solids, due to the decrease in the number
   of the phenol-formaldehyde terminals onto which resorcinol
   is grafted during PRF manufacture
2. The increase in molecular weight of the resin, which, by
   the need to decrease the percentage resin solids content
   to a workable viscosity, decreases the percentage of resorcinol
   on liquid resin (not on resin solids)

It is clear that in a certain sense a branched PRF will behave
as a more advanced, almost precured, phenolic resin. While the
first effect described is a definite advance on the road to better
engineered PRF resins, the second effect also can be obtained with
more advanced (reactionwise) linear resins. The contribution of
the second effect to the decrease in resorcinol is not less marked
than that of the first effect. It is however the second effect that
accounts for the differences in behavior between "branched" and
"linear" PRF adhesives.

III. DIFFERENT BRANCHING MOLECULES

Many chemicals can be useful to branch a phenol-formaldehyde or
a PRF resin during its manufacture. In principle, any chemical
possessing three or more sites capable of reacting with formaldehyde
should be able to branch the resin. This is true, however, only in

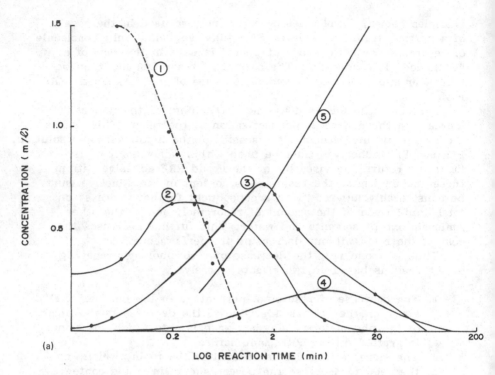

Fig. 7.1 Schematic representations of extrapolated concentration curves of phenol-formaldehyde reactions catalyzed by NaOH at (a) 93°C, above and (b) 30°C. Curves are as follows: (1) phenol, (2) monomethylol phenols, (3) dimethylol phenols, (4) trimethylol phenol, and (5) cumulative curve of condensation oligomers formed. [(b) From Megson, 1985.]

theory. Trimethylol phenol, for example, can in theory achieve such branching without any addition of extraneous chemicals to the phenol and formaldehyde used as reagents. Unfortunately, trimethylol phenol is formed in insignificant quantities under the reaction conditions used to form a PF resin useful for wood adhesives application [see Fig. 7.1(a)]. To use reaction conditions in which the formation of trimethylol phenol is maximized, such as high formaldehyde proportions, would still not solve the problem because the PF resin formed would be quite unstable and unsuitable for the preparation of balanced phenolic wood adhesives. Trimethylol phenol could of course be prepared separately and then added to the PF reaction mixture, but this would be far too troublesome and expen-

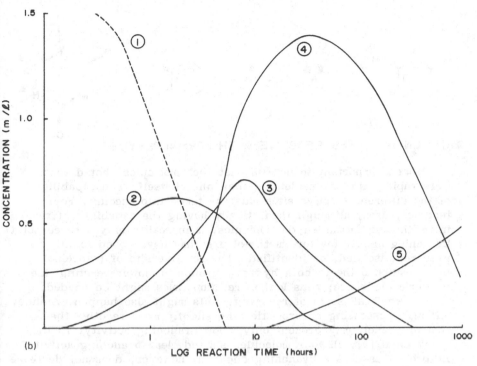

Fig. 7.1 (Continued)

sive. It is interesting to note however, that when a PF resin is curing, tridimensional crosslinks form; hence some phenols start to present three methylene bridges. This is part of the curing process; hence the statement that to branch the resin in some manner also means to advance it to a more progressed curing stage than usual. It is also clear, then, that linear PF resins, upon considerable increase of their molecular weight when progressing toward curing, also become "naturally" branched. This branching of course is much lower than would be obtained by deliberately inserting branched structural characteristics during the preparation of the resin, while still maintaining the resin in liquid and usable form.

The more useful branching chemicals, and simpler to use in the water/polar solvent reaction environment in which PF resins are usually prepared, are simple molecules such as resorcinol, melamine, urea, and aniline. Many others can be thought of, but the former were the ones tried successfully in a recent study [9].

HO　　　OH

H₂N - CO - NH₂

NH₂

H | O=C-N-C=O / N-C-N \ H H

$$H_2N - CO - NH_2$$

* POTENTIAL SITES REACTIVE WITH FORMALDEHYDE

What is important to note is that such molecules should react
more rapidly with formaldehyde than phenol itself. The capability
of the different reactive sites must be taken into account. For
instance, urea, although theoretically having the capability of reacting with four formaldehyde molecules, is in reality only a three-point
branching agent, for tetramethylol urea has never been found,
prepared, isolated, or identified. The reactive site of resorcinol
in a position ortho to both hydroxy groups is slower reacting due
to steric hindrance; thus longer reaction times might be needed,
during which all sorts of rearrangements might also happen. Aniline,
although possessing an aromatic ring slightly more reactive than
that of phenol, may present only a low branching activity. In the
study cited [9], all four molecules showed clear branching activity.
Resorcinol used as a branching chemical, however, does not decrease
the total resorcinol amount in the resin and thus would not decrease
the cost as much as wanted. Melamine and urea, in molar amounts
equivalent to the optimum determined for resorcinol, not only gave
better results but are also considerably cheaper. The results obtained
with the PRF resins branched with resorcinol, melamine, and urea
are shown in Table 7.1. The results obtained for ambient temperature
curing of the PRF resins produced using melamine and urea are
clearly superior to those using resorcinol, particularly for the percentage wood failure results, and satisfy the requirements of international standard specifications for close-contact, cold-setting
adhesives for wood. As regards urea, it must be clearly pointed
out that the percentage of urea on total liquid resin is only 1.8%,
increasing to 2.5% in more sophisticated formulation and to 4.8%
on total resin solids. Similar weight percentages are used for melamine
(3.6 and 9.6%, respectively). This is stated to clearly indicate the
lack of any effect on joint durability by the indeed very small amounts
of urea or melamine added. The resins prepared using urea as a
brancher have initially an amber/greenish tint that evolves after
aging for 5-7 days in storage to a deep inklike blue color: hence
the nickname of "blue glue" used to quickly identify these resins

Table 7.1 Comparison of Results Using Three Different
Branching Chemicals

	Melamine	Urea	Resorcinol
Brancher (g)	4.50	2.18	3.9
Solids content (%)	41.67	40.44	37.65
Parts by mass			
liquid resin	16.67	16.18	18.83
paraformaldehyde	4.50	4.37	5.08
wood flour	1.28	1.28	1.6
pH	8.30	8.61	8.20
Pot-life (hr)	143	154	120

| | Curing at hot and ambient temperatures | | | | | |
| | Melamine | | Urea | | Resorcinol | |
	Hot	Ambient	Hot	Ambient	Hot	Ambient
Strength (kN)						
dry	3.72	2.82	3.69	2.73	3.09	1.75
soak	3.05	3.14	2.87	2.89	2.68	2.15
boil	3.15	2.94	2.71	3.02	2.11	1.82
Failure (%)						
dry	99	29	100	29	100	12
soak	99	87	99	75	100	57
boil	98	90	94	95	93	32

[9]. When hardener is added, the glue mix of these blue glues assumes a grayish color. When cured, however, they are dark brown and indistinguishable from traditional PRF resins. A point of importance is that in the urea-branched PRF resin in Table 7.1, the percentage of resorcinol on liquid resin (43% solids) is already as low as 9.5%.

In the branching reaction, the branching time, the amount of brancher, and the time of addition of the brancher during PF resin preparation present optimum values. For example, in PRF resins prepared completely under base-catalyzed conditions, the relationships among resin solids content, terminal resorcinol content on resin solids, reflux time before and after branching and before addition of final resorcinol, and equivalent amount of branching molecules are shown in Table 7.2.

Table 7.2 Initial Variation of Parameters of Preparation for Branched PRF Adhesives

Parameter	Commercial control	Adhesive preparation number									
		1.1	1.2	1.3	2.1	2.2	2.3	3.1	3.2	3.3	4.1
Initial reflux time (hr)		1	2	3	1	2	3	1	2	3	1
Amount of resorcinol as brancher (%)[a]	—	0.3	0.3	0.3	1.4	1.4	1.4	2.3	2.3	2.3	2.8
Reflux time for branching (hr)		3	2	1	3	2	1	3	2	1	3
Amount of resorcinol as terminal (%)[a]	16.5	16.2	16.2	16.2	15.1	15.1	15.1	14.2	14.2	14.2	13.7
Reflux time for termination (hr)	1	1	1	1	1	1	1	1	1	1	1
Final solids content at 1000 cP (%)	58	47	59	48	55	44	50	49	38	53	43
Strength (kN)											
dry	3.28	3.12	3.32	3.40	3.62	2.81	3.23	3.06	3.09	2.49	2.75
soak (24 hr)	3.21	2.63	3.10	2.96	3.02	2.54	2.49	2.46	2.68	2.67	3.00
boil (6 hr)	2.99	2.11	2.53	2.40	2.03	2.45	2.17	2.34	2.11	2.09	2.21
Wood failure (%)											
dry	98	100	95	100	100	100	100	100	100	100	97
soak (24 hr)	100	100	99	100	100	100	97	100	100	100	84
boil (6 hr)	100	100	97	100	99	100	100	95	93	99	94

[a]On liquid resin.

IV. BRANCHED PRF CHARACTERISTICS

The characteristics of branched PRF adhesives are in many respects similar to but in certain others very different from those of traditional PRF adhesives. Curing temperature experiments at 25, 20, and 18°C showed no marked decrease in strength and percentage wood failure [9], a behavior very similar to that of traditional PRF resins. Branched PRF resins of 41, 39, and 35% solids content do not show any decrease in strength and percentage wood failure of the bonded joints. This is quite different from the behavior of traditional PRF adhesives, which generally operate at 52-54% resin solids content and show marked decrease of strength and wood failure in joints bonded with them when their resin solids contents are markedly decreased. Another difference in behavior is the marked shortening in the pot-life of the branched resins with decreasing resin solids content.

The main difference between branched and traditional PRF adhesives is the resin solids content of application of the two types of PRF. Branched PRF resins of 35-43% resin solids content give strength and wood failure results comparable to those obtainable with traditional PRF resins of 52-57% resin solids content. The need for the lower solids content is due to the higher viscosity at equal solids content and the need to adjust the branched resin (before addition of hardeners and fillers) to a handable viscosity (400- to 1000 cP) by decreasing the solids content.

Other properties are also different. For instance, although the pH of branched PRF resins can easily be adjusted upward by addition of caustic soda solutions (up to 40% concentration), it appears impossible to lower the pH using 10% acid solutions or more concentrated acids. Sulfuric acid, acetic acid, formic acid, phosphoric acid, and hydrochloric acid were used diluted in water without any success. As soon as these solutions were added to the resin, small insoluble white particles formed, indicative of nearly instantaneous localized precipitation from solutions of the phenolic resin. This is understandable, since phenolic resins of such a degree of advancement as well as high molecular weight are likely to be very sensitive to any type of water addition. Attempts with acid solutions of 30% methanol/70% water also failed; addition to the hot resin of acids in any concentration, at the end of the reaction, also failed. However, a 40% acetic acid in methanol (no water) appears to lower the pH successfully, without any noticeable side effects, at ambient temperature.

Sometimes, when hardener is added to a branched PRF, especially when the resin is in the solids content range 35-38%, the polymer condenses and precipitates, eliminating water. This effect can be

observed during application, since the resins whitens and can become almost puttylike. This occurrence is not too common and is
indicative of a resin with a high degree of polymerization, which
has already taken just about as much water as it can. By increasing
the resin pH or by reducing the reaction time after addition of
the branching chemical, this occurrence is considerably reduced.
The use of less volatile solvents also appears to improve the situation.

This apparent defect indicates how important the control of the
end point of the reaction is. The problem just outlined is caused
by improper control of the end point of the reaction, namely the
time of addition of resorcinol at the end of the reflux time of the
branching reaction. These formulations are very sensitive to timing
in this respect. Five minutes overreaction is enough to cause problems. Ten minutes underreaction causes the resin to be at the 43%
solids content level rather than at the 35-37% solids content, which
is more economical due to lower resorcinol and lower solids. A 43%
solids content branched PRF resin is at 9.5-10% resorcinol content
still vastly more economical than a traditional PRF resin, which
generally has a solids content of 52-54% and a resorcinol content
of 16.5-25%.

Open and closed assembly times have been found to be of
importance only for tests on beech strips as specified by relevant
standard specifications [10,11]. Open assembly times of 15 minutes
and closed assembly times of 30 minutes have been found to give
good results.

Table 7.3   Strength and Wood Failure Results Using 43% Solids
Content Urea-Branched PRF Adhesive (blue glue)

Parameters[a]	Finger joints	Beams
Strength[b]		
dry	3.91	14.71
soak	3.76	8.36
boil	3.19	6.96
Wood failure (%)		
dry	98	94
soak	96	98
boil	96	95

[a]Ambient curing at 23°C.
[b]Given in kilonewtons for finger joints and in megapascals for
beams.

Table 7.4 Comparative Results of "Honeymoon" Fast-Set Adhesives Finger Joints Prepared with Branched and Linear PRF Adhesives—Results of Bending Tests

Curing time	Blue glue/ blue glue		Blue glue/wattle extract		Commercial PRF/ commercial PRF		Commercial PRF/ wattle extract	
	Strength (MPa)	Wood failure (%)	Strength (MPa)	Wood failure (%)	Strength (MPa)	Wood failure (%)	Strength (MPa)	Wood failure (%)
15 min	5.63	5	9.87	5	—	—	—	—
30 min	8.49	0	10.19	5	18.1	30	20.5	41
1 hr	16.99	15	26.18	20	29.6	64	28.0	60
2 hr	29.02	45	28.30	70	37.0	98	33.0	95
4 hr	31.42	25	32.41	70	43.4	93	35.6	83
24 hr	43.17	80	47.41	65	47.3	90	45.1	87

This is identical to what has already been observed for any adhesive resin of unusually high molecular weight, such as cold-set adhesives based on wattle tannin [12]. It appears that the high molecular weight and bulkiness of the polymers require a certain time for "grip" when the timber glued is as tight-grained as beech.

Gap-filling properties of branched PRF resins are also good and comparable with what is obtainable with traditional PRF adhesives [9]. Finger joints and glulam beams bonded with these new resins give, on testing, results comparable to those with the PRF resins of commerce (Table 7.3).

Branched PRF adhesives can also be used to prepare separate application "honeymoon" fast-setting adhesives of two types. The first consists of fast sets in which both components A and B are composed of the branched PRF resins, with the PRF of component A containing both paraformaldehyde hardener and fillers and the PRF resin of component B being without hardener and filler but with pH adjusted to 12 or higher. The second type consists of fast sets in which component A is the same as for the first case and component B is a commercial 50% tannin extract solution with pH adjusted to 12-13 [3,13]. This latter type is of interest because

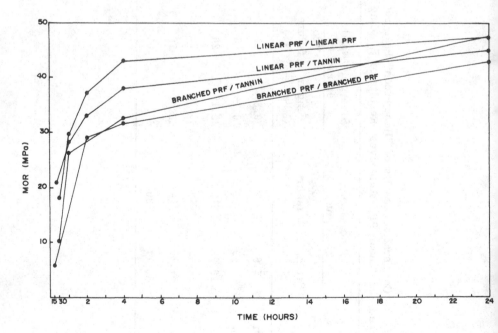

Fig. 7.2 Modulus of rupture (MOR) versus time for "honeymoon"-glued finger joints using urea brancher.

the resorcinol content of the whole adhesive system drops to only
4.5-5% on liquid resin. Excellent weather- and boil-resistant joints
are obtained with these fast-setting adhesives (Table 7.4). In comparison to fast-setting adhesives prepared from traditional commercial
PRF resins, those prepared from branched resins appear to be
initially slightly slower, due to the high molecular weight and possibly
lower mobility of the resin, although the 20 MPa in bending requirement is reached nearly as fast, and the final results are comparable
(Table 7.4, Fig. 7.2).

V. BASE- AND ACID-CATALYZED FORMULATIONS; POT-LIFE AND SHELF-LIFE

Branched PRF resins can be prepared by using a base-catalyzed
or an acid-catalyzed route during the manufacture of the PF intermediate. There is no difference in this respect from traditional
PRF adhesives. In the case of the branched resins, however, the
two routes give different results.

In the case of branched resins where the PF intermediate is
prepared by base catalysis, the shelf-lives of the resins in storage
tend to be shorter than those obtained by an acid route. Base-
catalyzed formulations have a shelf life of only 2 months when the
phenol/formaldehyde molar ratio is 1:1.3. A reduction in the amount
of formaldehyde in the PF resol preparation does considerably
lengthen the resin shelf-life. In accelerated shelf-life tests at 50°C,
the shelf-life of base-catalyzed resins of phenol/formaldehyde ratios
of 1:1.3, 1:0.73, and 1:0.66 were respectively, 2, 4-5, and 6-7
days [9]. A traditional PRF resin, under these conditions, has
generally a shelf-life at 50°C of 18-21 days. The strength and
percentage wood failure of joints bonded with a branched PRF having
a phenol/formaldehyde ratio of 1:0.73 are acceptable. This is not
the case for 1:0.66 ratios, for although the joint strength is unaffected, the percentage wood failure starts to be marginally affected
(Table 7.5). The decrease of the phenol/formaldehyde ratio does
lengthen the condensation time of phenol and formaldehyde during
the preparation of the PF intermediate. Figure 7.3 indicates the
relation between phenol/formaldehyde ratio and the condensation
time of phenol and formaldehyde needed to obtain comparable strength
results.

A far more effective solution of the problem is to use a PF
intermediate prepared in acid environment, and to refrain from
adjusting the pH to the alkaline range until just before the addition
of the branching chemical. This approach improves dramatically
the shelf-life in storage of the resin to levels comparable to and
in some cases even better than those of freshly prepared commercial

Table 7.5   Beech Strip Results Obtained at Different Initial
Phenol/Formaldehyde Ratios

	Initial P/F ratio	
Parameters	1:0.73	1:0.66
Solids contents at 1000 cP	3%	45%
Storage life at 50°C	4-5 days	6-7 days
pH	9.4	9.2
Pot-life	35 min	50 min
Curing of beech strips		
Ambient strength (kN)		
dry	2.44	2.60
soak	2.60	2.14
boil	2.92	2.84
Wood failure (%)		
dry	10 (dry glue line)	55
soak	80	88
boil	92	84

PRF resins. According to the preparation of the resin, results
for 50°C accelerated shelf-life tests of 10-28 days are obtained.
    As regards the pot-life of base-catalyzed resins, it is necessary
to point out that the length of the reaction times as well as the
final viscosities after addition of the urea brancher and of the last
resorcinol influences the pot-life of the finished resin drastically.
Overreaction at the stage after last resorcinol addition causes precuring and shortens the finished resin pot-life, while overreaction
of the branching stage after urea addition causes extensive pre-
condensation. The problem is then one of overreaction affecting
both the pot-life and the shelf-life of the resin, when the latter
is short because the phenols/formaldehyde ratio in the finished
resin is too high. Shelf-life, on the other hand, seems to be particularly dependent on the viscosity of the resin obtained after last
resorcinol addition; nothing can be done to improve the shelf-life
of the resin once the resin has been overreacted after the last
resorcinol addition. The pot-life instead can be lengthened by (a)
the reduction of the paraformaldehyde hardener in the glue mix,
(b) the lowering in the lower fifties of the percentage resin solids
content by further dilution with 30% industrial methylated spirits

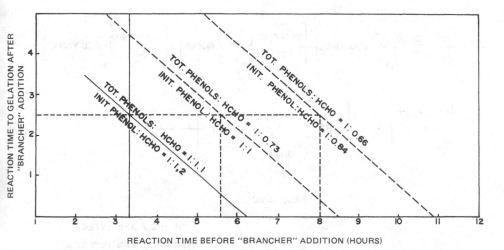

Fig. 7.3 Relation between P/F ratio and condensation time of components, showing proportions needed to obtain comparable strength results.

in water, and (c) the addition of ethylene glycol (±2% on liquid resin) to the glue mix.

Under these conditions, the addition of an acid-catalyzed PF intermediate route rather than the base-catalyzed one solves the shelf-life problem in a more satisfactory manner.

It is also important to point out that the amount of resorcinol used in branched PRF resins is fairly critical. A percentage that is too high while not adding anything to the performance of the adhesive causes a noticeable increase in cost. Too low an amount, apart from extreme cases where gelation occurs, causes further condensation of the resin, with the following effects: (a) decreasing the number of the resorcinol reactive sites giving cold-set characteristics, to the detriment of resin reactivity in the glue mix, (b) increasing further the average molecular weight of the resin, which may cause problems with resin grip of the substrate and lowering of percentage wood failure, and (c) producing resins of higher viscosities at the high solids of preparation, but with an increase of solids content, hence an increase in cost, at the standard viscosity used for the finished adhesives. A possible explanation for the latter phenomenon could be the formation of "super" linear polymers of very high molecular weight, composed of branched polymers as its units and of the following tentative structure.

VI. FORMULATION

The following formulation can be used as a starting point for experimentation in the preparation of branched PRF cold-set adhesives.

Base-Catalyzed PRF Formulation (short shelf-life)

Component	Parts by mass
Phenol	41.01
Water	16.90
Methanol	4.17
Paraformaldehyde, 96% powder	17.15
NaOH, 40% solution	2.80
Urea	1.48
Resorcinol	16.49-14.47 (lower resorcinol amounts can be used on lowering paraformaldehyde)

Bring the mixture of phenol, water, methanol, paraformaldehyde, and caustic soda solution to reflux at atmospheric pressure under continuous mechanical stirring. Reflux for 2 hours, then add urea and reflux for another 2 hours; add resorcinol, and reflux for 1 hour. Cool the resin and dilute to 37-38% solids according to viscosity with a water/methanol mixture. Adjust pH with 40% NaOH solution to a value between 8.4 and 9.0 according to balance of strength and pot-life wanted. The glue mix used 27% paraformaldehyde on resin solids and 3% wood flour on liquid resin. For beech strips, use only 15 minutes open and 30 minutes closed assembly time for best results.

To lengthen shelf-life, decrease total phenol/formaldehyde ratio (hence decrease formaldehyde) in the preparation of the resin.

VII. CONCLUSION

Branched PRF cold-setting adhesives are possibly a new generation of engineered phenol-resorcinol-formaldehyde resins of lower resorcinol content and lower solids content than traditional PRF cold-set adhesives. They appear to be more economical than traditional resins of this type.

REFERENCES

1. Dinwoodie, J. M. (1983). "Properties and Performance of Wood Adhesives." In *Wood Adhesives Chemistry and Technology*, A. Pizzi, Ed. New York, Dekker, Chapter 1, pp. 1-58.
2. Kreibich, R. E. (1984). "Exposure of Glue Lines to Weather." In *Wood Adhesives: Present and Future*, A. Pizzi, Ed., Applied Polymer Symposium No. 40, pp. 1-18.
3. Pizzi, A., Rossouw, D. Du T., Knuffel, W., and Singmin, M. (1980). "Honeymoon Fast-Setting Adhesives for Exterior-Grade Fingerjoints," *Holzforsch. Holzverwer.*, 32(6), 140-151.
4. Pizzi, A., Vosloo, R., Cameron, F. A., and Orovan, E. (1986). "Self-Neutralizing Acid-Set PF Wood Adhesives." *Holz Rohet. Werkst.*, 44, 229-234.
5. Cameron, F. A., and Pizzi, A. (1984). "Acid-Setting, Cold-Setting Phenolic Adhesives for Glulam: A Controversial Issue." In *Wood Adhesives: Present and Future*, A. Pizzi, Ed., Applied Polymer Symposium No. 40, pp. 229-234.
6. Pizzi, A., Orovan, E., and Cameron, F. A. (1988). "Cold-Set Wood Adhesives of Low Resorcinol Content." *Holz Rohst. Werkst.*, 46, 70.
7. Van der Klashorst, G. H., Cameron, F. A., and Pizzi, A. (1985). "Lignin-Based Cold-Setting Wood Adhesives—Structural Fingerjoints and Glulam," *Holz Rohst. Werkst.*, 43, 477-481.

8. Pizzi, A., Horak, R. M., Ferreira, D., and Roux, D. G. (1979). "Condensates of Phenol, Resorcinol, Phloroglucinol, and Pyrogallol as Flavonoids A- and B- Ring Model Compounds with Formaldehyde." *J. Appl. Polym. Sci.*, 24, 1571-1587.
9. Pizzi, A., and Vermeulen, S. (1987). "Blue Glue—A Low Resorcinol PRF Cold-Set Adhesive." In preparation.
10. British Standard BS1204. (1965). Specification for Synthetic Resin Adhesives for Wood. Part 2: Close Contact Joints.
11. South African Bureau of Standards. (1981). Standard Specification for Phenolic and Aminoplastic Adhesives for the Laminating and Fingerjointing of Timber, and for Furniture and Joinery. SABS 1349-1981.
12. Pizzi, A., and Roux, D. G. (1978). "The Chemistry and Development of Tannin-Based Weather- and Boil-Proof Cold-Setting and Fast-Setting Adhesives for Wood." *J. Appl. Polym. Sci.*, 22, 1945-1955.
13. Pizzi, A., and Cameron, F. A. (1984). "Fast-Set Adhesives for Glulam," *For. Prod. J.*, 34(9), 61-64.

8

Hotmelts for Wood Products

NEVILLE E. QUIXLEY* / *Expandite (PTY) Ltd.*
Isando, South Africa

I. INTRODUCTION

This chapter deals with the practical application of hotmelt adhesives
for the woodworking industry. It describes the application areas
for woodworking hotmelts and hotmelt types, as well as their formulation and manufacture. Included, too, is an in-depth discussion
of requirements and problems commonly associated with the use
of these materials and their bonding substrates.

II. BACKGROUND

The use of hotmelt adhesives in the timber conversion market has
increased substantially since the introduction of these materials
in the mid-1950s to early 1960s.

Hotmelt growth in this industry can be attributed largely to
the trend away from solid wood and boards to the cheaper particleboard and blockboard developments. Mass production of these materials required a quick and effective method of attaching decorative
surface veneers. It is in this area, and more specifically edge
veneering, where the bulk of hotmelt adhesives are employed.

The type of veneer material over the years has also seen a
noticeable change from the more traditional natural wood veneers
and lippings to less expensive synthetic decorative foils, particularly
polyvinyl chloride (PVC) and melamine.

In keeping with these developments, veneer application equip-

---

Present affiliation: T.A.C. National (PTY) Ltd., Benoni South,
South Africa

ment has now reached a high degree of sophistication to cater to all requirements. Hotmelt adhesive development, too, has had to keep pace with these trends. It is thus important to ensure the availability of adhesive systems that are compatible with application requirements and equipment, coupled with acceptable overall performance as a bonding medium for the materials in question.

Hotmelt adhesives are rarely used for structural bonding in the woodworking industry. Although a few exceptions do exist, these products are generally unable to match the tensile strengths of the more traditional structural adhesives (e.g., polyvinyl acetate cold glues, or thermosets, such as urea-formaldehyde resin, tannin, or resorcinol formaldehyde). Hence their more traditional role as veneer splicing adhesives or edge veneer adhesives, where forces are predominantly peel in nature, coupled with their ability to bond well to plastics. Other important reasons for employing hotmelt adhesive systems for veneer applications are as follows.

1. Ease of application via high speed equipment
2. Formation of strong, permanent, and durable bonds within a few seconds of application
3. No environmental hazard and minimal wastage because of 100% solid systems
4. Ease of handling
5. Absence of highly volatile or flammable ingredients
6. Excellent adhesion to both wood and plastics
7. Wide formulation possibilities to suit individual requirements (e.g., color, viscosity, application temperature, and performance characteristics)
8. Cost effectiveness

III. HOW DO HOTMELTS WORK?

Hotmelts are 100% solid thermoplastic materials that are supplied in pellet, slug, block, or irregular shaped chip form. They require heating via appropriate application equipment, which usually is fairly sophisticated in order to control the required temperature and coverage rate. Upon application, the heat source is removed and the thermoplastics set immediately (within a few seconds). Hotmelts are thus well suited to high speed, continuous bonding operations.

IV. APPLICATION AREAS

A. Veneer Splicing

Particleboard with a decorative wooden veneer surface usually employs a hotmelt adhesive to effectively mate veneer edges down

the length of the joint. Polyamide hotmelt adhesives are widely
used for this veneer splicing process. The adhesive is more often
than not supplied as a thread, and positioned as such in a zig-zag
configuration as shown in Fig. 8.1. A heated press is employed
to activate the adhesive followed by rapid cooling and setting. The
rapid set required is best achieved with polyamide hotmelt adhesives,
since their setting temperatures are much higher than, for example,
ethylene-vinyl acetate (EVA) types, and since range between application and setting temperatures achievable with polyamides is narrow.

Another important feature of polyamide resins in the context
of this application is the low melt viscosities achievable, thus ensuring
rapid spreading and wetting of the molten film.

B. Edge Veneering and Edge Banding

Edge veneering and edge banding constitute by far the main area
employing hotmelt adhesives, which are predominantly based on
EVA copolymer resins. For some applications, however, formulated
polyamide hotmelts are also used, particularly where exceptional
heat resistance of the bond is required.

Modern materials such as decorative surface board products,
used in the manufacture of furniture components, require exposed
edges to be covered with suitable edging materials.

Most laminated surface board products consist of a decorative
melamine or PVC layer bonded to a chipboard substrate. These
board products have their own performance characteristics, which
may influence the edge bond.

The choice, application, and fabrication method of edging plays
a very important part in the manufacturing and final application
of the furniture produced. The choice of correct edging selection,
therefore, depends very much on performance requirements and
aesthetic value.

The different types of edging available are as follows.

1. *Edge-veneering* (veneer < 1 mm thick)
   a. Aminoplastics (i.e., high pressure decorative laminates,
      e.g., Formica; continuous decorative laminates, e.g.,
      Deccon)
   b. Decorative PVC
   c. Amino resin paper foils
   d. Wood
2. *Edge-banding* (strips usually 3-10 mm thick)
   a. Wooden strips
   b. Aluminum strips
   c. Extruded rigid PVC strips

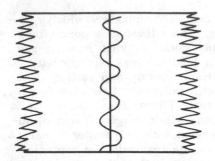

Fig. 8.1   Polyamide hotmelt thread for veneer jointing.

C.   Important Considerations

1.   *Surfacing Materials*

Materials are designed to meet specific requirements. Choose an
edging material that is compatible with the surface laminate.

Some laminates are designed to have very high temperature
resistances (e.g., 130°C). The heat resistance of the edging material
as well as the edge bond performance (adhesive) should therefore
match these criteria.

2.   *Adhesives*

Select a suitable hotmelt adhesive that:

    Is compatible with application equipment.
    Is suitable for bonding chosen substrates.
    Will perform under final application requirements/conditions.

3.   *Edging*

The melamine layer of aminoplastic decorative laminate edges is
hydrophilic. This allows moisture contained in the edge to evaporate
at elevated temperatures with low humidities. As the edge dries
out, stresses are built up in the edging material, imparting additional
strain to an already softened glue line.

If the heat resistance of the hotmelt adhesive is too low, the
strain generated in the edging material will cause delamination.

Edging materials will behave differently during bonding in
accordance with their individual characteristics.

    Aminoplastic decorative laminated edges (Formica and Deccon)
        are thermosetting plastic materials and will not soften when
        exposed to high temperatures.

Amino resin paper foil edging (melamine foil) is produced with a urea-formaldehyde acrylic resin, resulting in a soft, flexible edge.

PVC edging is thermoplastic and will therefore repeatedly soften with heat

Soft and flexible edging materials have little influence on the bond strengths of edged components at elevated temperatures.

V. ETHYLENE-VINYL ACETATE HOTMELTS FOR EDGING

A. Physical Characteristics

1. Viscosity

Application equipment for edge veneering requires a hotmelt adhesive that is relatively high in viscosity at application temperatures (usually around 200°C). Reasons for this are as follows.

The adhesive must have sufficient body to prevent flowing from vertical surfaces after application.

It must not penetrate too deeply into particleboard, causing glue starvation.

It must have easy spreading and excellent wetting characteristics.

Viscosities of these hotmelts are in the order of 50-60,000 mPa·s (cP) at 200°C.

Viscosity is achieved through the correct selection of EVA copolymer grades, coupled with the quantity and type of reinforcing filler that is added to the system.

The ball and ring softening point is an early indication of the degree of heat resistance of a particular hotmelt. The softening point is influenced by the combination of ingredients, but to a large extent by the grade and quantity of EVA copolymer and tackifying resin contained in the system. Using a 5.1-g lead ball, average softening points are between 90 and 105°C.

2. Wettability

For optimum adhesion, wetting characteristics (of the hotmelt to substrates during application) are vital. Proper wetting is related to viscosity but is again largely influenced by resin selection and quantity.

3. Stability

During prolonged periods at elevated temperature while contained
in the hotmelt applicator, the hotmelt must resist oxidation and
thermal breakdown of components. This often leads to discoloration,
charring, and inferior bonds. Nozzle blockages, as a result of
charred material, can also be encountered.

B. Formulation Considerations

EVA hotmelts consist basically of the following:

> 1. EVA Copolymer (e.g., Elvax, a duPont product)
> 2. Tackifying and adhesion-promoting resins (e.g., hydrocarbon,
>   rosin esters, coumarone-indene, terpene resins)
> 3. Fillers—usually barium sulfate (barytes) or calcium carbonate
>   (whiting)
> 4. Antioxidants

1. EVA Copolymer

EVA copolymer is the main binder in the system and largely influences the following:

> Viscosity and rheology characteristics
> Cohesive strength
> Flexibility
> Adhesive strength

A variety of EVA grades are available, allowing the formulator a
choice of varying vinyl acetate contents coupled with varying
viscosities (melt index).

Higher VA contents generate greater adhesion to plastics coupled
with increased flexibility. The higher the VA content, however,
the higher the cost.

Broadly speaking, EVA-based edge-veneering hotmelts utilize
grades averaging 28% vinyl acetate, and formulations usually contain
around 40% binder.

2. Resins

A certain percentage of resin is almost always incorporated into
formulations, with resin content varying from 8 to 25%. Hydrocarbon
resins are most commonly used, but rosin esters, terpenes, and
indene resins, which are more heat stable, are also common. Resins
provide better flow, hot-tack, adhesion, and wetting characteristics.

3. Fillers

The heavy fillers such as barytes are used at levels of up to 50% by weight, but more commonly at around 35-40%.

The filler imparts cohesive strength and body to the adhesive, and naturally also reduces cost considerably.

Barium sulfate is the chosen filler in most cases because of its high density and hence low pigment volume concentration. Barium sulfate grades vary from beige to dark brown, and this assists in formulating specific opaque colors to match veneer color requirements. Finely ground calcium carbonate is sometimes also used as a filler where very light colors are required. Titanium dioxide pigment is commonly used as toner at levels of 2-5%.

4. Antioxidants

Antioxidants are added to protect the organic components, especially resins, from oxidation/discoloration at high temperatures. A large choice exists. These materials are usually added at levels of 0.2-0.5%.

5. Typical EVA Edge-Veneering Hotmelt Formulas

Following are formulas for hotmelts in various applications. Ingredients are listed parts by weight.

General-Purpose Hotmelts for Both Wood and Plastic Veneers

	White	Natural	Brown
Hydrocarbon resin, 90°C m.p.	5.50	5.50	5.50
Rosin ester, 85°C m.p.	8.00	8.00	8.00
Coumarone Indene resin, 105°C m.p.	5.00	5.00	5.00
Butylated hydroxytoluene (BHT) antioxidant	0.20	0.20	0.20
Elvax 250	14.00	14.00	14.00
Elvax 210	10.00	10.00	10.00
Elvax 150	15.00	15.00	15.00
TiO_2 pigment	4.50	0.80	—
Superfine light barytes	37.80	41.50	—
Pink barytes	—	—	42.30
	100.00	100.00	100.00

Low Cost Hotmelt for Wooden Veneer Only, Natural Color

Hydrocarbon resin, 90°C m.p.	12.00
Hydrocarbon resin, 100°C m.p.	10.00
Elvax 250	30.00
BHT antioxidant	0.20
Superfine light barytes	47.20
TiO_2	0.60
	100.00

Hotmelt for Difficult Plastic Surfaces (e.g., Deccon, Natural Color)

Polyterpene resin, 115°C m.p.	30.00
BHT antioxidant	0.20
Elvax 260	10.00
Elvax 250	35.00
$CaCO_3$ (15 µm)	24.80
	100.00

6. *Production Technique and Equipment*

Because of their relatively high melt viscosities, the EVA hotmelts need special manufacturing equipment. For example, a Z-blade mixer such as a Baker Perkins or Winkworth with oil-heated jacketing is required. Mix temperatures are kept as low as possible (±110°C) to keep bulk thick.

The high viscosity kneading action ensures rapid dissolution of EVA copolymer and resin. Fillers are easily dispersed and a homogeneous mix rapidly achieved with this type of agitation. Upon completion, the molten product is extruded into ropes approximately 6 mm in diameter, which are cooled through a chilled water trough and then granulated into pellet form. Alternatively, hotmelt slugs are supplied where application equipment utilizes this form.

It is essential to ensure that any residual moisture picked up during the cooling process is eliminated via an air-drying cyclone before packing.

VI. POLYAMIDE HOTMELTS FOR EDGING

The polyamide hotmelts are high performance systems and are used selectively where good heat resistance is required. Their high cost, relative to EVA types, makes them rather unattractive for general use.

Polyamide resins offer high tensile strengths and high initial tack, often without the need for additional formulating. Their higher melt points ensure good heat-resistance qualities and are also responsible for rapid setting on cooling. Their two main drawbacks are cost and the tendency to char easily if kept at high temperatures.

Hot-melt polyamide resins are obtained by the reaction of diamines with diacids. While in their simplest form polyamides are the reaction of a particular diamid with a particular diamine, most of the polyamides used in adhesive formulations are complex reaction products obtained by combining several diacids and diamine in order to obtain the particular, required properties.

The most common diacid used is a dibasic acid obtained by polymerizing oleic or linoleic acid or other unsaturated fatty acids. This acid can be represented as HOOC-R-COOH where R is a hydrocarbon residue of 34 carbon atoms and of indeterminate configuration. Commercial forms of this dimeric diacid also contain preparations of products obtained by polymerization of three or more molecules of unsaturated fatty acids, and thus contain varying quantities of trimeric acids and of higher homologues. Monomeric forms are also present. The most used diamine for this type of adhesive is ethylene diamine $H_2N-CH_2CH_2-NH_2$, but other diamines are also used, responding to a general formula

$$
\begin{array}{cc}
X & X \\
| & | \\
HN-R-NH
\end{array}
$$

where X and Y can be H or other chemical groups. Polyamides are then formed according to the schematic reaction.

$$
\begin{array}{cc}
X \quad\quad X & \quad\quad X \quad\quad X \\
| \quad\quad | & \quad\quad | \quad\quad | \\
HOOC-R-COOH + HN-R^1-NH \longrightarrow HOOCRCONR^1-NH + H_2O
\end{array}
$$

or simply

$$
\begin{array}{cc}
X \quad\quad Y & \quad\quad X \quad\quad Y \\
| \quad\quad | & \quad\quad | \quad\quad | \\
n\ HOOC-R-COOH + nHN-R^1-NH \longrightarrow \ \overline{\ OCRCONR^1-N}\ \overline{\ }_n
\end{array}
$$

The reaction occurs with the elimination of water to form amide groups. The high polarity of the amide groups contribute to give, by formation or interchain hydrogen bonds, the characteristic polymer strength and adhesive properties to the polyamides.

As far as edge veneering polyamide hotmelts are concerned, the basic resins need some form of modification to achieve:

Suitable application viscosities
Flexibility
Reduction in cost if possible

Suitable polyamide resins (those of the more flexible variety) are thus frequently modified by the addition of EVA copolymer, such as Elvax 265 (high viscosity, high melt point grade). The amount of EVA that can be added is restricted to a maximum of 25% in most cases because of compatibility problems. The blend is then further modified with selected tackifying resin addition and small quantities of filler, to reach an optimum balance of performance properties. To achieve maximum adhesion, it is common for polyamide hotmelts of this type to be used in conjunction with a polyamide resin solution primer system for edging material. The primer is invariably a dilute solution of the base polyamide resin.

VII. EDGE-BONDING TECHNIQUES

The correct use of edge-bonding hotmelts and application equipment requires knowledge and understanding of environmental condition requirements, materials, and adhesive and machine technology in application and fabrication methods. Important considerations in each of these areas are outlined next.

A. Environmental Condition Requirements

1. The ambient temperature and materials to be bonded should not be below 15°C.
2. Avoid draughts, which would prematurely chill the glue line, resulting in poor wetting of the edge.
3. To improve the situation in cold conditions, fit preheating infeed fences, blow heaters, and condition edging (Fig. 8.2).
4. Keep humidity of materials between 8 and 12%. High humidities result in a moisture vapor barrier that will interfere with the bond. Lower humidities are generally a less serious problem but can cause curling of certain veneers.

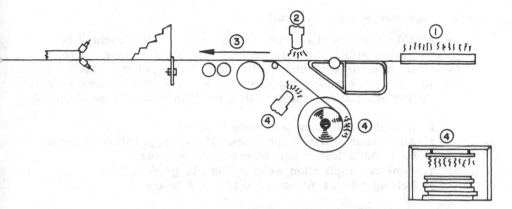

Fig. 8.2 Improving the bond strength in cold conditions: (1) pre-heat infeed fences, (2) heat space between applicator and pressure rollers, (3) increase machine speed, and (4) preheat edge materials.

B. Materials

1. Check squareness of board components. Out-of-square components will result in a pressure decrease on edges, leading to delamination.
2. Check evenness of board edges. Saw scoring marks will lead to delamination in these areas.
3. Check squareness of edging. Curled edging will result in tension during and after the pressure cycle, with possible delamination.
4. Note that poor density or open core of the substrate will reduce the bond strength.

C. Adhesive

1. Always follow supplier's instructions as to correct use of adhesive, with particular reference to application temperatures, viscosity, and open time.
2. Do not change adhesives until manufacturing processes have been properly checked or new adhesive extensively tested.
3. Do not overheat hotmelt adhesives or allow excessive periods during which the adhesive remains at operating temperature. Reduce temperature to 150°C during any interruption of work.

D. Adhesive Application Guidelines

In general, one should ensure that operation of the machine is
in accordance with the manufacturer's instructions, being sure
to set the machine according to the adhesive supplier's specifications
for line speed, operating temperature, and adhesive coating weight.
During application, the following guidelines should be observed.

1. Adhesive reservoir temperature:204°C
2. Application roller temperature: 191°C (application roller to
   be 12-13°C lower than reservoir temperature)
3. Adhesive application weight: 200-250 g/m^2
4. Melting time of adhesive: 1-1/2 to 3 hours

Correct application weight and spread of the hotmelt can be
checked by bonding a transparent PVC strip and applying at a
pressure of 2-4 kg/cm^2. If the correct pressure has been applied,
the pattern caused by the applicator wheel on the adhesive should
disappear, with little or no squeeze out at the edges.

The adhesive reservoir must be filled completely and the lid
kept in place, to avoid heat loss. It is necessary to clean gluepots
thoroughly at regular intervals, as well as filters, nozzles, and
glue lines (every 2-4 weeks). Suitable solid hotmelt cleaners, available from adhesive suppliers, have largely replaced the traditional
use of cleaning solvents.

The applicator wheel or roller is used according to the following
procedure.

1. Start roller only after hotmelt becomes sufficiently fluid.
2. Set applicator roller to protrude 2 mm over the panel edge.
3. Always use a knurled gravure applicator roller.
4. Replace worn applicator rollers as this affects spreading/
   application weight.
5. Ensure correct angle of applicator roller.
6. See also Section E, "Gravure Applicator Wheel Technology."

For pressure rollers, the following considerations apply.

1. Pressure rollers must be clean at all times. Any foreign
   material on the rollers will affect the pressure and subsequently the bond, as well as possibly damaging the edging
   material.
2. Pressure exerted by the rollers must be set in accordance
   with the manufacturer's instructions: usually 2-4 kg/cm^2.
3. High pressure will cause slipping of the edging under the
   head. Low pressure will cause poor contact of the edging
   to board edge.

Trimming units should be handled as follows.

1. Set trimming units per manufacturer's instructions.
2. Ensure that all trimming blades are sharp and in good condition. Blunt blades will cause chipping and breaking of edges.
3. Clean adhesive buildup on blades at regular intervals, using a suitable solvent.
4. Set air pressure on trimming units in proportion to feed speed.

Chamfering units have the following requirements.

1. Set chamfering units per manufacturer's instructions (bevel normally set at 25° angle).
2. Ensure that knives are sharp and in good condition.
3. Clean adhesive buildup on blades regularly with a suitable solvent.
4. Ensure correct setting of tracing wheels and clean wheels regularly.

3. Gravure Applicator Wheel Technology

1. Application Process

The gravure applicator wheel is the most important individual component on an edgebander. Adhesive transfer to the substrate takes place when the substrate comes in contact with the gravure wheel applicator, which should rotate at the same speed as the moving track. Fresh adhesive is resupplied to the applicator wheel via the doctor blade in less than one revolution of the wheel.

By adding another doctor blade and reversing the direction of the rotation of the gravure applicator wheel, much more adhesive can be driven onto the substrate. This may be necessary when edging substrates with a low density, wide edges, or edges that require a lot of gap-filling (plywood).

Worn gravure applicator wheels should be replaced immediately. On replacement, a change in machine performance will take place and equipment adjustments should be carried out.

2. Heating

Where a cartridge heater is mounted in the center of the applicator wheel shaft, a high temperature grease must be used as a heat-conducting medium between cartridge heater and applicator wheel, otherwise there will be rapid cartridge heater burnout due to over-heating.

Adhesive temperature at the applicator wheel should be 12-13°C
cooler than the adhesive temperature in the reservoir, to increase
cartridge heater life.

3.   Behavior of Hotmelt Adhesives on Gravure Wheel Applicators

As mentioned earlier, the single most important component on an
edgebander is the gravure wheel applicator. All other components
of the machine support the performance of the gravure wheel applicator.

It is the gravure wheel application that controls the amount
of hotmelt adhesive that is applied to the substrates, which, in
turn, determines the quantity of calories of heat present to keep
the hotmelt liquid until the time of bond formation.

Of the hotmelt adhesive that is located between the peaks and
the doctor blade, only 1% is transferred to the substrate; the balance
becomes nothing more than squeeze-out. The volume of adhesive
that is found in the groove area is the actual material that is transferred to the substrate edge.

The purpose of the doctor blade is not to act as an adjustment
to increase or decrease adhesive transfer to the substrate; rather,
it serves to replace in the grooves the adhesive that has transferred
to the substrate. An incorrect doctor blade setting will either cause
excess squeeze-out or insufficiently fill the grooves, which will
result in less than maximum adhesive transfer.

The volume of adhesive that is transferred to the substrate is
the single major controlling factor in determining the open time of
the hotmelt adhesive. The volume of adhesive that does transfer
is preset at the factory and is determined by the actual dimensions
and geometry of the groove area.

If the adhesive is too cohesively strong (too cold) or too cohesively weak (too hot), the adhesive will break out of the top
of the groove in the gravure applicator wheel, reducing the adhesive
transfer. This is exactly why the open time of the hotmelt adhesive
is determined by the amount of adhesive transferred, not by raising
the application temperature. If the shear force is too low to physically
move the adhesive from the gravure applicator wheel (slow line
speed), adhesive transfer is reduced from maximum; too high a
shear force (fast line speed) also reduces transfer. Shear force
affects the cohesive strength of the hotmelt exactly like temperature.

The land in the gravure applicator wheel pattern assists in
creating a constant fracture point in the hotmelt adhesive when
the gravure applicator wheel comes into contact with the substrate.
Theoretically, the best adjustment for groove refilling is to have
the doctor blade at a 90° angle with the plane of the land (Fig. 8.3).

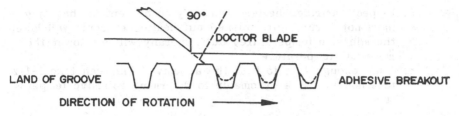

Fig. 8.3 Gravure applicator wheel.

F. Edgebander Setup and Fault Detection Procedures

1. Directions for Use of Clear PVC for Edgebander Setup

By using clear PVC in the edgebander setup, the operator can quickly and easily adjust the machine to maximize adhesion and heat-resistance properties of the specific adhesive being used. This method is easily adaptable into a simple quality control method that is capable of almost totally eliminating delamination complaints, with respect to the edgebanding process.

This method is also highly useful to accurately determine the best product to suit the application. The following equipment is necessary to institute this method.

1 Roll of clear PVC (or 0.4-0.5 mm Melinex or Genolon)
1 Pyrometer (digital temperature recorder) with immersion and
    surface probes (for measuring adhesive temperature)
1 Tachometer (for determining line speed)
1 Magnifying glass (~ 10 × magnification)

Adhesive reservoir flushing is required before the actual adjustment procedure, which consists of the following steps.

1. Remove all old adhesive from reservoir with a wide square-edged metal spatula. (Exercise safety precautions.)
2. Fill reservoir completely with fresh adhesive. (Running 5-10 boards through the machine should remove any old adhesive from the applicator wheel section and associated reservoir feed lines.)
3. Keep the reservoir and applicator wheel free from char, which reduces the transfer rate of heat to the adhesive. Char also can clog the grooves on the applicator wheel, thus reducing the amount of adhesive transferred to the substrate.
4. Keep all instruments, indicator lights, and instrument read-outs in functioning order.

5. Inspect cartridge heaters routinely to be certain that they have not burned out. Burned-out heating elements will lower the adhesive temperature, but generally will not lower the reservoir temperature.
6. Allow enough time to heat the adhesive. This can take 1-1/2 to 4 hours. Use a pyrometer to determine adhesive temperature.

2. *Setup Procedure*

The edgebander procedure is set up as follows.

1. Set reservoir temperatures to 204°C—do not change.
2. Set applicator wheel temperature to 191°C—do not change.
3. Measure the adhesive temperature in the reservoir close to the applicator wheel with a pyrometer. Do not make machine adjustments until the temperature is between 191 and 204°C.
4. Measure the surface temperature of the applicator wheel to ensure that it is controlling properly at 191°C. A temperature readout of 100-125°C will indicate a cartridge heater burnout.
5. Ensure that the nip roll pressure is moderate.
6. Set line speed to 20 m/min with a tachometer—do not change.
7. Set applicator wheel with respect to infeed guiderail (± 2 mm). It is necessary to maintain sufficient applicator wheel pressure against the substrate edge without causing the applicator wheel to bounce.
8. Replace existing edging on the machine with clear PVC edging.
9. Edge test pieces with clear PVC edging.
10. Inspect the glue line with a magnifying glass for voids or any sign of applicator wheel pattern.

If *no* voids or wheel pattern detected in the glue line, and a minimal to zero amount of squeeze-out is exhibited, the machine is adjusted properly. If voids or wheel pattern *are* observed in the glue line, the gap between the doctor blade and the applicator wheel must be increased. Run a second board through machine and check the results. Repeat these steps until the voids or the wheel pattern disappears from the glue line.

If difficulty is still encountered in applying sufficient adhesive to the substrate, the following procedure should be used.

1. *Do not* increase or change equipment temperature settings. (This will actually decrease the amount of adhesive transferred to the substrate, thus shortening the adhesive open time.)
2. Examine adhesive transfer on good quality, high density particleboard, which is less than 18 mm thick, and at a

substrate temperature greater than 18°C. (Substrate thickness, density, and temperature have a drastic effect on the open time.) This step is taken to determine applicator wheel assembly problems.

3. If the problem is caused by substrate thickness or density, an adhesive with a longer open time or lower heat resistance (if one exists) should be tried. Failing that, the doctor blade/applicator wheel assembly should be modified.

4. If the adhesive transfer problem is caused by low substrate temperature (< 18°C), install additional heating sources.

5. If the problem still persists, inspect the doctor blade/applicator wheel assembly. The doctor blade should be flush and square with the applicator wheel. It may be necessary to replace the applicator wheel with another wheel that has wider or deeper grooves.

Because of manufacturing tolerances or wear characteristics, a given applicator wheel may be suitable for use with a given combination of edging material, hotmelt adhesive, and substrate material, but not for others.

VIII. HEALTH AND SAFETY

Suitable protective clothing should be worn when handling adhesives in the molten state. If burns do occur, the following steps should be taken.

Immerse the affected area in clean cold running water for 20 minutes.
Do not attempt to remove the cold adhesive from the skin.
Cover the affected area with a wet compress and seek medical advice immediately.

Any vapor fumes emitted during normal operation are of relatively low toxicity, but they should not be repeatedly inhaled, and suitable ventilation should be provided where necessary.

Refer to adhesive manufacturers' specifications for information about the properties and characteristics of the adhesive, including safety instructions during use.

ACKNOWLEDGMENTS

The author thanks E. Hafner and J. Lubbe of Laminate Industries, Johannesburg, South Africa, for their valuable contribution to this chapter, with particular emphasis on the practical application of edge-veneering hotmelt adhesives.

9

Fast-Setting Adhesives for Finger-Jointing and Glulam

A. PIZZI and F. A. CAMERON* / Council for Scientific
and Industrial Research, Pretoria, South Africa

I. INTRODUCTION

Finger joints are commonly used to produce long boards from short
length timber. Such joints are acceptable in structural timber and
in laminations for glulam. The adhesives normally used for finger-
jointing are melamine/urea/formaldehyde and phenol/resorcinol/
formaldehyde, which require lengthy periods to set. There is therefore a 1-day delay between finger-jointing and further processing
and dispatch, which is inconvenient and interferes with production
flow.

Separate application adhesives capable of setting faster than
conventional adhesives were developed in the United States [1,2]
and in other countries [3,4] to glue large components where presses
were impractical [5-8]. Kreibich [8] describes these "honeymoon
systems" as follows: component A is a slow-reacting resin with
a reactive hardener and component B is a fast-reacting resin with
a slow-reacting hardener; when A and B are mated, the reactive
parts of the components react within minutes to form a joint that
can be handled and processed further. Full curing of the slow-
reacting part of the system takes place with time.

In South Africa the basic "honeymoon" gluing system was considerably modified by the use of tannin-based rather than synthetic
phenols for both components A and B and by the elimination of
the hardener from component B.

As indicated in Fig. 9.1, component A is a wattle/resorcinol/
formaldehyde cold-setting adhesive to which paraformaldehyde
hardener has been added. Coconut shell and wood flour fillers may

---

*Present affiliation: Industrial Laminates (PTY) Ltd., Alrode, South
Africa

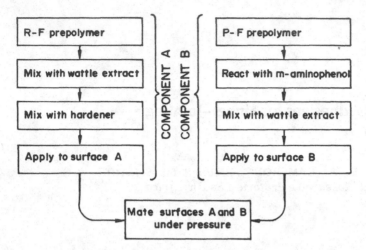

Fig. 9.1   Schematic representation of fast-set adhesive system using wattle tannin extract and m-aminophenol.

be added or omitted from the glue mix. Component B is instead composed of a small amount of a phenol/m-aminophenol/formaldehyde resin added to a 50-55% aqueous solution of commercial wattle extract at high pH, hence high reactivity, and containing no hardener. This system has been proved to be successful at both laboratory and plant levels. While finger joints manufactured industrially with this adhesive proved to be excellent, both in speed of curing and durability, the availability and high price of m-aminophenol chemical presented a problem.

    This chapter describes alternative phenolic-based and tannin-based fast-setting adhesives not containing m-aminophenol such as phenolic/phenolic, tannin/tannin, and tannin/phenolic adhesive systems, which can be used for the same purpose. All these systems have now been used extensively in forestry in Southern Africa since 1981.

II.   COMPONENTS A AND B

The adhesives used for components A and B of the "honeymoon" systems are shown in Table 9.1.

    All the combinations of components A and B (10 combinations) were used to manufacture finger joints.

    To better understand the functions of the different A and B components of the different adhesive systems presented, it is necessary to give some background about the resins of which they are composed.

Table 9.1   Adhesive Components A and B Used for "Honeymoon" System

Component A	Component B
System A1. Commercial wattle/ resorcinol/formaldehyde cold-setting laminating adhesive + hardener + fillers (pH = 7.5)[a]	System B1. Commercial wattle/ formaldehyde adhesive fortified with phenol/*m*-aminophenol/ formaldehyde resin; no hardener; pH = 9.0[b]
System A2. Commercial phenol/ resorcinol/formaldehyde cold-setting laminating adhesive + hardener + fillers (pH = 8.0)	System B2. Commercial wattle/ resorcinol/formaldehyde cold-setting laminating resin; no hardener[a]; pH adjusted to 11.4
	System B3. Pine tannin extract; no hardener; pH adjusted to 12.4
	System B4. Wattle tannin extract[c] no hardener; pH adjusted to 12.6
	System B5. Commercial phenol/ resorcinol/formaldehyde resin[d]; no hardener; pH adjusted to 11.4

[a]Supplied by Bondtite Adhesives (PTY) Ltd.
[b]F101 supplied by Hüttenes-Albertus-Falchem (South Africa) (Pty) Ltd.
[c]Mimosa extract supplied by Natal Tannin Extract Company, Ltd.
[d]Cascophen RS12 supplied by General Chemical Corporation Ltd.

A. Components A

1. Wattle extract/resorcinol/formaldehyde cold-setting laminating adhesive. This adhesive is prepared by simultaneous synthesis of resorcinol/formaldehyde and flavonoid/formaldehyde condensates; namely, the reactions involved are as follows:

$$-CH_2-\ \text{flavonoid}-\ CH_2-\text{condensates}$$
$$+$$
uncombined resorcinol and flavonoid polymers

This mixture can be cured at ambient temperature by addition of paraformaldehyde 96% powder, at a pH of 7.5–7.9, according to the paraformaldehyde grade added. The proportion of wattle extract solids, resorcinol, and formaldehyde (hardener excluded) used in the preparation of the resin is generally 100:50:6 mass/mass. The exact formulation for this resin has already been reported [3]. This adhesive is weather- and boil-proof.

2. Phenol/resorcinol/formaldehyde laminating adhesive. This is one of the many resorcinol-terminated phenolic resins available on the market for cold-setting laminating applications. These adhesives are weather- and boil-proof.

Components B

1. Wattle extract, 50–55% solution, to which is added a phenol/*m*-aminophenol/formaldehyde resin. The *m*-aminophenol-terminated phenol/formaldehyde resin is prepared according to the following reactions.

Resin I is added to a 50–55% wattle extract solution and the pH adjusted to 9. The proportion of wattle extract solution (55% solids) to resin I to a 40% solution of sodium hydroxide is 100:42:25, mass/mass. The exact formulation of this resin has already been reported [3,4].

I

2. Wattle extract/resorcinol/formaldehyde cold-setting laminating adhesive prepared as component A1, with no hardener or fillers and with pH adjusted to 11.4.

3 and 4. Pine bark tannin extract and wattle bark tannin extract. The phenolic nature and consequent behavior of pine and wattle tannins are evident from the schematic structure of their main monomeric constituents.

Wattle tannin   Pine tannin

These schematic wattle and pine tannin structures represent the two major types of flavonoid tannin known, wattle tannins being composed of mainly resorcinolic A rings and pine tannins by phloro-

glucinolic A rings [3,9-11]. The pyrogallic and catecholic B rings
do not take part in the reaction with formaldehyde to any noticeable
extent (up to pH = ± 10) due to their low nucleophilic character
[3,9-11]. A further difference between these two tannins is the
mode of linkage of the flavonoid units: 4,6 in the case of wattle
tannins and 4,8 in the case of pine tannins.

The stronger nucleophilicity of the pine tannins' phloroglucinolic
A rings causes pine tannin to be considerably faster reacting, with
formaldehyde, at comparable pH values, than the resorcinolic A
rings of the wattle tannins. As a consequence, pine tannin extract,
with the pH adjusted to 12.4, was a more logical choice as a component B of the "honeymoon," system as it should allow faster curing,
hence shorter time delay before finger-joint machining, than wattle
tannin extract. Untreated pine tannin extract was thus given closer
attention and was more extensively experimented with.

5. Phenol/resorcinol/formaldehyde without any hardener and
with pH adjusted to 11.4.

The bark extracts of the black wattle tree (*Acacia mearnsii*,
formerly *mollissima*), commercially available and containing 70-80%
tannins, and of the pine species *Pinus patula*, pilot plant produced
by ambient temperature countercurrent methanol extraction were
used for this study.

The influence of the following parameters on the performance
of the different adhesive systems and of the finished finger joints
was evaluated: (a) wood density, (b) wood species, and (c) tolerance
to quantity variations, namely, variations of A/B components were
investigated at ratios of 2:1, 1:1, and 1:2.

To show that the curing speed of "honeymoon" systems is considerably faster than conventional adhesive systems, finger joints
were also prepared by using on both profiles a commercial phenol/
resorcinol/formaldehyde resin + hardener and also a commercial
wattle/resorcinol/formaldehyde resin + hardener with unadjusted
pH. Thus, the components A1 and A2 were used to glue both sides
of the finger joints.

III. ADHESIVE RESINS PREPARATION

The quantities of chemicals as well as the processes of manufacture
of the different adhesives used have already been reported, namely,
resins A1 [3,9], B2 [3,9], and B1 [4]. The phenol/resorcinol/
formaldehyde resin used was a commercial product. However, any
of the numerous phenol/resorcinol/formaldehyde resins commercially
available can be used with similar results. Resins A1, B1, and B2,
as well as resins A2, B3, B4, and B5, are described in Table 9.2.

Table 9.2   Glue Mixes of Adhesive Systems

Glue mixes[a]	Parts by mass	
System A1. Wattle-based liquid resin (52% solids)	100	
Paraformaldehyde 96%, powder	10	15
Fillers (200 mesh wood flour; coconut shell flour 50:50 by mass)	10	
pH:	7.45-7.55	
Pot-life:	2.5 hours at 25°C	
The addition of fillers in this glue mix is optional.		
System A2. Liquid resin	100	
Hardener + fillers, powder	20	
pH:	8.0	
Pot-life:	1.5-3.0 hours at 25°C	
System B1. Wattle extract solution (55% solids)	100	
Phenol/m-aminophenol/ formaldehyde resin, liquid	42	
pH:	9.0	
Pot-life:	indefinite, as no hardener is added	
System B2. Wattle-based liquid resin (52% solids)	100	
pH:	11.4	
Pot-life:	indefinite, as no hardener is added	

(continued)

Table 9.2 (continued)

Glue mixes[a]	Parts by mass
System B3. Pine extract solution (50% solids)	100
pH:	12.4
Pot-life	indefinite, as no hardener is added
System B4. Wattle extract solution (50-55% solids)	100
pH:	12.6
Pot-life:	indefinite, as no hardener is added
System B5. Liquid resin	100
pH:	11.4
Pot-life:	indefinite, as no hardener is added

[a]All pH values adjusted with 30-40% NaOH solution.

IV. FINGER—JOINTING

Boards of South African pine (*Pinus patula, P. radiata, P. pinaster, P. taeda,* and *P. elliotii*) of dimensions 5000 mm × 120 mm × 37 mm and of *Eucalyptus saligna* of dimensions 5000 mm × 75 mm × 25 mm were finger-jointed, using the various adhesive systems presented. All the adhesive systems were used to produce finger joints to be tested at 0.5, 1.0, 2.0, 4.0 and 24.0 hours, and 7 days, after gluing to evaluate the rate of strength increase with time.

The series of finger joints manufactured consisted of the following:

1. *General.* Finger joints were prepared using medium density pine (approximate density range: 0.450-0.550 g/cm^3) using the adhesive systems presented.

2. *Effect of wood density.* Finger joints were prepared using high density pine (approximate density range: 0.650-0.750 g/cm^3) using the adhesive systems A1/B2 and A1/B3.

3. *Tolerance to quantities and viscosity variations.* Since it is practically impossible, under factory production conditions, to have identical quantities of adhesive on both profiles of each finger joint, it was deemed necessary to investigate the tolerance of the adhesive systems to variation of quantities. This was achieved by varying the glue mix viscosity of the two adhesive components to have approximate proportions by weight of A/B of 2:1, 1:1, and 1:2. The viscosities used were expressed in centipoise as follows:

Component A	Component B	Approximate mass
3800	2000	2:1
1900	1900	1:1
1900	3500	1:2

The A/B adhesive systems used were A1/B2 and A1/B3. Gel times at 60°C of mixtures of different mass ratios of component A to component B, namely 25:75, 33:67, 50:50, 67:33, 75:25, and 100:0 were also measured. The gel time results for the A/B adhesive system A1/B1 are shown in Table 9.3.

4. *Control.* Finger joints in which both profiles were glued with component A1 or component A2 were also prepared to compare the rate of strength increase of the "honeymoon" adhesive systems with that of more traditional cold-setting adhesives.

Table 9.3 Gel Time at 60°C of Adhesive A1/B1 Containing Different Ratios of Components A and B

A (parts by mass)	B (parts by mass)	Mean gel time (sec)
25	75	50.0
33	67	39.3
50	50	37.3
67	33	39.3
75	25	45.6
100	0	320.0

Table 9.4   Results of Finger Joints Accelerated Weathering Tests

| Adhesive system | South African pine | | | | | | Average density (gr/cm³) |
| | Dry | | 24 hr cold soak | | 6 hr boil | | |
	Strength, tension (N)	Wood failure (%)	Strength, tension (N)	Wood failure (%)	Strength, tension (N)	Wood failure (%)	
A1/A1	4772	72	2466	66	2613	87	0.545
A2/A2	3720	96	2132	76	2170	87	0.500
A1/B1	4501	94	3332	77	2978	73	0.493
A1/B2	4249	95	2814	56	2803	67	0.466
A1/B3	3996	97	2686	65	2711	73	0.461
A1/B4	4237	92	2432	46	2725	56	0.535
A2/B1	–	–	–	–	–	–	–
A2/B2	4977	95	3777	88	3350	90	0.512
A2/B3	5252	90	3474	85	3358	89	0.531
A2/B5	4060	94	2658	94	2824	92	0.531
High wood density							
A1/B2	5580	97	3661	61	3472	64	0.711
A1/B3	5296	95	3231	48	3410	59	0.711
Variation of proportions							
A1/B2 2:1	4539	99	2796	44	2472	74	0.539
A1/B2 1:1	4846	94	2859	50	2844	63	0.539
A1/B2 1:2	4757	94	2488	38	2584	52	0.539
A1/B3 2:1	4174	96	2811	61	3381	82	0.546
A1/B3 1:1	4500	91	2150	40	2040	79	0.546
A1/B3 1:2	4216	99	1846	49	2102	68	0.546

Adhesive system	Eucalyptus saligna						
	Dry		24 hr cold soak		6 hr boil		Average density (g/cm³)
	Strength, tension (N)	Wood failure (%)	Strength, tension (N)	Wood failure (%)	Strength, tension (N)	Wood failure (%)	
A1/A1	—	—	—	—	—	—	—
A2/A2	—	—	—	—	—	—	0.513
A1/B1	4687	85	3543	58	3803	82	0.490
A1/B2	4516	72	2215	31	2691	40	0.490
A1/B3	3927	75	3008	58	3160	53	0.490
A1/B4	—	—	—	—	—	—	—
A2/B1	5548	98	3465	75	3368	89	0.478
A2/B2	—	—	—	—	—	—	—
A2/B3	—	—	—	—	—	—	—
A2/B5	3985	88	4182	78	3538	80	0.481
High wood density							
A1/B2	—	—	—	—	—	—	—
A1/B3	—	—	—	—	—	—	—
Variation of proportions							
A1/B2 2:1	—	—	—	—	—	—	—
A1/B2 1:1	—	—	—	—	—	—	—
A1/B2 1:2	—	—	—	—	—	—	—
A1/B3 2:1	—	—	—	—	—	—	—
A1/B3 1:1	—	—	—	—	—	—	—
A1/B3 1:2	—	—	—	—	—	—	—

V. TESTING

The finger joints were tested to failure in bending, using three-point loading (span 914.4 mm) after time lapses of 0.5, 1.0, 4.0, and 24.0 hours, and 7 days after gluing and assembling. Percentage wood failure of the broken finger joints was assessed according to the South African Bureau of Standards (SABS) specification 096-1976 for structural finger joints [12]. The SABS specification requires a minimum of 20 MPa in bending. Several specimens for each time span were tested. For all the finger-joint series prepared with the different adhesive systems, accelerated weathering tests according to SABS 096-1976 were carried out 7 days after gluing and assembling. The accelerated weathering tests consisted of testing in tension finger-joint specimens of 250 mm × 25 mm × 5 mm, dry, after 24 hours of cold water soaking and after 6 hours of boiling. The SABS 096-1976 requirements for exterior-grade structural finger joints after 24 hours of cold water soaking and/or 6 hours of boiling are on a sliding scale depending on failing load as well as wood failure:

Minimum failing load (newtons)	Wood failure (%)
1400	90-100
1700	70-89
2100	50-69
2500	30-49
2800	10-29

The average shear strength and percentage wood failure results obtained for the different adhesive systems and finger-joint series are shown in Table 9.4.

VI. STATISTICAL ANALYSIS

Because the total experiment involved nearly 1000 finger joints, meaningful statistical analysis could be applied to the results. In this type of experiment the most useful information concerns the relationship between the strength of a joint and the period of time after gluing and assembling. The curves of strength versus time are of considerable importance, since they show how long after gluing and assembling the finger joint has enough strength to be machined or dispatched without damage. This relationship normally can be expressed in some form of regression equation or graphic curve.

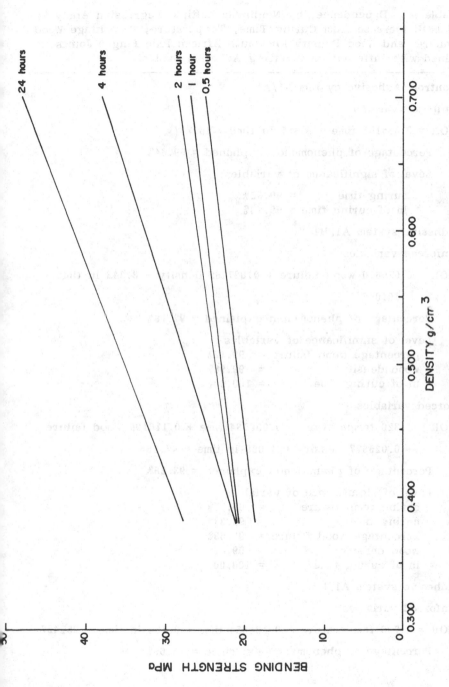

Fig. 9.2 Bending strength vs. timber density of pine fingerjoints bonded at ambient temperature with fast-set adhesives after 30 min, 1 h, 2 h, 4 h, and 24 h curing at ambient temperature.

Table 9.5 Dependence, by Nonlinear Multiple Regression Analysis, of MOR Increase from Curing Time, Temperature, Percentage Wood Failure, and Wood Density for South African Pine Finger Joints Glued with Different Fast-Setting Adhesive Systems

---

Control: Adhesive system A1/A1

Unforced variables

MOR = 0.065181 time + 3.157 ln time + 19.707

    Percentage of phenomenon explained = 99.64%

    Level of significance of variables:

        curing time   = 99.05%
        ln of curing time = 99.74%

Adhesive system A1/B1

Unforced variables

MOR = 0.070610 wood failure + 0.027162 density + 3.342 ln time

    + 10.092

    Percentage of phenomenon explained = 92.78%

    Level of significance of variables
      percentage wood failure = 99.12%
      wood density   = 99.98%
      ln of curing time   = 100.00%

Forced variables

MOR = 1.320 temperature - 0.058784 time + 0.116196 wood failure

    + 0.025677 density + 3.690 ln time + 47.886

    Percentage of phenomenon explained = 93.28%

    Level of significance of variables
      curing temperature   = 68.75%
      curing time   = 92.33%
      percentage wood failure = 93.58%
      wood density   = 99.83%
      ln of curing time   = 100.00%

Adhesive system A1/B2

Unforced variables

MOR = 1.796 temperature + 0.134854 time + 3.153 ln time - 25.427

    Percentage of phenomenon explained = 73.69%

Level of significance of variables
curing temperature = 99.99%
curing time   = 99.92%
ln of curing time = 99.99%

Forced variables
percentage wood failure = 32.20%
wood density   = 48.58%

Adhesive system A1/B3

Unforced variables

MOR = 2.773 temperature + 5.285 ln time − 48.308

Percentage of phenomenon explained = 79.00%

Level of significance of variables
curing temperature = 100.00%
ln of curing time = 100.00%

Forced variables

MOR = 2.119 temperature + 0.047170 time + 0.075485 wood failure

   + 0.029009 density + 4.026 ln time − 50.583

Percentage of phenomenon explained = 80.95%

Level of significance of variables
curing temperature   = 98.79%
curing time   = 75.58%
percentage wood failure = 88.79%
wood density   = 93.24%
ln of curing time   = 100.00%

The experimental results were analyzed according to two different approaches. In the first method, one adhesive system was selected and simple regression analysis was used to establish the relationship of strength as a function of the density of the timber. Several regressions were performed, and each equation corresponded to a unique gluing time (viz., for 0.5, 1.0, 2.0, 4.0, and 24.0 hours, and for 7 days).

The graphs of strength versus timber density were plotted for the 0.5-4.0 hour period, which was the more interesting time span. A certain density was then selected and the corresponding strength value was determined from the strength-density curves for each time span. These strength-time span values were then plotted, and the procedure was repeated for the rest of the adhesive systems. Numerous strength-timber density graphs were generated, but only one is presented (Fig. 9.2). All the relevant strength-time span curves are shown (Figs. 9.3-9.9, below).

In the second approach it was decided to use multiple regression analysis to obtain, for a certain adhesive system, an equation that would describe strength in terms of the other variables. The independent variables considered were the following:

Curing temperature (°C)
Curing time (hours)
Percentage wood failure
Wood density (g/cm^3)

The dependent variable was the modulus of rupture (MOR) expressed in megapascals (MPa). Since the points obtained appeared to follow a logarithmic law for the curing time, it was decided to use both the curing time and its natural logarithm to obtain more elastic and pliable equations that were better able to describe the phenomenon. The multiple regression analysis was performed with forced and unforced variables. The multiple regression equations with the relative coefficients for the statistically significant independent variables and the percentages of the phenomenon explained for the different adhesive systems for which there were enough data to perform a multiple regression analysis are shown in Table 9.5.

VII. INITIAL INDUSTRIAL APPLICATION

Table 9.6 shows the strength-versus-time test results of finger joints manufactured in a 3-day factory trial using the adhesive systems A1/B1, A1/B2, A1/B4, A2/B1, A2/B4, and A2/B5. Table 9.7 gives accelerated weathering test results for four of these systems. To determine whether planing of the finger joints 0.5 hour after manufacture would damage the glue line and decrease its strength, finger joints were tested rough, as-manufactured, and after planing. From the bending and tensile strength results, it appeared that planing of the joints 0.5 hour after manufacture has no bearing on the strength of the joint. The joints were tested 0.5, 1.0, 2.0, and 24.0 hours, and 7 days, after manufacture. All the components A and B of the adhesive system used had been industrially manufactured. The viscosities used for various adhesive systems were the following:

Table 9.6 Development of Bending Strength Versus Time: Industrial Production Finger-Joints Trial

Curing time (hr)	Treatment	Mean values[a]	Adhesive system, at ambient temperature					
			A1/B1, 23°C	A1/B2, 24°C	A1/B4, 21°C	A2/B1, 25°C	A2/B4, 21°C	A2/B5, 24°C
0.5	Planed	MOR (MPa)	16.8	13.6	14.4	18.9	8.6	9.6
		Density (kg/m³)	537	544	581	577	538	555
		Wood failure (%)	15	3	2	7	2	1
	Rough	MOR (MPa)	17.0	14.0	12.7	180	85	12.1
		Density (kg/m³)	522	509	568	564	527	531
		Wood failure (%)	13	2	3	9	1	0
1.0	Planed	MOR (MPa)	19.8	13.6	16.8	21.5	12.5	11.5
		Density (kg/m³)	553	570	629	506	617	612
		Wood failure (%)	19	6	10	32	2	5
	Rough	MOR (MPa)	17.8	14.8	16.9	20.8	12.9	13.4
		Density (kg/m³)	526	532	629	497	598	584
		Wood failure (%)	19	6	11	26	2	2
2.0	Planed	MOR (MPa)	22.5	19.0	16.4	26.0	18.6	16.8
		Density (kg/m³)	573	541	548	561	505	501
		Wood failure (%)	31	25	25	39	26	18
	Rough	MOR (MPa)	20.2	17.4	15.0	24.9	17.1	16.2
		Density (kg/m³)	533	538	545	558	510	480
		Wood failure (%)	37	17	28	35	19	18
24.0	Planed	MOR (MPa)	30.4	32.5	31.8	35.1	35.4	34.3
		Density (kg/m³)	548	579	607	591	527	640
		Wood failure (%)	81	80	32	67	100	70
	Rough	MOR (MPa)	28.6	28.4	36.4	32.6	35.1	33.0
		Density (kg/m³)	542	541	579	572	512	633
		Wood failure (%)	84	72	35	66	91	47

[a]Mean taken of 5 specimens.

Table 9.7 Accelerated Weathering Test of Finger Joints[a] Manufactured Industrially: Tension Test of 150 mm × 25 mm × 5 mm after 7 Days

Adhesive system	Dry		24-hr cold soak		3-hr boiling	
	Failing load (N)	Wood failure (%)	Failing load (N)	Wood failure (%)	Failing load (N)	Wood failure (%)
A1/B1						
planed	5585	84	3305	75	3454	80
rough	4783	79	3082	75	3451	84
A1/B2						
planed	5770	73	3033	35	2982	25
rough	5030	65	2860	46	3086	50
A1/B4						
planed	5135	87	3189	77	3066	51
rough	5492	79	3144	65	3034	59
A2/B1						
planed	6120	90	3853	28	3430	44
rough	5788	96	3984	22	4042	44
A2/B5						
planed	5741	97	3467	25	3946	47
rough	5846	97	3833	34	4247	54

[a]Each value presented in this table is the mean taken of 25 specimens.
Source: South African Bureau of Standards.

Viscosities (Pa·sec)		
Adhesive system	Component A	Component B
A1/B1	4.7	4.85
A1/B2	2.55	2.35
A1/B4	2.45	3.05
A2/B1	3.4	3.7
A2/B4	3.4	3.2
A2/B5	3.8	3.7

The South African Bureau of Standards accepted the new adhesive systems as suitable for finger-jointing timber of grades 4, 6, and 8.

The results shown in Figs. 9.2-9.9 and Tables 9.6-9.7 indicate that all the adhesive systems performed sufficiently well to allow fast bonding of finger joints at ambient temperatures of ±25°C. Notwithstanding the differences in curing rates among them, they all developed strength fast enough to allow a short delay between assembly and machining or dispatch if glued at temperatures of ±25°C. However, it was expected that the curing rate would be considerably slower at lower temperatures. A follow-up experiment was therefore necessary to determine the behavior of these adhesives at temperatures down to 5°C.

The shapes of the development of strength-versus-time curves of the A1/B1, A1/B2, and A1/B3 systems in Fig. 9.3 indicate that the setting and curing characteristics of these three systems are considerably different. The A1/B1 system has the fastest initial rate of setting due to the presence of the highly reactive *m*-aminophenol, whose reaction with formaldehyde is also violently exothermic, causing an even faster initial rate of condensation. However, once the *m*-aminophenol has reacted and the initial *m*-aminophenol/formaldehyde exotherm has been exhausted and the temperature has subsided, the development of strength with time is slower than in the A1/B2 and A1/B3 systems. The A1/B2 system shows initial curing rate nearly as fast as the A1/B1 system, while the A1/B3 resin is somewhat slower. The long-term results (24 hours and 7 days) are good for all three resin mixtures. Considering the minimum bending strength requirements for machining finger joints without damage as lying in the 10-15 MPa range, the A1/B1, A1/B2 and A1/B3 adhesive systems allow finger-joint machining approximately 5, 10, and 15 minutes after assembly. The minimum final bending strength requirements according to SABS 096-1976 are a 20-MPa average, with the lowest value not less than 12 MPa. The 20-MPa level is reached in 15, 30, 42, and 30 minutes for the A1/B1, A1/B2, A1/B3, and A1/B4 adhesive systems respectively (Figs. 9.3 and 9.4). The apparently anomalous result of the A1/B4

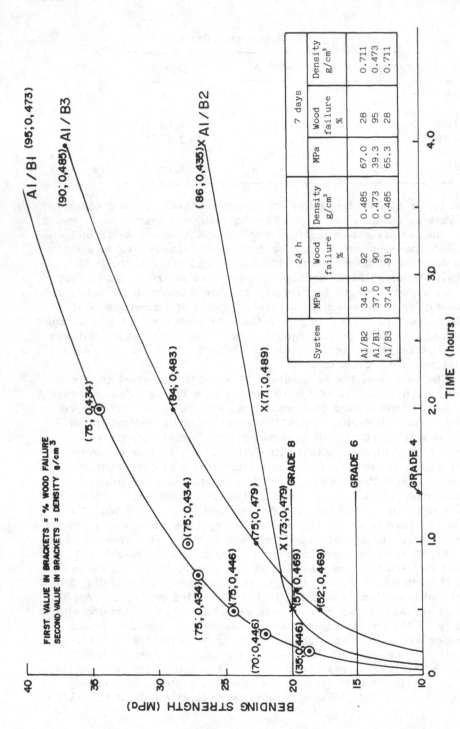

Fig. 9.3 Bending strength vs. curing time of pine fingerjoints bonded at ambient temperature with fast-set adhesives types A1/B1, A1/B2, and A1/B3.

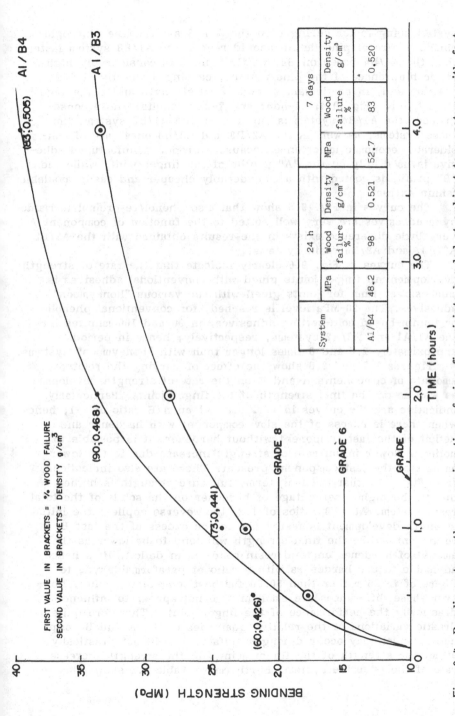

System	24 h			7 days		
	MPa	Wood failure %	Density g/cm³	MPa	Wood failure %	Density g/cm³
A1/B4	48.2	98	0.521	52.7	83	0.520

Fig. 9.4 Bending strength vs. curing time of pine fingerjoints bonded at ambient temperature with fast-set adhesives A1/B3 and A1/B4.

system being faster curing than the A1/B3 system (the phloroglucinolic nature of pine tannin should render the A1/B3 system faster
than the A1/B4 based on B4, wattle tannin) is considerably higher
steric hindrenace of the pine tannins, causing a slightly slower
reaction and, in particular, a lower final strength and percentage
wood failure (Fig. 9.4; 24-hour and 7-day results). As a consequence, the A1/B4 system is superior to the A1/B3 system. Both
these systems, as well as the A2/B3 and A2/B4 ones, are of considerable economic importance because a proper manufactured adhesive is used only on the "A" profile of the finger joint, while the
"B" profile is coated with a considerably cheaper and easily available
tannin extract.

The curves in Fig. 9.5 show that also phenol/resorcinol/formalde-
hyde adhesives are very well suited to the function of component A.
Very little difference exists in the results obtained with the A2/B2,
A2/B3, and A2/B5 resin systems.

The curves in Fig. 9.6 clearly indicate that the rate of strength
development in finger joints glued with conventional adhesives is
much slower than for joints glued with the various "honeymoon"
adhesives. The 20-MPa level is reached, for conventional phenolic
or tannin-based cold-setting adhesives, in 90 and 100 minutes for
the A1/A1 and A2/A2 systems, respectively; hence in periods
approximately 2.5 and 6 times longer than with "honeymoon" systems.

Figures 9.7 and 9.8 show the effect of varying the relative
amounts of components A and B on the rate of strength development
as well as on the final strength of the finger joints. Particularly
indicative are the curves in Fig. 9.8. At an A/B ratio of 2-1, hence
when there is excess of the slow component with hardener and
deficit of the fast component without hardener, it is possible to
notice a slower initial rate of strength increase due to the lower
amount of the fast component present. There are also indications,
from Fig. 9.7, that the long-term, full-cure strength is higher
due to the higher percentage of hardener on the solids of the total
resin system. At A/B ratios of 1:2, the reverse applies: the initial
strength development is faster due to the excess of the fast B
component, while the final strength will tend to be lower as the
amount of hardener on total resin solids is in deficit. It is hence
advisable to use hardeners with a ratio of paraformaldehyde to
fillers of 75:25 rather than 50:50 for best long-term results. However these differences are small and do not appear to influence
drastically the performance of the finger joints. Therefore quite
drastic variations of the relative quantities of the A and B components, which may occur during manufacture, will not drastically
affect the strength of the finger joint and the strength increase
rate of the adhesive system. Furthermore Table 9.3 shows that gel

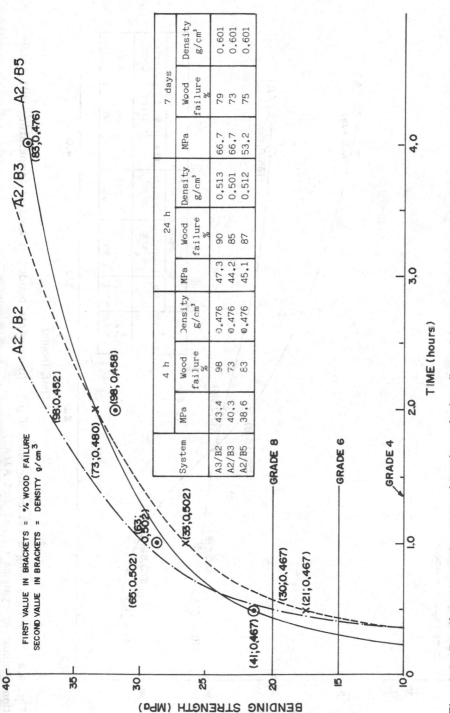

Fig. 9.5 Bending strength vs. curing time of pine fingerjoints bonded at ambient temperature with fast-set adhesives A2/B2, A2/B3, and A2/B5.

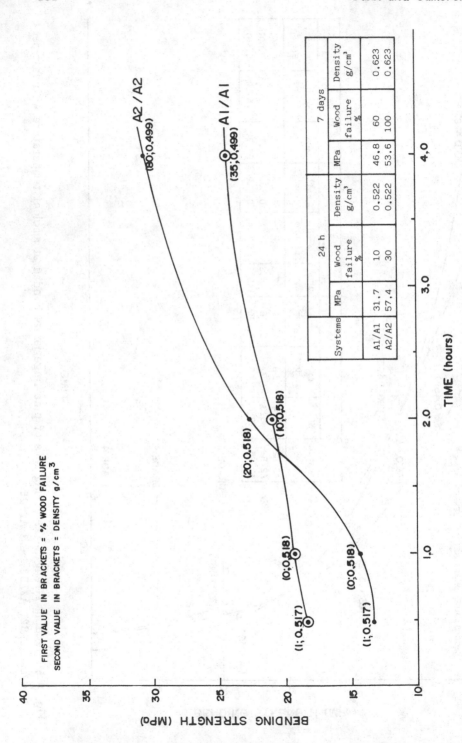

Fig. 9.6 Bending strength vs. curing time of pine fingerjoints bonded at ambient temperature with tradi-tional adhesives systems (A1/A1 and A2/A2).

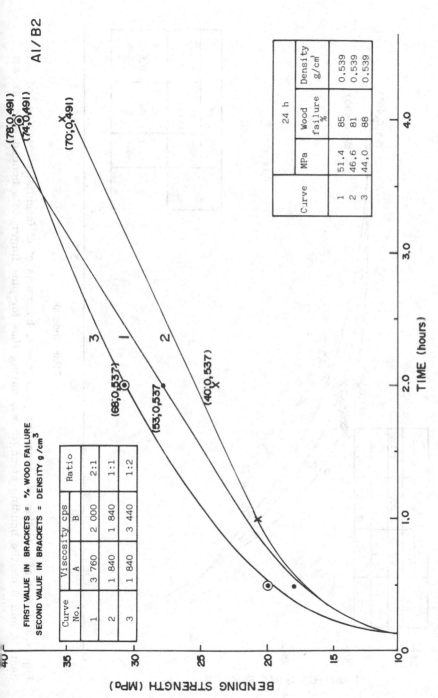

Fig. 9.7 Effect of variation of respective proportions, by means of differential viscosities, of components A and B on the bending strength vs. curing time for pine fingerjoints bonded with fast-set adhesives Al/B2.

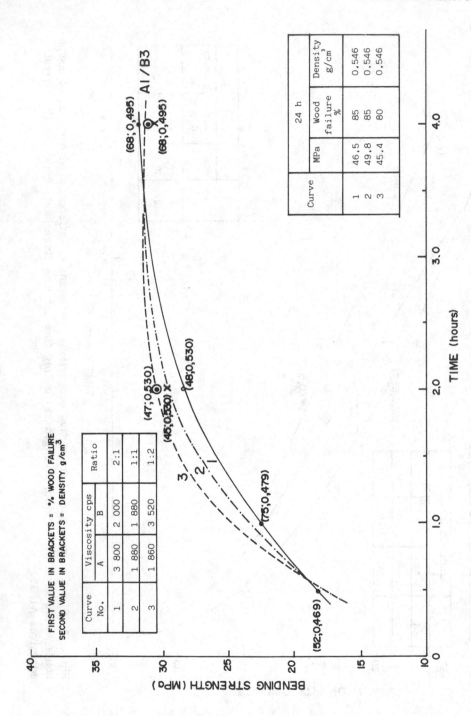

FIRST VALUE IN BRACKETS = % WOOD FAILURE
SECOND VALUE IN BRACKETS = DENSITY g/cm³

Curve No.	Viscosity cps		Ratio
	A	B	
1	3 800	2 000	2:1
2	1 880	1 880	1:1
3	1 860	3 520	1:2

Curve	24 h		
	MPa	Wood failure %	Density g/cm³
1	46,5	85	0,546
2	49,8	85	0,546
3	45,4	80	0,546

Fig. 9.8 Effect of variation of respective proportions, by means of differential viscosities, of components A and B on the bending strength vs. curing time for pine fingerjoints bonded with fast-set

times increase only slightly when the proportions of components A and B are changed from 1:1 to 1:2 or 2:1. This confirms that a considerable deviation from the intended components (1:1) is permissible. Since it is felt that the inequality of the mixing proportions would, in practice, not exceed 2:1, the adhesive systems should be highly suitable for industrial application.

Figure 9.9 shows the curves of strength increase versus time of finger joints manufactured with *Eucalyptus saligna* using the same adhesive systems. The trends shown are the same as those of pine finger joints.

The results of the accelerated weathering tests showed that all finger joints produced with the various "honeymoon" adhesive systems satisfy the SABS 096 requirements for weather- and boil-proof finger-jointed structural timber even when high density timber is used (Table 9.4: 0.711 g/cm^3). However, the unduly high fall-off in wood failure upon soaking with water and boiling suggests that the durability of these adhesives could be suspect. Further testing, especially outdoor weather exposure tests, was therefore carried out to prove the durability beyond doubt.

The results of the accelerated weathering tests are not significantly affected by the variation of the A/B ratio from 1:1 to 2:1 and 1:2.

It is also interesting to compare the equations correlating the modulus of rupture (MOR) in bending with other independent variables obtained by nonlinear multiple regression analysis for the A1/A1, A1/B1, A1/B2, and A1/B3 adhesive systems. This comparison gives an insight on the differences in intrinsic characteristics of the various resins used as these are related to their molecular structure. The independent variables that mostly contribute to the MOR values are those appearing in the unforced variable equations. It is noticeable that the MOR of the conventional control adhesive, which is quite slow curing, depends mainly on the length of time the adhesive is left to cure. It is seen that the MOR of the A1/B1 adhesive system containing the very reactive *m*-aminophenol is instead quite independent of the curing time, since high MORs are achieved very quickly after assembly of the joint. The latter adhesive system is, however, strongly dependent on the density of the wood used (both the percentage wood failure and the wood density variables appear in the equation). Such behavior has often been noticed in industrial practice, but no logical explanation for it has been proposed.

The A1/B2 system appears to be a somewhat better binder than the A1/B1 adhesive. It is not very dependent on wood density (the level of significance of the wood density variable is only 48%) but only on the curing time and the curing temperature. The slightly more marked dependence on the curing temperature,

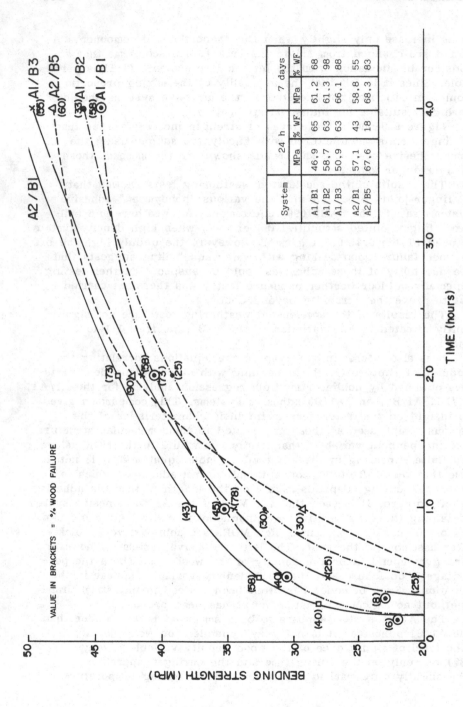

VALUE IN BRACKETS = % WOOD FAILURE

System	24 h			7 days		
	MPa	%	WF	MPa	%	WF
A1/B1	46.9	65		61.2	68	
A1/B2	58.7	63		61.3	98	
A1/B3	50.9	63		66.1	88	
A2/B1	57.1	43		58.8	55	
A2/B5	67.6	18		68.3	83	

Fig. 9.9 Comparison of bending strength vs. curing time of pine fingerjoints bonded with 5 different fast-set adhesives.

with respect to the A1/B1 system, is due to the absence of the very reactive *m*-aminophenol present in the latter. The A1/B3 (and similarly A1/B4) system appears to be a case of "in between" the A1/B1 and A1/B2 systems; less dependent on wood density for the A1/B3 system, obtained by the different route of single linear regression analysis, as shown in Fig. 9.2. In it, as expected, it is noticeable that the MOR value is more dependent on the wood density after longer curing times, this being due to the more cured, hence stronger, adhesive.

VIII. VARIATIONS IN PARAFORMALDEHYDE CONTENT AND pH

A. Increase in the Paraformaldehyde Concentration in Component A and pH Adjustment in Component B

The level of paraformaldehyde in the A component was increased from approximately 18% to 30% on total resin solids, which means from 11% to 18% on liquid resin. To maintain a 5:1 mass ratio of liquid resin to hardener, the amount of filler in the blend was reduced. Glue mixes are shown in Table 9.9.

The pH values of the B component were lowered as follows:

B1 from 11.6 to 10.5
B2 from 12.5 to 11.5
B3 from 10.0 to 9.0

Beech strips were assembled and clamped, 10 at a time. Mean results are shown in Tables 9.13 and 9.14.

Series 1, 2, and 3 in the table were cured at 50°C for 4 hours; series 4 was the same as 1 but was cured at 25°C for 7 days.

B. Further Testing with Increased Level of Paraformaldehyde in Component A

Further testing was restricted to three adhesive systems: A1/B4, A2/B1, and A2/B2.

The glue mixes used for the A components are shown in Table 9.9.

The A1 component had too low a viscosity for application with 2 parts by mass of wood flour. It was therefore necessary to add 1.5 parts by mass more wood flour than to the A2 system.

To match the viscosities of the A1/B4 system, 3.5 parts by mass of wood flour was added to 100 parts by mass of the B component.

Table 9.8   Glue Mixes for Component A with
Increased Paraformaldehyde

| | Parts by mass | |
Ingredient	A1	A2
Resin, liquid	100	100
Wood flour 200 mesh	1.25	2.5
Coconut shell flour 200 mesh	1.25	—
Paraformaldehyde, 96%	18	18

Thirty beech strips were assembled for each system: 15 were
cured at 25°C for 7 days and 15 were cured at 50°C for 4 hours.
Mean results dry, after 6 hours in boiling water or 24 hours in
cold water are shown in Tables 9.10 and 9.11.

C.   Increased Paraformaldehyde in Component A
     and Lower pH in Component B

From the results (Table 9.12) it is apparent that lowering the pH
of the component B has no effect on the cold water resistance of
the adhesive systems.
    The increased level of paraformaldehyde in component A improved
the bond sufficiently in the case of the A2/B5 and A2/B4 systems
to meet the requirements of BS 1204 [13].

D.   Increased Level of Paraformaldehyde Only:
     Further Testing

The mean results of further tests carried out on three selected adhesive
systems (Tables 9.13 and 9.14) satisfy the British standard requirements (BS 1204 [13]) at both 25 and 50°C curing temperatures.

Table 9.9   Glue Mixes for Component A with
Increased Paraformaldehyde

| | Parts by mass | |
Ingredient	A2	A2
Resin	100	100
Wood flour 200 mesh	3.5	2
Paraformaldehyde 96%	18	18

Table 9.10 Results for Beech Strips and Finger Joints Cured at 50°C for 4 Hours

Adhesive system A/B	Finger joints[a]								Beech strips[a]							
	Pine				Saligna				10 strips				5 strips			
	24-hr cold soak		6-hr boil		24-hr cold soak		6-hr boil		24-hr cold soak		6-hr boil		24-hr cold soak		6-hr boil	
	N	%	N	%	N	%	N	%	N	%	N	%	N	%	N	%
A1/B4	2195	10	8,275	14	2180	22	9850	0	2612	21	3102	23	2787	53	2590	28
A1/B1	2655	24	5,625	15	1693	5	6167	0	2026	4	2130	32	2125	3	2310	33
A2/B5	1238	21	10,100	6	2467	27	8000	0	2620	69	2955	44	3140	63	*2155	95
A2/B4	2718	81	6,175	39	2997	17	9433	22	2714	67	2310	55	2760	28	*1935	95
A2/B1	1917	30	8,975	24	2453	23	6400	10	1820	4	1946	8	1893	2	*1210	33

*Asterisk indicates suspect result.
[a]Specimens were tested for strength in tension; these results appear in the columns headed N(ewtons). Wood failure (%) is also given.

Table 9.11 Mean Results for Beech Strips and Finger Joints Cured at 25°C for 7 Days

Adhesive system A/B	Finger joints[a]								Beech strips[a]							
	Pine				Saligna				10 strips				5 strips			
	24-hr cold soak		6-hr boil		24-hr cold soak		6-hr boil		24-hr cold soak		6-hr boil		24-hr cold soak		6-hr boil	
	N	%	N	%	N	%	N	%	N	%	N	%	N	%	N	%
A1/B4	920	8	5,525	23	1233	0	8,100	0	2056	7	2448	28	1807	2	2390	5
A1/B1	1170	21	5,375	4	1660	2	4,700	0	1346	1	2034	21	1377	0	1850	8
A2/B5	1750	23	13,333	17	1323	7	7,833	17	1558	0	2466	35	1793	0	2850	60
A2/B4	2110	32	8,600	55	2737	27	10,200	58	1586	0	2630	40	1873	3	2655	45
A2/B1	897	3	7,600	2	2150	20	9,300	15	1272	2	2004	6	1320	0	2060	33

[a]Specimens were tested for strength in tension; these results appear in the columns headed N. Wood failure (%) is also given.

Table 9.12 Mean Results for Beech Strip Tests with Increased Paraformaldehyde (CHOH) in A and Reduced pH in B[a]

Adhesive system A/B	A: Increased HOHC B: Normal				A: Normal B: Lower pH				A: Increased HOHC B: Lower pH				A: Increased HOHC B: Normal			
	24-hr cold soak		6-hr boil		24-hr cold soak		6-hr boil		24-hr cold soak		6-hr boil		24-hr cold soak		6-hr boil	
	N	%	N	%	N	%	N	%	N	%	N	%	N	%	N	%
A1/B4	2502	41	2268	18	1914	2	2292	10	2580	48	2446	39	2660	38	2258	31
A1/B1	2092	11	2296	3	1330	1	2016	8	2154	14	2172	3	2066	23	2250	26
A2/B5	3166	88	2826	97	2620	30	2840	35	2906	83	2780	76	3080	95	2634	90
A2/B4	2830	45	2773	94	2472	20	2766	65	2944	32	2834	87	2682	80	3068	78
A2/B1	2326	26	2330	41	2258	24	2658	35	2428	14	2650	19	2438	9	2374	52

Cured at 50°C for 4 hours

[a]Specimens were tested for strength in tension; these results appear in the columns headed N. Wood failure (%) is also given.

Table 9.13 Mean Results of Beech Strip Tests with Increased Paraformaldehyde in Component A: Specimens[a]

Adhesive system A/B	Cured at 50°C for 4 hours					
	Dry		24-hr cold soak		6-hr boil	
	N	%	N	%	N	%
A1/B1	3468	88	3006	92	2706	88
A2/B5	3686	99	2898	99	2544	97
A2/B4	3814	97	3172	91	2690	91

[a]Strength and failure results given as explained in note *a*, Table 9.12.

IX. FACTORY TRIALS

Factory trials have been carried out using the A2/B5 system.

The mechanically assembled finger-joint samples of high density (grades 8 and 10) South African pine 152 mm × 38 mm in the first trial and high density *E. saligna/grandis* 100 mm × 25 mm in the second trial were divided into three groups of 10 samples each. The first group of samples was left unplaned, the second group was planed after 30 minutes, and the third group was planed after 1 hour. The temperature during the trials was 22°C

Samples from each group were tested to destruction on a TRU stress grader after 30 minutes and after 1, 2, 3, and 24 hours.

Results of the first trial (South African pine: Table 9.15 and Fig. 9.10) show the increase in strength versus time from 1 to 3

Table 9.14 Mean Results of Beech Strip Tests with Increased Paraformaldehyde in Component A: Specimens[a]

Adhesive system A/B	Cured at 25°C for 7 days					
	Dry		24-hr cold soak		6-hr boil	
	N	%	N	%	N	%
A1/B1	3756	91	2900	83	2474	90
A2/B5	4057	92	3158	94	2588	97
A2/B4	3534	83	2574	90	2146	76

[a]Strength and failure results given as explained in note *a*, Table 9.12.

hours of testing. The percentage wood failure results are inserted on the graph. The picture presented is very similar to that previously reported [14].

Results of the second trial (eucalyptus) (Table 9.16 and Fig. 9.11) show the buildup in strength versus time. It may be noted that the stress grading machine was unable to break the planed sample beams after 24 hours.

In addition, since there was some doubt that the factory operators could handle a two-component system, during the second trial, when all the experimental finger-joint samples had been assembled, the balance of the adhesive was handed over to the operators for un-supervised production. Of the 24 beams (114 mm × 25 mm) that were produced, four were selected at random by the South African Bureau of Standards for full evaluation (Fig. 9.12). The tests showed that the samples drawn in this manner produced satisfactory strength and wood failure results in the wet and dry tension tests.

X. VARIATION IN THE PROPORTION OF A AND B COMPONENTS

A possible disadvantage of the "honeymoon" adhesive system in the manufacture of finger joints in industry is that it is not possible under factory conditions to apply exactly the same amount of adhesive to each finger-joint profile.

The B component of the "honeymoon" system does not contain any hardener. If the solids of the B component are in excess of the resin solids plus hardener of the A component, there may be insufficient hardener to bring about curing of the adhesive.

If the variation in proportion of the components is brought about purely on the basis of weight, which has been done up to a ratio of A to B of 1:10, then the excess resin runs away and no real variation in proportion occurs.

Since viscosity is related to the solids content of the resin, and since it is the possibility of an excess of resin solids of the B component over the percentage hardener in the A component that is important, the variation in proportion has been brought about by difference in the viscosity of the two components in the ratio of A to B of 1:3, 1:4, and 1:5, based on viscosity variations. Therefore, the viscosities (cP) were as follows:

Component A	Component B	Approximate mass
2200	6,000	1:3
2200	8,000	1:4
2200	10,000	1:5

Table 9.15 Factory Trial Results: High Density South African Pine Finger Joints (A2/B5)

	Unplaned			Planed 1/2 hr			Planed 1 hr		
	Strength (N)	MOR (MPa)	Wood failure (%)	Strength (N)	MOR (MPa)	Wood failure (%)	Strength (N)	MOR (MPa)	Wood failure (%)
Grade: 10 time (hr)									
1	3090	17.2	10	3090	25.7	35	1960	16.5	30
2	5420	30.4	65	4480	38.1	85	3720	30.9	90
3	5420	32.0	70	—	—	—	3810	32.4	85
24	7915	49.6	100	5820	49.2	55	5590	46.1	95
Grade: 8 time (hr)									
1	3720	22.3	65	3780	31.4	45	2400	19.9	50
2	—	—	—	—	—	—	—	—	—
3	—	—	—	—	—	—	—	—	—
24	7450	44	95	5420	46.4	95	3510	30.0	98

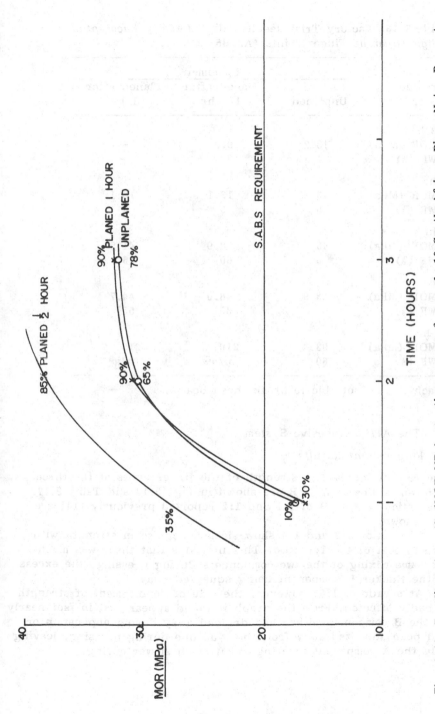

Fig. 9.10 Bending strength (MOR MPa) vs. time curves for grade 10 South African Pine. Note: Percentage wood failure results are inserted on the graph.

Table 9.16   Factory Trial Results: High Density *Eucalyptus Saligna/Grandis* Finger Joints (A2/B5)

		Specimen	
Stressed after	Unplaned	Planed after 1/2 hr	Planed after 1 hr
1/2 hr			
MOR (MPa)	13.2	9.3	—
WF (%)	1	0	—
1 hr			
MOR (MPa)	13.2	18.1	18.1
WF (%)	0	0	0
2 hr			
MOR (MPa)	45	38.9	41.5
WF (%)	0	60	5
3 hr			
MOR (MPa)	43.9	48.0	44.7
WF (%)	50	35	55
24 hr			
MOR (MPa)	93.3	210	210
WF (%)	60	n/a[a]	n/a[a]

[a]Machine was not able to break these beams.

A.   The A1/B2 Adhesive System

1.   *Finger-Joint Results*

The rate of increase in strength of the finger joints at the three different ratios of A to B are shown in Fig. 9.13 and Table 9.17. The ratios of A to B of 1:1 and 1:2 reported previously [14] are also shown.

The ratios 1:3 and 1:4 show similar increase in strength with time to that of the 1:1 ratio. This indicates that there was much the same mixing of the two components during pressing, the excess of the thicker B component being squeezed away.

At a ratio of 1:5, however, the rate of development of strength is badly affected. From the graph it would appear that in fact nearly all the B component either had dripped away during application or had been squeezed away from the glue line during pressing, leaving only the A component to bring about much slower curing.

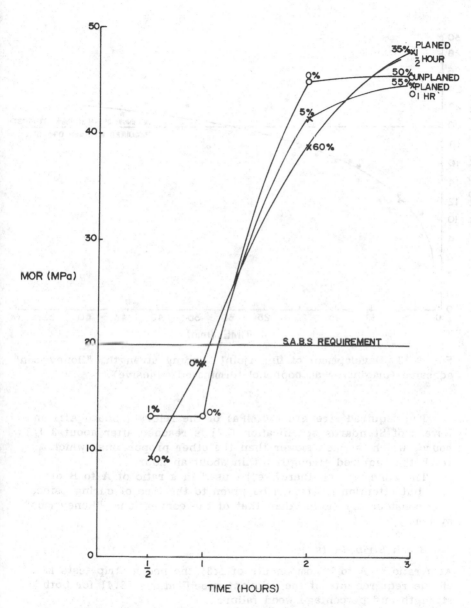

Fig. 9.11 Bending strength (MOR MPa) vs. curing time curves
for high density E. E. Saligna/Grandis. Note: Percentage wood
failure results are inserted on the graph.

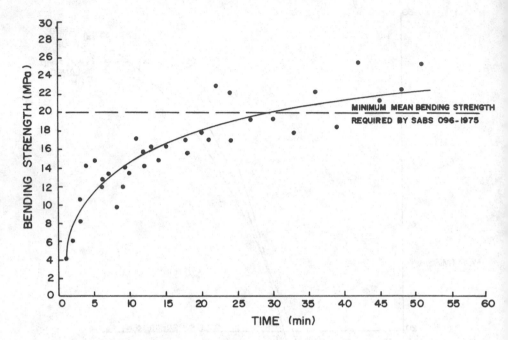

Fig. 9.12 Development of fingerjoint bending strength, "Honeymoon" adhesive (original m-aminophenol-terminated adhesive).

The required strength (20 MPa) of the relevant South African Bureau of Standards specification [12] is reached after about 3-1/2 hours, which is much slower than the other proportions, which reach the required strength within about an hour.

The adhesive can therefore be used in a ratio of A to B of 1:5, but attention must then be given to the time of curing, since it is considerably slower than that of the conventional "Honeymoon" system.

2. Beech Strip Tests

At a ratio of A to B components of 1:3, the beech strip tests meet all the requirements of the relevant specifications [3,4] for both strength and percentage wood failure.

At a ratio of A to B components of 1:4, the system fails to meet the strength and wood failure requirement of the standard specification [3,4] after 24 hours of soaking in cold water, and at a ratio of 1:5, this system fails to meet the weather- and boil-proof requirements of the specification completely.

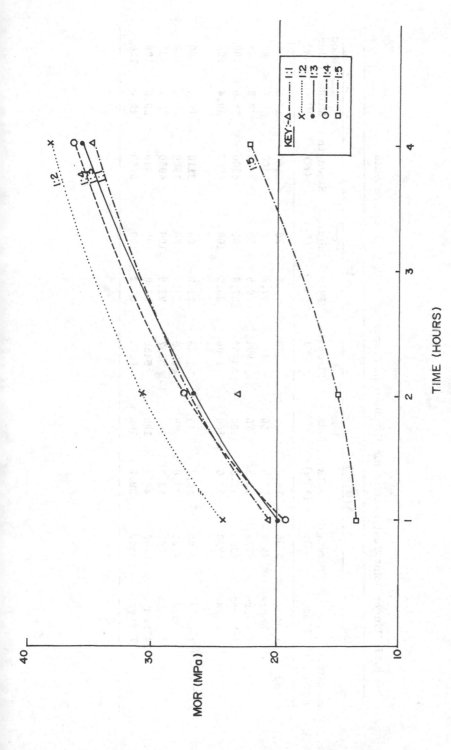

Fig. 9.13 Bending strength (MOR, MPa) vs. curing time curves for variations in proportion of the A1 to B1 "Honeymoon" components.

Table 9.17 Finger-Joint Test Results for Variation in Proportion of A to B Components

Adhesive system	Time	A/B = 1:3			A/B = 1:4			A/B = 1:5		
		Strength (N)	MOR (MPa)	Wood failure (%)	Strength (N)	MOR (MPa)	Wood failure (%)	Strength (N)	MOR (MPa)	Wood failure (%)
A1/B2	1 hour	3195	19.8	10	3100	19.2	26	2360	14.6	5
	2 hours	4293	26.6	50	4410	27.3	10	2400	14.9	40
	4 hours	5778	35.8	75	5891	36.5	85	3600	22.3	45
	24 hours	4890	30.3	100	4450	27.6	100	4300	26.6	100
	7 days	5810	36.0	100	5406	33.5	100	4906	30.4	100
A2/B5	1 hour	2490	15.4	15	3500	21.7	50	3110	19.3	80
	2 hours	4200	26.0	75	3810	23.6	80	3900	24.2	100
	4 hours	4720	29.2	100	3857	23.9	100	4950	30.7	100
	24 hours	7100	44.0	100	5450	33.8	100	5450	33.8	100
	7 days	4800	29.7	100	6000	37.2	100	6780	42.0	100

From these results it would seem that the excess of the B component did mask the effect of the A component completely in the
case of the beech strips, since both the strength and wood failure
results deteriorated. Without hardener in the B component, there
was therefore very little crosslinking.

It is apparent that at the high ratio of A to B components,
the end pressure applied during finger jointing is sufficient to
cause the "squeeze" away of most of the excess B component, and
the finger joints are able to cure sufficiently, after a slightly longer
time, to meet the required strength [12].

However, the type of jointing involved in the more critical beech
strip test is not able to bring about the same results.

The A1/B2 "honeymoon" adhesive system should not be used
at a variation of A to B component greater than 1:3 for safety.
However a higher ratio appears to still give acceptable results on
finger joints.

It is unlikely that a variation in proportion greater than this
will occur, as at weight variations represented by the increased
solids contents created by higher viscosity above 6000 cP the adhesive is very difficult to apply to the finger-joint profile. Furthermore
so much waste occurs that it would not be practical to use this type
of system at such a high level of disproportion.

B. The A2/B5 System

1. Finger-Joint Results

The rate of development of strength of these adhesive systems at
the three different ratios of A to B components are shown in Fig.
9.14. Results for 1:1 and 1:2 are not available.

The A-to-B component ratios of 1:3 and 1:4 show what would
be expected from the variation in proportion. At a ratio of 1:3,
a rapid initial increase in strength due to the excess of fast-reacting
B component is noticeable. The adhesive, however, maintains a
high strength value after 4 hours. At a ratio of 1:4, instead, a
high initial strength value is noticeable, but lower strength values
are obtained after 4 hours, due to a deficiency in hardener.

The A/B ratio of 1:5, however, differs very little from the
normal rate of development of strength of this adhesive system
[14,15]. This implies that enough of the thicker B component drips
away during application or is squeezed away during pressing to
return the ratio to approximately 1:1.

2. Beech Strip Results

There is a slight reduction in percentage wood failure dry, after
24 hours of soaking in cold water and 6 hours in boiling water at
an A/B ratio of 1:5. However at all three ratios, 1:3, 1:4, and 1:5,

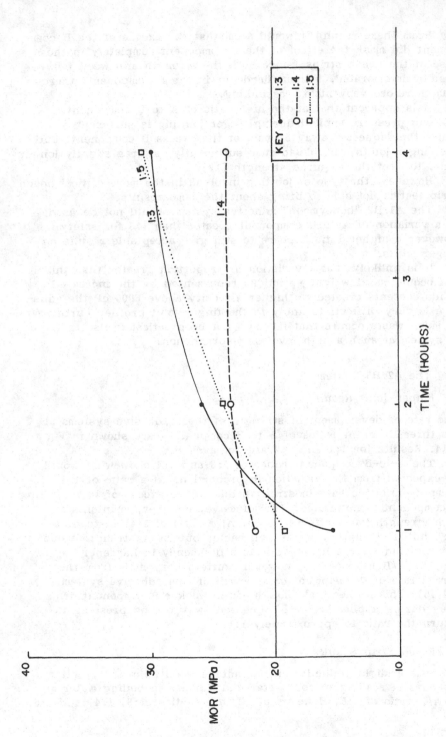

Fig. 9.14 Bending strength (MOR MPa) vs. curing time curves for variations in proportion of A2 to B2 "Honeymoon" components.

all the requirements for weather- and boil-proof wood adhesive of the relevant standard specification [13] are quite comfortably met.

The A2/B5 system is unaffected by the extreme variation of proportion (Table 9.19), but again the amount of waste incurred during application and the difficulty in applying the adhesive at viscosities above 6000 cP would prevent such a variation in proportion from occurring in practice.

From the point of view of practical use of the "honeymoon" systems in industry, it is highly unlikely that a variation in proportion of the A to B components above 1:3 will occur. Above this the viscosity of the B component is such that it cannot easily be applied to the finger-joint profile. This would result in slowing down of the operation. The excessive waste incurred in applying the B component would also prevent the operator from continuing to use a B component that is so much thicker than the A component.

XI. MINIMUM CURING TEMPERATURES

It is accepted at present that a cold-setting adhesive will cure to attain maximum bond strength in a specified time at temperatures between 18 and 30°C.

Most experimentation has been at temperatures of 21°C and above. This series of experiments, therefore, involves the curing of the adhesive at 20°C and below.

It has been proved that the beech strip (*Fagus sylvatica*) test described in both the British standard BS 1204 specification [13] and the SABS specification is suitable for the assessment of wood adhesives. In fact, if the adhesive passes all the requirements of the British standard specification BS 1204, and the South African Bureau of Standards [16,17], then a good structural joint is assured. This test is therefore the main criterion in deciding whether an adhesive system is suitable or unsuitable for use at the chosen temperature.

To study the effect of lower temperature on the rate of development of strength in finger joints, finger-joint specimens were prepared using low to medium density South African pine. These were tested to destruction in midpoint bending after 1, 2, 4, and 24 hours, and 7 days. Finger joints were tested for each adhesive system, at the time intervals specified, at all except the 10°C test temperature.

The minimum conditioning time at the required temperatures was 24 hours for adhesive plus hardener and/or fillers, beech strips, and finger joints.

The assembled test samples, beech strips, and finger joints, were kept at the required conditions of temperature and humidity until the time of testing.

Table 9.18 Beech Strip Test Results for Variation in Proportion of A to B Components

A/B ratio	Dry		24-hr cold soak		6-hr boil	
	Strength (N)	Wood failure (%)	Strength (N)	Wood failure (%)	Strength (N)	Wood failure (%)
A1/B2						
1:3	2888	75	2400	79	2497	85
1:4	2852	21	1924	12	2366	75
1:5	2242	5	1062	0	1680	13
A2/B5						
1:3	3782	100	2938	99	2662	99
1:4	2894	100	2770	100	2578	100
1:5	4460	77	2988	91	2502	91

Initial conditioning was at 20°C with the relative humidity set to give an equilibrium moisture content (emc) of the wood at 12%.

This was followed by conditioning at 15°C and 12% emc for all the adhesive systems.

The adhesive systems that passed the British Standard specification BS 1204 and the South African Bureau of Standards provisional specification at 15°C were tested at 10°C.

Because a smaller conditioning cabinet had to be used, only the beech strips and adhesive components were conditioned. The emc was altered to 9%. One of the adhesive systems passed the beech strip test according to the standard specification [13] and was therefore tested at 5°C.

Since a conditioning cabinet was not available, a refrigerator was used in which there was no humidity control, and the temperature varied between 5 and 7°C.

The cold-setting adhesives used are the wattle-tannin based A1 component and the phenolic A2 component. The A1 or A2 components were applied to both beech strips to be joined and to both halves of the finger joint. This represents the traditional system for timber laminating and finger-jointing.

The "honeymoon" system involves the application of the A component to one of the beech strips or one half of the finger joint and the B component to the other beech strip or other half of the finger joint to be joined.

The adhesives were used in combinations A1/B2, A2/B5, A1/B1, and A2/B4.

The glue mixes used for the A components are shown in Table 9.19.

Table 9.19 Glue Mixes for the A Component (parts by mass)

Ingredient	Cold-setting		Fast-setting ("honeymoon")	
	A1	A2	A1	A2
Resin, liquid	100	100	100	100
Paraformaldehyde, 96% Degussa	11	—	18	—
Wood flour 200 mesh	4.5	—	2	—
Coconut shell flour 200 mesh	4.5	—	2	—
Commercially blended hardener RSX-1	—	20	—	—
Commercially blended hardener RSX-2	—	—	—	20

Note the increase in the level of paraformaldehyde in the "honey-moon" system from about 18% on total resin solids to 30% on total resin solids as reported previously.

The glue mixes were prepared just before use after conditioning of the resin, hardener, and/or fillers at the required temperature for a minimum of 24 hours.

The pH adjustment to the B components was made with 30-40% NaOH before conditioning at the required temperature, for a minimum of 24 hours.

The viscosities of the "honeymoon" A and B components were matched at 20°C and were not adjusted at the lower temperatures.

Beech strip glued specimens were assembled as outlined in the British standard specification BS 1204 and the South African Bureau of Standards specification after conditioning at the required tempera-ture and emc.

They were clamped 15 at a time and after 16-24 hours in the conditioned environment they were removed from the clamp and returned to the required conditions for 7 days.

The test strips were then divided into three groups of five strips. One group was tested dry, one group after 24 hours soaking in cold water, and one group after 6 hours in boiling water.

Boards of South African pine (*Pinus patula*) with densities between 0.42 and 0.57 g/cm^3 and dimensions of 400 mm × 114 mm × 38 mm were finger jointed, using the various adhesive systems after conditioning of the wood at the required temperature and emc for 24 hours.

The end pressure applied in the finger-joint press was 4.0 MPa.

To assess the rate of increase in strength, the finger joints were tested to failure in midpoint bending after 1, 2, 4, and 24 hours, and 7 days, using a universal testing machine.

The assembled finger joints were kept in the required conditions of temperature and emc up to the time of testing.

Beech strip test results are shown in Table 9.20 and finger-joint test results in Table 9.21.

A. Cold-Setting Adhesives

1. *Beech Strip Tests*

At 20°C both the wattle-tannin based A1/A1 and phenolic A2/A2 systems just fail to pass the strength requirement of the British standard specification BS 1204 [13] and the South African Bureau of Standards specification [16,17] in the 24 hour cold water soak test (Tables 9.21 and 9.22).

The percentage wood failure results in both 6-hour boil test and the 24-hour cold water soak test are good, and the strength

Table 9.20 Seven-Day Beech Strip Test Results at Different Temperatures[a]

Adhesive system	20°C Dry		24-hr cold soak		6-hr boil		15°C Dry		24-hr cold soak		6-hr boil	
	N	%	N	%	N	%	N	%	N	%	N	%
A1/A1	2020	75	1900	75	2520	80	1814	0	1636	2	2516	23
A2/A2	2244	75	2034	78	2690	92	2008	0	1836	15	2744	72
A1/B2	3752	99	2908	97	2600	99	1598	71	2056	18	2710	11
A1/B1	3756	91	2900	83	2474	90	2684	11	1852	75	1594	86
A2/B5	3272	100	2900	97	2126	90	3346	83	2345	75	2312	90
A2/B4	3518	87	2498	75	1942	99	3244	78	2462	75	2842	100

Adhesive system	10°C Dry		24-hr cold soak		6-hr boil		5°C Dry		24-hr cold soak		6-hr boil	
	N	%	N	%	N	%	N	%	N	%	N	%
A1/A1	Not tested: failed at 15°C											
A2/A2	Not tested: failed at 15°C											
A1/B2	Not tested: failed at 15°C											
A1/B1	2318	0	1716	7	1726	9	Not tested: failed at 10°C		Not tested: failed at 10°C			
A2/B5	3678	86	2824	97	2482	99	3444	96	2942	86	2592	100
A2/B4	2748	50	2432	17	1934	90	Not tested: failed at 10°C		Not tested: failed at 10°C			

[a]Specimens were tested for strength in tension; these results appear in the columns headed N(ewtons). Wood failure (%) is also given.

Table 9.21   Finger-Joint Midpoint Bending Strength Results at Different Temperatures

Adhesive system	Testing time	20°C		15°C		5°C	
		MOR (MPa)	Wood failure (%)	MOR (MPa)	Wood failure (%)	MOR (MPa)	Wood failure (%)
A1/A1[a]	1 hour	14.3	0	11.8	5	Not tested:	
	2 hours	18.0	20	9.9	15	failed beech	
	4 hours	18.0	5	13.6	5	strip test	
	24 hours	21.4	75	26.0	10	at 15°C	
A2/A2[a]	1 hour	12.4	0	8.1	0	Not tested:	
	2 hours	12.4	0	9.9	0	failed beech	
	4 hours	14.3	5	12.4	5	strip test	
	24 hours	22.3	95	24.8	50	at 15°C	
	7 days	40.9	80	31.0	90		
A1/B2	1 hour	16.9	0	12.4	0	Not tested:	
	2 hours	24.4	5	20.4	5	failed beech	
	4 hours	31.6	10	27.9	10	strip test	
	24 hours	45.9	100	40.9	95	at 15°C	
	7 days	31.0	95	31.0	100		
A1/B1	1 hour	22.1	55	18.6	10	Not tested:	
	2 hours	30.3	75	24.8	15	failed beech	
	4 hours	31.5	95	31.0	20	strip test	
	24 hours	33.3	90	42.8	100	at 15°C	
	7 days	39.0	95	44.0	65		
A2/B5	1 hour	16.7	10	15.3	15	14.3	5
	2 hours	20.0	95	19.3	20	19.0	5
	4 hours	25.6	85	24.4	40	23.4	95
	24 hours	37.2	100	26.0	100	25.5	95
	7 days	24.8	100	34.7	90	28.8	95
A2/B4	1 hour	16.1	20	16.9	5	Not tested:	
	2 hours	26.6	25	28.5	15	failed beech	
	4 hours	26.0	80	34.7	75	strip test	
	24 hours	26.0	100	34.7	90	at 10°C	
	7 days	24.8	100	36.5	95		

[a]Traditional cold-set adhesive.

after boiling is well above specification. For this reason the adhesives were tested at 15°C.

At 15°C both adhesives fail to meet the strength or wood failure requirements of the relevant specifications after 24 hours of soaking in cold water, and the percentage wood failure requirement after 6 hours in boiling water.

It is clear from the results that these adhesives are able to meet the requirements of the British standard specification BS 1204 and the South African Bureau of Standards specification 1349 only for close contact adhesives at temperatures above 20°C.

2. Finger Joints

Although it was not necessary to test the rate of development of strength in the case of these two adhesives, this was done, to assess the effect of the reduction in temperature on the rate of curing.

The results show a reduction in strength development at the lower temperature.

Both adhesives build up enough strength after 24 hours at both 20 and 15°C to meet the minimum strength requirement (20 MPa) of the South African Bureau of Standards code of practice for structural finger joints but not SABS adhesive specification no. 096-1976.

B. Fast-Setting "Honeymoon" Adhesive Systems

1. Beech Strip Test Results

a. The A1/B2 System: This system comfortably meets the strength and percentage wood failure requirements of the relevant specifications at 20°C for both the 24-hour cold water soak test and the 6-hour boil test (Tables 9.20 and 9.21).

At 15°C, the strength after boiling increases, but the percentage wood failure is too low. In addition, the strength and percentage wood failure after 24 hours in cold water do not meet the requirements of the relevant specifications.

These adhesive systems are therefore suitable for use only at temperatures above 15°C.

b. The A1/B1 System: This was the only system badly affected by the increase in viscosity at low temperatures. Whereas in the case of the other systems the increase in viscosity affected both components almost equally, in the A1/B1 system the B component was noticeably more viscous than the A component at 15°C, and at 10°C it could hardly be spread.

At 15°C, this system fails the strength requirement after 24 hours of cold water soaking, but since it had good wood failure results, it was tested at 10°C.

At 10°C there is a sharp drop in percentage wood failure (Tables 9.20 and 9.21), and although the boil strength value is higher than the specification requirements, it is lower than that obtained for any other adhesive system at any of the temperatures.

This adhesive system would be suitable for use only at temperatures above 15°C.

c.   The A2/B5 System: This system shows excellent strength and percentage wood failure results at 20, 15, 10, and 5°C. The results indicate that from 20°C and below, temperature has little effect on the bond strength of the adhesive.

All the results of the various temperatures satisfy the requirements of the British standard BS 1204 and SABS specifications for exterior graded wood adhesives.

d.   The A2/B4 System: This adhesive system meets all the requirements of the British standard specification BS 1204 and the South African Bureau of Standards specification 1349 for both the 24-hour cold water soak and the 6-hour boil, for strength and percentage wood failure, at 20 and 15°C. The sharp increase in boil strength at 15°C is unusual (Tables 9.21 and 9.22).

At 10°C, although both boil and soak test strengths are within specification, there is a drop to values below the specification requirements in the percentage wood failure of the cold water soak test results.

This adhesive is suitable for use at temperatures above 10°C.

2.   Finger Joints

There has been a decrease in the rate of development of strength in the case of all except the A2/B4 system. An explanation for this exception would appear to be in the percentage wood failure results, which are higher for low strength values in the case of the 20°C test, which indicates weak wood, and lower for high strength values in the case of the 15°C, test indicating stronger wood.

The depression in the rate of strength development in most cases means that instead of attaining the minimum strength requirement of the South African Bureau of Standards code of practice [12] within an hour at 25°C, it would take 1.5-2 hours, in the range of temperatures tested, for this strength to develop.

Figure 9.15 shows the development of strength versus time for the A2/B5 system and includes results previously obtained at 25 and 22°C.

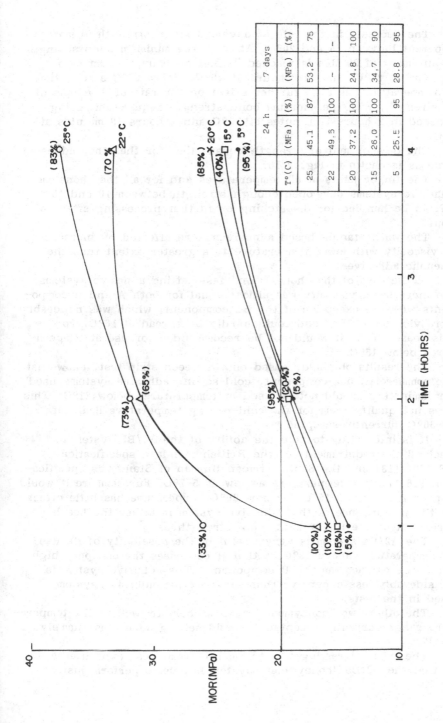

T°(C)	24 h		7 days	
	(MPa)	(%)	(MPa)	(%)
25	45.1	87	53.2	75
22	49.5	100	–	–
20	37.2	100	24.8	100
15	26.0	100	34.7	90
5	25.5	95	28.8	95

Fig. 9.15 Bending strength vs. curing time curves for adhesive system: A2/B5 at five different temperatures (decreasing). Note: Percentage wood failure results are shown in brackets.

The decrease in the rate of development in strength is most apparent between 25 and 20°C. At 25°C the minimum bond strength requirement of 20 MPa is reached in half an hour, in just over an hour at 22°C, and in 2 hours at 20°C. Below 20°C a reduction in temperature has little further effect on the rate of development of strength, with the minimum bond strength requirement being reached in 2 hours 12 minutes at 15°C and 2 hours 18 minutes at 5°C.

The development of strength versus time for the other adhesive systems is shown in Fig. 9.16.

The finger joints at all temperatures and for all the "honeymoon" adhesive systems developed enough strength between 10 and 15 MPa to be handled for dispatching or further processing after an hour.

The wattle-tannin based adhesives were affected by increase in viscosity with lower temperature to a greater extent than the phenolic adhesives.

In the case of the "honeymoon" fast-setting adhesive systems, the increase in viscosity was almost equal for both A and B components with the exception of the B1 component, which was noticeably more viscous at 15°C and could hardly be spread at 10°C. For this reason alone it would not be recommended for use at temperatures below 15°C.

The results obtained, based on the beech strip test, show that the commercial, one-component, cold-setting adhesive systems used for laminating should not be used at temperatures below 21°C. This does not qualify them for the cold-setting temperature limits of 18-30°C currently accepted.

It is interesting to note the ability of the A2/B5 system to meet all the requirements of the British standard specification BS 1204 [13] and the South African Bureau of Standards specification [16,17] at a temperature as low as 5-7°C. Furthermore it would appear from Fig. 9.15 that below 20°C temperature has little effect on the performance of this adhesive system in either the beech strip test or the finger-joint bond strength.

The A2/B4 system is very useful in the possibility of its use at temperatures above 10°C in that it comprises the cheaper, high pH wattle extract, as the B component. This adhesive system is considerably less expensive than all the other adhesive systems used in the tests.

The other two "honeymoon" systems perform well in the temperature range currently accepted for cold-setting adhesives, namely 18-30°C.

The most interesting fact to emerge from these experiments is that the A2/B5 "honeymoon" system is able to perform just as

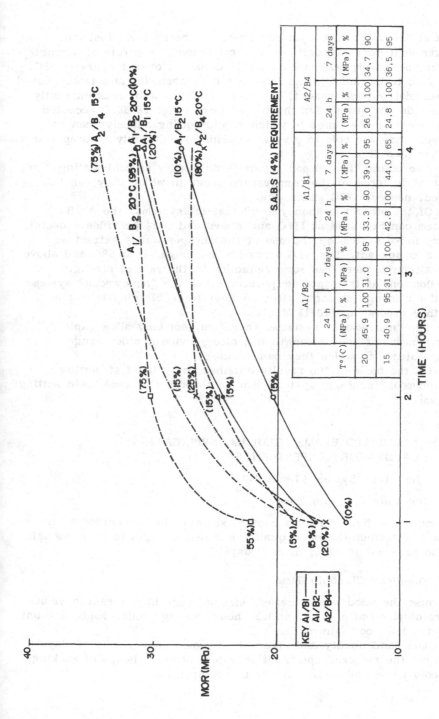

T°(C)	A1/B2				A1/B1				A2/B4			
	24 h		7 days		24 h		7 days		24 h		7 days	
	(MPa)	%	(MPa)	%	(MPa)	%	(MPa)	%	(MPa)	%	(MPa)	%
20	45.9	100	31.0	95	33.3	90	39.0	95	26.0	100	34.7	90
15	40.9	95	31.0	100	42.8	100	44.0	65	24.8	100	36.5	95

S.A.B.S (4%) REQUIREMENT

Fig. 9.16 Bending strength vs. curing time curves for three "Honeymoon" systems at two different temperatures. Note: Percentage wood failure results is shown in brackets.

well at 5°C as at 20°C with regard to the beech strip test and
finger-joint bond strength, with a reduction in the rate of strength
development from half an hour at 25°C to just over 2 hours at 5°C.

Also to be noted is the fact that both phenolic- and tannin-based
commercial, one-component, cold-set laminating adhesives currently
in use do not qualify for the temperature range 18–30°C accepted
at present as the range in which a cold-setting adhesive can be
cured. They are able to give satisfactory results only at temperatures
higher than 20°C.

Two of the "honeymoon" systems do qualify as cold-setting
systems in terms of the temperature range in which they can be
cured, namely, 18°C and above.

Of the other two "honeymoon" fast-set systems, the A2/B4
system can be cured at 11°C and above and thus provides a useful,
inexpensive system by the use of the cheaper wattle extract as
the B component. The A2/B5 system can be cured at 5°C and above.

Although there was some reduction in the rate of strength
development at the lower temperatures, all the "honeymoon" systems
built up enough strength after an hour to enable dispatching or
further processing (10–15 MPa).

The "honeymoon" systems, therefore, not only allow rapid
development of bond strength but also provide a wider range of
temperatures in which they can be used.

On the basis of the results obtained, only the fast-setting
"honeymoon" adhesive systems can be classified as true "cold-setting"
adhesives.

XII. LAMINATED BEAMS (GLULAM) APPLICATION: LABORATORY EXPERIMENTS

A. The A1/B2 System (Table 9.22)

1. One-Hour Clamping Time

Although the beam built up enough strength to be handled after
4 hours, neither the bond quality nor the strength is good enough
to be resistant to water after 7 days.

2. Two-Hour Clamping Time

Because the wood was extremely strong, very high strength values
were obtained after 2, 4, and 24 hours curing, which might account
for the low wood failure results.

The bond quality and strength are good enough after 7 days
to meet the required specification values after 24 hours of soaking
in cold water and after 6 hours in boiling water.

Table 9.22 Block-Shear Test Results for Small Beams Made with
A1 and B2 Components

| Testing times | Clamping time | | | |
	1 hour	2 hours	4 hours	24 hours
1 hour				
shear strength (MPa)	0.5	—	—	—
wood failure (%)	0	—	—	—
2 hours				
shear strength (MPa)	2.2	7.0	—	—
wood failure (%)	0	40	—	—
4 hours				
shear strength (MPa)	5	10.1	7.4	—
wood failure (%)	5	40	30	—
24 hours				
shear strength (MPa)	9.7	11.7	12.9	8.1
wood failure (%)	80	80	95	90
7 days				
24-hr cold soak				
shear strength (MPa)	3.6	5.0	5.5	4.5
wood failure (%)	25	95	95	100
6-hr boil				
shear strength (MPa)	1.1	6.0	5.0	4.4
wood failure (%)	10	90	80	85

3. Four-Hour Clamping Time

There is no improvement in bond strength and quality relative to
the 2-hour clamping time, and all the requirements of the relevant
specification have been passed.

4. Twenty-Four Hour Clamping Time

Although the dry strength and wood failure results meet the requirements of the relevant SABS specification [16] after 24 hours, the
strength is lower than the required value of 5 MPa in the 7-day
samples after the 24-hour cold water soak and 6 hours in boiling
water. The excellent percentage wood failure results indicate that
the wood itself was weaker than 5 MPa after water treatment, and
this is in fact the reason for the low strength values, not the adhesion.

Table 9.23 Block-Shear Test Results for Small Beams Made with A2 and B5 Components

Testing times	Clamping time			
	1 hour	2 hours	4 hours	24 hours
1 hour				
shear strength (MPa)	2.1	—	—	—
wood failure (%)	0	—	—	—
2 hours				
shear strength (MPa)	4.5	8.4	—	—
wood failure (%)	15	0	—	—
4 hours				
shear strength (MPa)	8.0	10.9	8.6	—
wood failure (%)	50	25	55	—
24 hours				
shear strength (MPa)	9.7	14.1	10.5	8.1
wood failure (%)	85	70	90	100
7 days				
24-hr cold soak				
shear strength (MPa)	5.0	6.9	5.0	4.1
wood failure (%)	85	100	95	100
6-hr boil				
shear strength (MPa)	6.3	6.5	6.4	4.3
wood failure (%)	90	80	75	100

B. The A2/B5 System (Table 9.23)

The pattern of results for the A2/B5 system is very similar to the A1/B2 system except that the 1-hour clamping time has sufficient bond quality and strength to be resistant to water attack after 7 days.

The sample beam that was clamped for 24 hours, like the A1/B2 system, shows insufficient bond quality to meet the required strength values after water treatment. This is because the sample beams were cut from the same piece of timber as used for the A1/B2 system. They therefore also fail because of weakness in the wood. As in the case of the A1/B2 system, the percentage wood failure results are excellent.

The block-shear test strength requirement of the South African Bureau of Standards specification for glued laminated timber structural members [16] is 5 MPa.

If the beams are removed from the clamps after only 1 hour, it takes 4 hours for the bond to reach the required strength for both adhesive systems. In addition, the A1/B2 system did not cure sufficiently to meet the strength or percentage wood failure requirements in the accelerated weathering test, after 7 days.

The 2-hour clamping time is the optimum for both adhesive systems, since although the A2/B5 system passes all tests after only 1 hour in the clamp, it takes 4 hours for enough strength to develop. After 2 hours in the clamp, the strength values of 7.0 MPa for the A1/B2 system and 8.4 MPa for the A2/B5 are already greater than required. Furthermore, the strength after 24 hours is better than any of the other clamping times for the A2/B5 (14.1-MPa) system and only slightly lower than the 4-hour clamping time after 24 hours for the A1/B2 system (11.7 MPa). Both systems satisfy the strength and percentage wood failure requirements after 7 days.

The failure of the beams that were clamped for 24 hours to meet the required strength value after 24 hours of cold water soaking and 6 hours in boiling water was due to weakness of the wood and not the bond, as previously discussed.

XIII. INDUSTRIAL PLANT APPLICATIONS

Following the successful development of two-component "honeymoon"-type adhesive systems for the manufacture of finger joints, the possibility of using this type of adhesive system in the glued-laminated timber industry, which would allow much shorter clamping time, was investigated.

Laboratory-scale, two-membered beams were assembled using two of the "honeymoon" systems. The results of block-shear tests done on samples cut at different times showed that after only 2 hours in the clamp, the beams had dry strength values in excess of the requirement of the relevant South African Bureau of Standards standard specification [16], and after 7 days (full curing), a fully durable bond had formed.

The advantage to the glulam industry of a 2-hour clamping time was quite obvious, but it was thought that some modifications might have to be made to the conventional methods of glulam assembly to reduce the open and closed assembly times, since just touching of the two components might be sufficient to cause areas of precure and therefore weakness.

The installation of an automated mechanical laminating press (Dimter high speed laminating press) at a South African sawmill provided the first opportunity to test the system under modified

manufacturing conditions [19]. The high speed press was used
for edge-jointing, and the "honeymoon" system used was a full
phenol/resorcinol/formaldehyde system (A2/B5). The press was
capable of applying the two adhesive components to the two opposite
wood faces by means of a one-operation, two-roller application
followed by rapid application of pressure and maintenance of pressure at elevated temperature for 11 minutes. The adhesive system
performed well in edge-jointing under these conditions.

The real usefulness of the fast curing of the "honeymoon" system,
however, is in the face-jointing of glulam, for which an automated
system is not as yet in use. Further laboratory investigations were
therefore carried out to establish what sort of open and closed
assembly times could be tolerated by the system. It was found that
just touching of the two surfaces for up to 45 minutes at 25°C
does not affect the bond quality in any way, so it remained to
be established what sort of assembly time could be obtained on
an industrial scale.

The opportunity to test the PRF "honeymoon" system arose
in the manufacture of three-layer beams 2.2 and 4.4 m long, for
post-office cross-arms. A vertical clamping system was used. Assembly time for the beams 2.2 m long was well within the 45-minute
limit at 22 minutes, while the longer beam just exceeded the limit
at 49 minutes. A clamping time of 3 hours was used, after which
the full length of each beam was cut into block-shear test samples.
Results indicated a drop-off in percentage and wood failure results
where the assembly time limit was exceeded. However, strength
results were excellent throughout, and where assembly times are
kept within the limit, no wood failure problems are experienced.
The fact that the 4.4-m beam was treated with copper/chromium/arsenic
(CCA) preservative also showed that this type of adhesive system
can be used to glue treated timber.

The opportunity to test this system under full factory conditions
arose when a factory trial to manufacture railway sleepers was held
at Mondi Timbers, Driekop Mill, Eastern Transvaal, South Africa.

A horizontal clamping system was used (L-clamps), and to assess
the effect of both long and short assembly times, a press load of
ten 3.1-m sleepers was made with the first beam at 40 minutes assembly time and the last at 6 minutes assembly time. A 3-hour clamping
time was used, at a temperature of 23°C. After thorough testing
it was quite apparent that the adhesive provided an exceptionally
strong and durable bond.

All the tests show that the PRF "honeymoon" system A2/B5,
which has been used for industrial finger jointing for some time
already, can quite successfully be used in the manufacture of glued-
laminated timber, in vertical or horizontal clamping, provided the

recommended assembly time limit is not exceeded (at ambient temperatures ≥ 20°C). This system allows clamping times of only 3 hours,
and from the results obtained, 2 hours would probably be quite
sufficient.

The adhesive system used in all the experiments was a phenol/
resorcinol/formaldehyde resin plus hardener, with a pot-life of
about 5 hours at 20°C (the "A" component) and a phenol/resorcinol/
formaldehyde resin with no hardener but at a pH of about 11.5,
with unlimited pot-life (component "B"). This system, A2/B5, has
been commercially available from 1981. Comparable systems A1/B2
and A2/B4 are also commercially available in South Africa. They
all meet the weather- and boil-proof requirements of the British
standard specification and the South African Bureau of Standards
standard specification [16].

The two-component system necessitates two separate glue
spreaders, since mixing of the two results in very rapid reaction
with resultant lumping of the resin.

A. Post-Office Cross-Arms: Beam Assembly

Medium density (400 g/cm^3) planed, South African pine timber,
60 mm wide and 23 mm thick, was used to manufacture three-layer
beams. The untreated timber beams were 2.2 m long and the CCA-
treated beams were 4.4 m long.

One press load was made, consisting of nine beams, of each
length. The adhesive was applied with paint brushes as follows:

Component A to one face of the outer beams
Component B to both faces of the center beams

In the case of the 2.2-m beam, as the adhesive was applied
to the timber, lay-up commenced. Application and lay-up took 14
minutes, and it took a further 8 minutes until final pressure (8.4
kg/cm^2) was applied. Total assembly time was therefore 22 minutes.

In the case of the 4.4-m beam, the laminates were coated with
adhesive first. This took 25 minutes. Lay-up in the press followed.
This took a further 7 minutes. Final application of pressure (8.4
kg/cm^2) took 17 minutes. Total assembly time was therefore 49
minutes.

The ambient temperature during the assembly was 23°C.

The full length of each beam was cut into block-shear test
samples and after 5 days, to allow full curing, was tested dry,
after 24 hours of soaking in cold water and 6 hours in boiling water.
The boil and soak specimens were subjected to 15 minutes of vacuum
(-0.06 MPa) and 1 hour pressure (+0.6 MPa) in an autoclave before
treatment. Each block-shear test sample had two glue lines.

Table 9.24 Summary of the Block-Shear Test Results for Three-Membered Beams for Post-Office Cross-Arms

	Dry			24-hr cold soak			6-hr boil		
Position along beam	Position along beam	Strength (MPa)	Wood failure (%)	Position along beam	Strength (MPa)	Wood failure (%)	Position along beam	Strength (MPa)	Wood failure (%)
				2.2-m untreated beam					
1	1	8.55	95	2	5.80	100	3	5.13	85
4	4	9.35	100	5	5.38	75	6	4.90	50
7	7	8.30	100	8	5.80	78	9	5.13	100
10	10	8.48	95	11	5.15	75	12	5.18	90
13	13	8.80	90	14	5.55	90	15	5.30	93
16	16	9.78	95	17	5.53	65	18	5.15	58
19	19	7.68	100	20	5.60	100	21	3.70	65
22	22	8.25	98	23	5.08	50	24	5.00	90
25	25	8.98	93	26	6.00	95	27	5.28	98
28	28	9.30	93	29	5.63	53	30	5.33	65
31	31	9.65	100	32	5.98	83	33	5.53	50
34	34	9.98	95	35	5.70	83	36	5.00	75
37	37	7.75	95	38	5.53	100	39	5.15	83
40	40	9.25	100						
Mean		8.86	96		5.59	80		5.06	77
SD		1.145	5.204		0.368	27.383		0.581	30.954
CV (%)		12.92	5.41		6.59	34.07		11.48	40.24
Entries		28	28		26	26		26	26
				4.4-m CCA-treated beam[a]					
1	1	8.75	85	2	5.08	90	3	4.53	90
4	4	8.35	100	5	4.70	70	6	4.15	83

7	8.68	83	8	5.05	88	9	4.63	88
10	6.45	100	11	4.55	55	12	4.70	63
13	8.83	83	14	4.98	50	15	4.48	55
16	10.75	100	17	3.30	43	18	5.28	85
19	7.15	95	20	5.38	95	21	5.08	95
22	9.05	85	23	4.98	75	24	4.58	70
25	8.40	100	26	5.30	100	27	4.65	95
28	9.35	95	29	5.28	85	30	4.53	75
31	6.58	90	32	5.25	90	33	4.73	70
34	9.78	70	35	5.25	38	36	5.05	43
37	7.35	85	38	5.53	48	39	4.68	100
40	9.68	90	41	5.30	73	42	4.60	90
43	8.88	80	44	5.38	65	45	4.85	80
46	8.13	95	47	5.35	30	48	4.80	48
49	9.45	83	50	4.48	35	51	3.83	45
52	9.15	73	53	4.30	48	54	4.60	3
55	10.08	45	56	5.15	18	57	4.38	10
58	11.03	53	59	5.50	40	60	4.83	55
61	10.08	85	62	4.93	35	63	4.88	15
64	11.28	63	65	5.15	25	66	4.30	50
67	10.60	85	68	4.83	30	69	4.25	45
70	9.80	73	71	5.05	18	72	4.80	28
73	10.15	80	74	4.73	43	75	4.05	65
76	10.75	70	77	4.50	38	78	4.78	60
79	10.53	63	80	5.13	30			
Mean	9.22	82		4.98	54		4.61	62
SD	1.643	20.764		0.573	29.814		0.485	33.177
CV (%)	17.82	25.42		11.51	55.52		10.52	53.51
Entries	54	54		54	54		52	52

[a]Treated with a copper-chromium-arsenic (CCA type B) solution (16 kg/m^3).

Table 9.24 shows the average results per block-shear sample.
The mean, standard deviation (SD), and coefficient of variation
(CV) shown, however, are based on the detailed results.

Figure 9.17 is a graphical representation of the block-shear
strength and percentage wood failure results for the 2.2-m beam.

Figure 9.18 shows the strength and percentage wood failure
results for the 4.4-m beam.

B. Railway Sleepers: Assembly

To test the capabilities of the "honeymoon" system at both long
and short assembly times, one full horizontal press load of 10 six-
membered railway sleepers of medium density South African pine
was assembled. The sleepers were 3.1 m long and 250 mm wide.

Ideally, two roller-spreaders are required for the application
of the two adhesive components. Only one spreader was available,
so the A component was placed in the two-side application trough
and the B component in the one-side application trough, in which
case the beams were passed through the spreader twice. This was
an extremely clumsy operation and resulted in runoff of the B component into the trough containing the A component, where rapid
reaction between the two components resulted in "lumping" of the
resin.

Pressure application was completed in 40 minutes, and the
laminates were left in the press for 3 hours at an ambient temperature of 23°C.

The chunks cut off from both ends of all the laminates were
used for duplicate percentage delamination testing [16].

Sleepers number 2 (36 minutes assembly time), 5 (21 minutes
assembly time), and 10 (6 minutes assembly time) were chosen for
thorough testing.

The whole 3.1 m length of sleepers 2 and 5 was cut into block-
shear test samples and tested dry, after 24 hours of cold soaking
and 6 hours in boiling water.

The results obtained were so good that only random samples
were tested from sleeper 10. The delamination test results are shown
in Table 9.25.

From the 250-mm width of each sleeper, four block-shear test
samples were cut. The position of the samples was recorded along
the length of the sleeper and the test samples were marked as follows:

Position 1: dry; soak; boil; dry
Position 2: soak; boil; dry; soak
Position 3: boil; dry; soak; boil

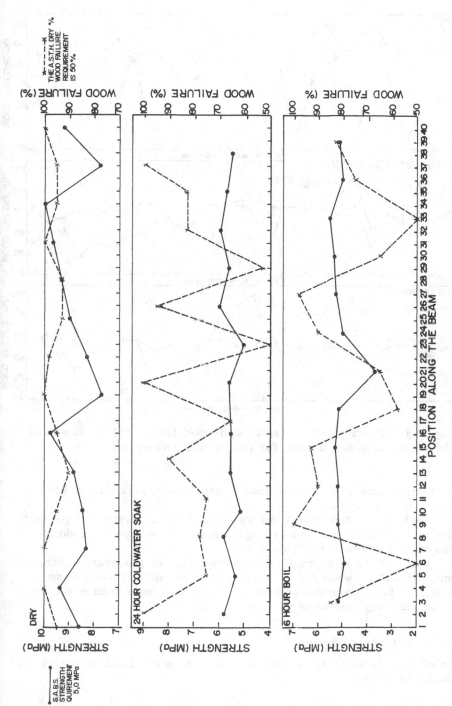

Fig. 9.17 Shear strength (MPa) and wood failure (%) results along the length of a 2.2 m beam for post-office cross-arms.

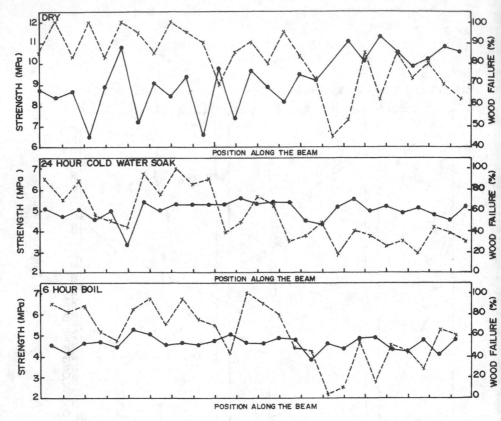

Fig. 9.18 Shear strength (MPa) and wood failure (%) results along the length of a 4.4 m beam for post-office cross-arms.

so that in each of three positions a duplicate dry, soak, or boil specimen was tested.

Each of the five glue lines was tested for strength and percentage wood failure, and the average results at each position along the beam are shown in Table 9.26.

Figure 9.19 is a graphical representation of the average strength and percentage wood failure results at each position along beam 2, and Fig. 9.20 gives the average strength and wood failure results along the length of beam 5.

C. Evaluation of Results

In all cases results are compared with values specified in appropriate standard methods.

Table 9.25   Percentage Delamination Test Results for Railway Sleepers

Beam number	Sample number	Delamination
1	1	0
	2	0
2	1	0
	2	0
3	1	0
	2	0
4	1	0
	2	0
5	1	0
	2	0
6	1	0
	2	0
7	1	0
	2	0
8	1	0
	2	0
9	1	0
	2	0
10	1	0
	2	0

The South African Bureau of Standards standard specification
[16] requires a minimum dry strength value of 5.0 MPa for softwood
timber of 400 g/cm^2 density. The adhesive must have met the weather-
and boil-proof requirements of the South African Bureau of Standards
standard specification for phenolic and aminoplastic resin adhesives [17].

The American standard method [20] requires a dry percentage
wood failure of 75%.

In addition to dry testing, however, the samples have been
subjected to accelerated weathering tests to assess long-term durability. These tests are very severe, involving:

Submerging the samples in cold water in an autoclave
Subjecting to vacuum at -0.06 MPa for 15 minutes
Subjecting to pressure at +0.6 MPa for 1 hour
Submerging the soak samples in cold water for 24 hours
Placing the boil samples in boiling water for 6 hours

Table 9.26 Summary of the Block-Shear Test Results for Industrial Manufactured Railway Sleepers: Mondi Timbers, Driekop Mill Plant Trial

			Beam number 2					
	Dry		24-hr cold soak		6-hr boil		Dry	
PAB[a]	Strength (MPa)	Wood failure (%)	Strength	Wood failure (%)	Strength	Wood failure (%)	Strength	Wood failure (%)
1	8.0	94	4.7	98	4.5	97	8.5	96
2	9.2	99	4.5	97	4.7	98	8.6	87
3	9.2	98	5.0	95	4.2	96	9.4	88
4	8.2	98	4.9	93	4.4	95	8.3	86
5	8.3	98	4.5	85	4.1	85	9.1	88
6	8.1	92	4.5	95	4.3	93	8.1	87
7	8.4	97	4.7	91	4.2	90	9.0	92
8	7.9	97	4.4	91	4.6	91	8.2	95
9	8.4	99	4.8	90	4.0	87	8.7	86
10	8.1	96	4.9	98	4.6	98	8.9	94
11	8.8	97	4.5	92	4.7	99	9.4	91
12	9.3	99	4.5	93	4.1	92	9.2	90
13	8.0	91	4.8	99	4.5	88	8.7	87
14	8.9	86	4.2	70	4.8	92	9.1	91
15	9.2	95	4.5	95	4.5	65	9.1	89
16	8.4	86	4.5	87	4.5	96	8.4	92
17	8.9	91	4.4	83	4.7	87	8.8	89
18	8.9	98	6.7	98	4.1	89	7.7	94
19	7.7	99	4.7	90	4.4	96	8.2	93
20	8.3	92	4.7	84	4.5	89	8.9	89
21	9.2	98	4.7	82	4.2	90	6.9	92
22	7.9	93	4.9	92	4.6	88	8.6	92
23	9.3	95	4.7	94	4.6	84	8.1	84
24	8.3	98	5.0	93	4.2	82	8.0	95
25	8.3	99	4.7	90	4.8	83	8.8	93
26	9.0	90	4.5	92	4.7	93	8.1	92
27	8.7	96	4.5	82	4.2	94	8.7	93
28	8.7	97	5.1	84	4.7	92	8.6	93
29	9.4	87	4.5	89	4.9	93	9.0	90
30	9.2	98	5.0	85	4.3	90	8.1	93
31	8.0	100	5.3	75	4.6	88	8.1	92
32	9.0	96	4.5	95	4.9	97	8.0	91
33	8.6	97	4.8	98	4.2	87	8.3	96
34	8.1	96	5.2	87	4.6	91	8.4	87
35	8.4	90	4.6	91	4.7	89	8.2	83
36	8.7	89	4.7	88	4.1	89	8.2	87
37	8.1	98	5.0	92	4.4	78	8.3	85
38	8.7	97	4.3	95	4.3	85	7.8	92
39	7.6	100	4.8	95	4.5	91	8.9	86
40	8.9	92	4.5	57	4.2	92	7.8	88
41	8.3	100	4.7	82	4.5	85	7.2	91
42	7.3	99	5.0	83	4.7	88	8.6	98
43	8.3	95	4.4	89	4.0	83	8.0	94
44	8.9	95	4.4	94	4.2	89	7.4	94

| Beam number 5 | | | | | | Beam number 10 | | | |
| 24-hr cold soak | | 6-hr boil | | Dry | | 24-hr cold soak | | 6-hr boil | |
Strength (MPa)	Wood failure (%)	Strength (MPa)	Wood failure (%)	Strength (MPa)	Wood failure (%)	Strength (MPa)	Wood failure (%)	Strength (MPa)	Wood failure (%)
4.0	95	4.2	88	9.2	90				
5.4	92	3.9	93						
4.5	94	4.5	93						
4.4	90	4.4	87						
4.2	88	4.2	98	8.0	93			4.4	93
6.0	84	4.2	88						
4.7	90	4.4	85					4.2	92[b]
4.6	89	4.2	93						
4.7	91	4.2	98						
5.8	91	4.5	93	9.7	88				
4.7	90	4.5	94						
4.5	87	4.3	94					4.1	97
3.4	95	4.7	95						
4.3	92	4.1	97			4.0	89		
4.7	91	4.2	85						
4.2	97	4.8	90						
4.3	80	4.2	94						
4.9	93	4.1	92	8.0	92	4.3	94[b]		
4.3	86	4.6	91						
4.2	84	4.2	87						
4.1	90	4.2	92						
4.4	87	4.7	87			4.6	96		
4.1	95	4.4	96						
4.8	93	4.0	94						
4.5	87	4.5	85						
4.3	92	6.0	94						
4.9	95	4.3	89						
4.6	94	4.4	91						
3.9	97	4.2	90					4.1	83
4.6	85	3.9	91						
4.7	92	4.4	98						
4.1	88	4.4	91	7.9	93				
4.9	89	4.1	88						
4.8	95	4.5	87						
4.2	89	4.6	95					4.0	90
4.7	83	3.9	92						
4.7	85	4.3	91						
4.2	82	4.6	93						
4.4	94	3.7	94						
4.9	89	4.2	96	7.5	97	4.5	85		
4.0	93	4.7	98			4.2	95		
4.1	89	3.9	90						
5.0	92	4.1	97						
4.3	94	4.6	81			4.9	92		

(continued)

Table 9.26 (Continued)

| | Beam number 2 | | | | | | | |
| | Dry | | 24-hr cold soak | | 6-hr boil | | Dry | |
PAB[a]	Strength (MPa)	Wood failure (%)	Strength (MPa)	Wood failure (%)	Strength (MPa)	Wood failure (%)	Strength (MPa)	Wood failure (%)
45	8.6	100	4.7	91	4.7	85	8.5	91
46	8.3	98	4.2	77	4.6	88	8.3	93
47	8.2	100	4.2	90	4.3	91	8.2	95
48	8.1	96	4.3	89	4.3	96	8.3	92
49	9.4	95	4.8	93	4.0	93	8.4	94
50	9.4	85	4.4	85	4.6	98	8.1	95
51	10.5	83	4.5	77	4.6	82	8.1	84
52	8.1	96	4.6	99	4.2	86	9.0	97
53	8.7	97	4.4	84	4.2	94		
54	7.7	98	4.3	80	4.2	87		
Mean	8.6	95	4.7	89	4.4	90	8.4	91
SD	0.583	4.258	0.380	7.857	2.248	5.878	0.523	3.623
CV (%)	6.78	4.48	8.08	8.83	5.64	6.53	6.23	3.98
Entries	54	54	54	54	54	54	52	52

[a]Position along the beam.
[b]Mean of two results.

There are no values for this type of test in any standard method; therefore results are compared with dry specification.

D.   Post-Office Cross-Arms

1.   The 2.2-Meter Beam

The average dry strength and percentage wood failure results are well above the requirements of the relative standard specification [16,00]. The lowest individual strength value is 6.75 MPa and the lowest individual wood failure result is 85%.

The average strength results for the 24-hour cold water soak samples is in excess of the dry strength requirement. Low individual

| Beam number 5 | | | | | | Beam number 10 | | | |
| 24-hr cold soak | | 6-hr boil | | Dry | | 24-hr cold soak | | 6-hr boil | |
Strength (MPa)	Wood failure (%)	Strength (MPa)	Wood failure (%)	Strength (MPa)	Wood failure (%)	Strength (MPa)	Wood failure (%)	Strength (MPa)	Wood failure (%)
4.7	90	3.9	91	8.0	90				
5.2	78	4.5	90			4.4	93		
4.1	90	4.8	93						
4.6	93	4.0	91						
4.8	96	4.3	97						
4.1	85	4.7	97						
4.5	98	3.8	94					4.0	90[b]
5.0	92	4.0	94						
				7.9	89				
4.6	90	4.3	92	8.3	92	4.4	92	4.1	91
0.459	4.446	0.360	3.911	0.745	2.878	0.259	3.615	0.167	4.224
9.98	4.94	8.37	4.25	8.98	3.13	5.89	3.93	4.07	4.64
52	52	52	52	8	8	8	8	8	8

percentage wood failure results recorded at positions 17, 23, and 29, in one seam only, brought the average results below the 75% dry requirement. In only one case, the low wood failure result was coupled with a low strength result. This was at position number 23, where the individual wood failure result was 10% and the strength value was 4.80 MPa. This was an isolated area of weakness—samples in positions 22 and 24 do not have low strength or percentage wood failure values.

In the case of the boil samples, the average result is above the standard requirements. In two cases, low individual strength results are coupled with low percentage wood failure results. This occurred at positions 6 and 21. As in the case of the soak samples,

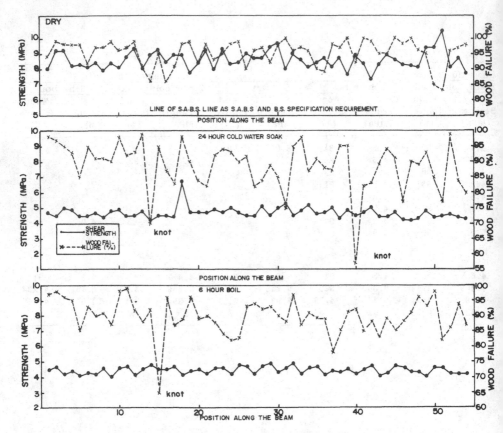

Fig. 9.19 Graphical representation of the average strength and percentage wood failure along the length of a railway sleeper 4.4 m long.

the "patches" of weakness are isolated—the adjacent samples do not reflect the same type of result.

The overall results are very good.

2.   The 4.4-Meter Beam

Although the overall average dry strength and percentage wood failure results are well within the specified values, there is a drop in percentage wood failure toward the end of the beam. The overall average percentage wood failure for the second 2.2-m length of the beam is 73% compared with 90% for the first 2.2-m length. The fact that the strength value is higher in the second 2.2-m section

of the beam than in the first (9.99 MPa versus 8.51 MPa) indicates
that there was sufficient pressure from the weight of the beams
to allow curing of the adhesive, but not enough pressure to penetrate the timber and the adhesive must have thickened somewhat
in the second half of it.

This is because the assembly time was 49 minutes, which is
longer than the recommended period (45 minutes). This would have
affected the section of the beam in which final pressure was applied
last. Curing might also have been more rapid because the assembly
of the beams was done in the open under the direct rays of the
sun, so the actual temperature at the glue line might have been
higher than the 23°C ambient temperature.

The overall average strength and percentage wood failure results
for the 24-hour soak and 6-hour boil samples are lower than for
the 2.2-m beam. This may be due to the effect of the CCA treatment

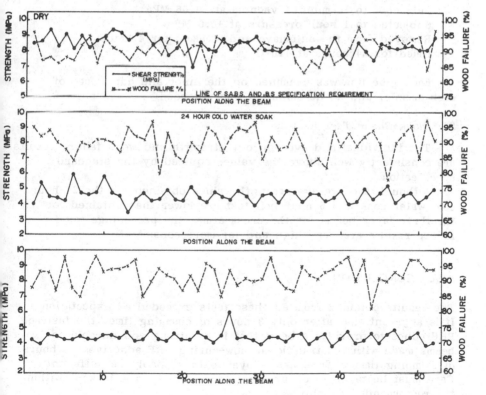

Fig. 9.20 Graphical representation of the average strength and
percentage wood failure along the length of another railway sleeper.

on the beams [21]. The percentage wood failure results reflect
the same lack of glue penetration or lack of pressure during curing,
insofar as the dry samples, with values for the soak samples of 71%
for the first half of the beam and 35% for the second, and for the
boil samples 79% for the first half of the beam and 42% for the second
half.

The results obtained, although satisfactory, illustrate the importance of the total assembly time when this adhesive system is used.

E. Railway Sleepers

1. Delamination Test Results

The delamination test was done on samples cut from each end of
all the railway sleepers made. The samples were treated as follows:

    Submerged in cold water in an autoclave
    Subjected to 15 minutes vacuum at -0.06 MPa
    Subjected to 1 hour pressure at +0.6 MPa
    Removed from the autoclave and steamed for 4 hours
    Dried for 24 hours at 110°C

Each glue line was examined on the inner and outer face of
each sample, and no delamination was recorded.

2. Block-Shear Test Results

The individual and average dry strength and wood failure results
are consistently well above the value required by the standard
specification.

Although the average strength values obtained for the 24-hour
cold water soak and 6-hour boil test are lower than obtained for
the three-layer beams for the cross-arms, the percentage wood
failure results are very high and far more consistent.

XIV. CONCLUSION

The results obtained from all these tests exceeded all expectations.
It is apparent that after only 3 hours of clamping time, the fast-set
PRF adhesive provides a beam that is as strong and durable as
beams made with the traditional slow-curing PRF adhesives; 2 hours
of clamping time at 20°C and above would probably be sufficient.
Care must be taken to ensure that the assembly time is kept within
the recommended 45 minutes.

It also appears that the fast-set system is more sensitive to assembly time in the case of vertical clamping (cross-arms), than for L-clamping.

Pine and saligna finger joints manufactured with these adhesive systems develop sufficient strength to allow further processing or dispatch within a very short period (5-30 minutes at 25-28°C). The "honeymoon" adhesive systems tested are capable of curing so fast at ±25°C that pine and eucalyptus finger joints comply with the requirements of the South African Bureau of Standards SABS 096-1976 specification for the manufacture of finger-jointed structural timber for bending strength (MOR) within 15-45 minutes, and for weather- and boil-proof tensile strength and wood failure within 24 hours to 1 week from assembly.

The main advantages of these adhesive are economical, as follows.

1. The delay between the manufacture of finger jointed timber and further processing or dispatch is decreased from an overnight period to anything between 5 and 30 minutes, depending on the adhesive system used. This eliminates the accumulation of a full day's production stock and therefore improves production flow. This is achieved without altering the existing equipment and method of production, and can therefore be easily implemented.

2. Components B3 and B4 are tannin extracts with adjusted pH. Component B4 is commercially available at a price of only 13-15% that of commercial phenol/resorcinol/formaldehyde adhesives. Component A1, a tannin-based cold-setting adhesive of proven reliability, is also commercially available from various sources at only 70% of the cost of commercial phenol/resorcinol/formaldehyde adhesives. Hence A1/B2, A1/B3, and A1/B4 are economically very attractive systems, with A1/B4 being the most economical and A1/B2 the best overall system of the three. The coupling of the A2 component, commercial phenol/resorcinol/formaldehyde adhesive, with the B2, B3, and B4 components also considerably decreases costs in adhesives for finger-jointing. The A2/B5 system, all synthetic, disproves the need for addition to fast-setting finger-joint adhesives of fast-reacting and expensive components, such as *m*-aminophenol, as claimed by previous authors.

3. There is increased independence from oil-derived synthetic adhesives and from a scarce, expensive and not easily available raw material such as *m*-aminophenol.

4. The same adhesive can be used for beam laminating and finger-jointing.

Disadvantages of these systems are as follows:

1. Preparation of the adhesive is complicated by the fact that it consists of more ingredients than conventional adhesives, namely two resins and one hardener, against the one resin and one hardener of conventional systems.

2. Complete coverage of both profiled faces is essential.

3. Mismatching of profiled faces after adhesive application would lead to slow curing or no curing at all.

These disadvantages, though, have been proved to be of little consequence in industrial production practice.

REFERENCES

1. Tiedeman, G. T., and Sanclemente, M. R. (1973). *J. Appl. Polym. Sci.*, 17, 1813.

2. Tiedeman, G. T., Sanclemente, M. R., and Smith, H. A. (1973). *J. Appl. Polym. Sci.*, 17, 1819.

3. Pizzi, A., and Roux, D. G. (1978). *J. Appl. Polym. Sci.*, 22, 1945.

4. Van der Westhuizen, P. K., Pizzi, A., and Scharfetter, H. (1978). *Wood Southern Africa*, 3, 7. Paper presented at I.U.F.R.O. Wood Gluing Working Party S5.04, Mérida, Venezuela, October 1977.

5. Baxter, G. F., and Kreibich, R. E. (1973). *For. Prod. J.*, 23, 1, 17.

6. Caster, R. W. (1973). *For. Prod. J.*, 23, 1, 26.

7. Ericsson, H. (1975). *Papper Och Trä*, 1, 19.

8. Kreibich, R. E. (1974). *Adhes. Age*, 17, 26.

9. Pizzi, A. (1979). *J. Appl. Polym. Sci.*, 23, 2777.

10. Pizzi, A. (1979). *J. Appl. Polym. Sci.*, 24, 1257.

11. Pizzi, A., and Scharfetter, H. O. (1978). *J. Appl. Polym. Sci.*, 22, 1745.

12. South African Bureau of Standards. (1976). SABS 096-1976. "Specification for the Manufacture of Finger-Jointed Structural Timber." Pretoria, South Africa, 18 pp.

13. "British Standards Institution (1979). "Synthetic Resin Adhesives (Phenolic and Aminoplastic) for Wood. British Standard BS 1204: Part 2: 1979 Specification for Close-Contact Adhesives.

14. Pizzi, A., Rossouw, D. Du T., Knüffel, W., and Singmin, M. (1980). "Fast-Setting Phenolic and Tannin-Based "Honeymoon" Adhesive Systems for Exterior-Grade Finger Joints." CSIR Special Report Hout 184, Pretoria. *Holzforsch. Holzwerwert*, 32(6), 140.

15. Cameron, F. A., and Pizzi, A. (1980). "Modified Fast-Setting Phenolic and Tannin-Based Adhesive Systems for Exterior Grade Finger Joints." CSIR Special Report, Hout 207, Pretoria.

16. South African Bureau of Standards. (1976). "Standard Specification for Glued Laminated Timber Structural Members." SABS 876-1976.

17. South African Bureau of Standards. (1981). "Standard Specification for Phenolic and Aminoplastic Resin Adhesives for the Laminating and Fingerjointing of Timber and for Furniture and Joinery." SABS 1349-1981.

18. Cameron, F. A., Pizzi, A., and Mosterd, D. (1981). "Fast-Setting Phenolic and Tannin-Based "Honeymoon" Adhesive Systems for Exterior-Grade Glued-Laminated Timber." CSIR Special Report HOUT 211, Pretoria, South Africa, February.

19. Pizzi, A. et al. (1983). "Honeymoon" Fast-Setting Adhesives for Timber Laminating." *Holz Rohst. Werkst.*, 41.

20. American Society for Testing and Materials. (1972). Standard Test Method," Standard Specification for Structural Laminated Wood Products for Use Under Exterior (Wet Use) Exposure Conditions." ASTM D2559-72.

21. Cameron, F. A., and Pizzi, A. (1983). "CCA-Treated Laminated Pine: The Effect of Preservative Treatment, Timber Density, Type of Phenolic Adhesive and Glue-Spread on the Adhesive Bond Quality." CSIR Special Report HOUT 280, Pretoria, South Africa, January.

22. Cameron, F. A., and Pizzi, A. (1981). "Minimum Curing Temperature of Cold-Setting and Fast-Setting 'Honeymoon' Laminating Adhesives." CSIR Special Report HOUT 212, Pretoria, South Africa, February.

23. Cameron, F. A., and Pizzi, A. (1981). "Variation in the Proportions of A and B Components of 'Honeymoon' Adhesive Systems Used for Exterior Grade Fingerjoints." CSIR Special Report HOUT 215, March.

24. Cameron, F. A., and Pizzi, A. (1983). "Fast-Setting 'Honeymoon' Adhesive System for Glulam Beams in Vertical and Horizontal Clamping Systems." CSIR Special Report HOUT 291, Pretoria, South Africa, May.

25. Pizzi, A., and Cameron, F. A. (1984). "Fast-Set Adhesives for Glulam." *For. Prod. J.*, September, 34.

10

Release of Formaldehyde by Wood Products

RAINER MARUTZKY / *Fraunhofer-Institute for Wood Research*
Braunschweig, Federal Republic of Germany

I. INTRODUCTION

In the past, the important criteria for assessing a wood adhesive
for a specified application were its technical properties, its gluing
behavior, and last but not least, its cost. During recent years,
a new set of criteria has attained more and more importance: the
environmental and health aspects of the adhesive itself.

Before using a wood adhesive, the manufacturers of wood
products must look at the following questions:

Is there any risk associated with the handling and application
of the adhesive?
Can the application of the adhesive harm the environment?
Can products bonded with the adhesive harm the end users?

The last question, in particular, is of importance in seeking
to avoid complaints and court actions brought by consumers and
construction product users.

The starting point of our considerations in this chapter is the
small and simple molecule HCHO emitted by many synthetic resins
that are used as adhesives for wood and wood products. Its official
name, given by the International Union of Pure and Applied Chemistry (IUPAC), is methanal, but it is much better known by its common
name, formaldehyde. Formaldehyde is the simplest aldehyde, closely
related to methane, methanol, and formic acid. It was first synthesized in 1859 by the Russian chemist Butlerov but not identified
with certainty until some years later by the German chemist Hofmann [1,2]. During the following decades the diverse properties

of formaldehyde were investigated in detail, and its high reactivity
toward a variety of other substances was determined. At the beginning of this century, the American chemist Baekeland combined
formaldehyde and phenol to synthesize the first polymers [3]. These
products later became known under the name "Bakelite." During
the years that followed, more formaldehyde-based polymers were
synthesized and studied with respect to their utility. Resins were
also developed by reaction with other phenols. The 1930s saw the
development of another important group of condensation products,
the aminoplasts. By the middle of the twentieth century, the aminoplasts started their successful penetration of the industrial wood
adhesives market. Aminoplasts are condensation products of formaldehyde with amido- and amino-group-containing compounds, especially
urea and melamine. The most advantageous properties of both aminoplasts and phenoplasts were their excellent performance as wood
adhesives and their high cost effectiveness. As a consequence,
urea-formaldehyde (UF) and phenol-formaldehyde (PF) resins obtained
a dominant position in wood gluing during the 1940s.

However, the use of formaldehyde-based resins also brought
some discomfort, due to the pungent odor of formaldehyde. Their
low stability, especially of urea-formaldehyde resins, results in
the liberation of formaldehyde by the glued wood products, a process
that can last over years. In 1962 Wittmann published the first studies
on the subsequent dissociation of formaldehyde from UF-bonded
particleboard [4]. Since then, some one hundred publications have
reported on the same theme.

Although formaldehyde complaints were sporadic during the
1960s, the problem increased dramatically after the oil crisis that
began in late 1973. The combination of draft-proof houses, low
air exchanges, and high board loading rates has given rise to a
considerable number of formaldehyde complaints. Sundin has shown
that the controversy on formaldehyde release was started by UF
foam releases, followed by waves of public debate about formaldehyde
health risks [5].

Regulations and standards for the release of formaldehyde by
particleboards and other wood products have now been specified
in many countries. The fast progress in formaldehyde research,
in industrial developments, and in official regulations makes it difficult to prepare a survey of this field that is not obsolete within
a few years. This chapter surveys some aspects of formaldehyde
release in wood products reflecting the literature up to the end
of 1987. More information is to be found in the original literature.
Detailed monographs describing formaldehyde release by wood products and other materials have also been published by Roffael in
1982 and by Meyer in 1986 [6,7].

II. FORMALDEHYDE: AN INTERESTING COMPOUND

A. Structure and Properties

Formaldehyde is a simple chemical compound containing carbon, hydrogen and oxygen, expressed as HCHO or CH_2O; the structure is shown in Fig. 10.1. At room temperature, formaldehyde is a colorless gas with a density slightly higher than that of air. Formaldehyde is inflammable and builds explosive mixtures with air at certain concentrations. It is soluble in water, alcohol, and other polar solvents, but nearly insoluble in unpolar solvents. Aqueous solutions of concentration between 35 and 50% are normally used for its transport and usage.

In solution, formaldehyde reacts with water, forming methylene glycol (Fig. 10.1). In solution, further reactions lead to the formation of oligomers structurally related to methylene glycol. Other forms of formaldehyde are the polymerization products trioxane and paraformaldehyde, which are both solid compounds (Fig. 10.1). At room temperature, paraformaldehyde decomposes slightly into formaldehyde; decomposition is faster at elevated temperatures. Paraformaldehyde is the other important form used for transport and usage of formaldehyde. Another formaldehyde-generating product is hexamethylenetetramine ("hexamine"), formed by reaction of 4 molecules of ammonia and 6 molecules of formaldehyde (Fig. 10.2). At elevated temperatures and in the presence of water, hexamine decomposes into monomeric formaldehyde and ammonia.

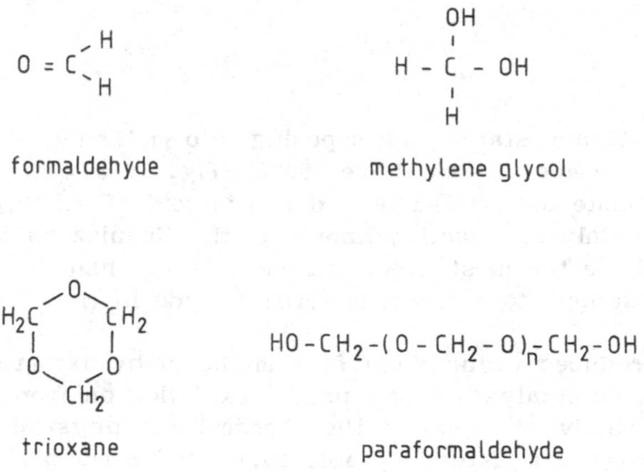

$$O = C \overset{H}{\underset{H}{\diagup}}$$

formaldehyde

$$H - \overset{\overset{\displaystyle OH}{|}}{\underset{\underset{\displaystyle H}{|}}{C}} - OH$$

methylene glycol

$$\overset{\displaystyle O}{\underset{\underset{\displaystyle CH_2}{O \diagdown \diagup O}}{H_2C \diagup \diagdown CH_2}}$$

trioxane

$$HO - CH_2 - (O - CH_2 - O)_{\overline{n}} CH_2 - OH$$

paraformaldehyde

Fig. 10.1 Structures of formaldehyde and formaldehyde polymers.

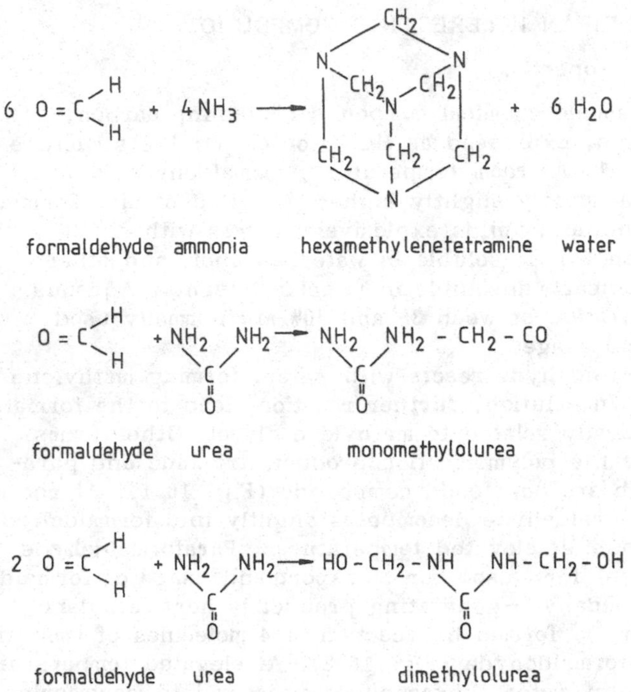

formaldehyde   ammonia   hexamethylenetetramine   water

formaldehyde   urea   monomethylolurea

formaldehyde   urea   dimethylolurea

Fig. 10.2   Reaction of formaldehyde with ammonia and urea.

Formaldehyde is not very stable, decomposing into methanol
and carbon monoxide at temperatures above 150°C (Fig. 10.3).
It can also disproportionate into methanol and formic acid (Fig. 10.3)
in an alkali- and acid-catalyzed reaction known as the Cannizzaro
reaction. This may well be the most important way, other than
oxidation and biodegradation, to decompose formaldehyde in the
natural environment.

Formaldehyde is produced industrially from methanol by oxidative
dehydrogenation on silver catalysts or by partial oxidation on iron
oxide/molbydate oxide catalysts. Some of the chemical and physical
properties of formaldehyde are listed in Table 10.1. In his monograph, Walker extensively describes the properties and the reactions
of formaldehyde [8].

(a) $2\ CH_2O \xrightarrow{>150°C} CH_3OH + CO$

(b) $2\ CH_2O + H_2O \xrightarrow{OH^-} CH_3OH + HCOOH$

(c) $n\ CH_2O \xrightarrow{OH^-} CH_3(OH) \cdot (CH(OH))_{n-2} \cdot CHO$

(d) $2\ CH_2O \xrightarrow{H^+} HCOOCH_3$

Fig. 10.3 Some typical reactions of formaldehyde: (a) thermal decomposition, (b) Cannizzaro reaction, (c) Aldol reaction, and (d) Tischenko reaction.

Table 10.1 Chemical and Physical Properties of Formaldehyde

Property	Value or description
Molecular weight	30.05
Form at 20°C	Gas
Boiling point	−21°C
Melting point	−92°C
Density	$0.815\ g/cm^3$
Relative density (air = 1)	1.03
Solubility	Water, alcohol, ether, acetone
Forms	Solution in water, paraformaldehyde, trioxane

B. Health Effects

First synthesized about one hundred years ago [1,2], formaldehyde has been produced synthetically for industrial purposes since the turn of the century. As a natural compound, it is much older. It has been discovered spectroscopically in intergalactic space, and hence we must assume that formaldehyde has been present in the universe since its beginning. Biochemists suppose that without formaldehyde, life on earth never would have begun. Later, living organisms lost their tolerance to formaldehyde. Being a highly effective poison for microorganisms, it is extensively used for medical and hygienic purposes. Formaldehyde is not as dangerous to human beings as to bacteria and other single-celled organisms, but many people may acquire sensitivity when coming in contact with formaldehyde gas or formaldehyde solutions.

Formaldehyde has a pungent odor and causes irritation of the eyes and of the mucous membranes of the respiratory tract. Human beings are able to detect formaldehyde at very low concentrations. The limit of awareness lies, depending on personal sensitivity, between 0.05 and 1 ppm (part per million; 1 ppm formaldehyde = 1.25 mg/m^3).

Other health risks of formaldehyde have also been extensively discussed. A long-term study of formaldehyde inhalation in rats resulted in a significant dose-related incidence of squamous cell carcinomas of the nasal cavities [9-11]. In mice, nasal tumors occurred only at the highest dose level (14.3 ppm) and at a statistically insignificant rate. There was no cancer formation in hamsters. As a consequence of these results and of some more genotoxic effects, formaldehyde was suspected to be a human carcinogen. Extensive discussions on the carcinogenic risk of exposure to formaldehyde [12-25] followed publication of the results of these studies. Wide-ranging studies among industrial workers exposed to formaldehyde have shown little evidence that mortality from cancer is associated with such exposure [26,27]. Nevertheless, formaldehyde was classified as a probable human carcinogen by a committee of the European Community (EC) and by the U.S. Environmental Protection Agency (EPA). It appears, however, that the actual risk of formaldehyde as a possible human carcinogen is negligibly small at concentrations to which people are likely to be exposed [25].

C. Occurrence of Formaldehyde in the Natural Environment

The atmosphere always contains small concentrations of formaldehyde [22,28-31]. It is continuously formed from other natural organic compounds like methane by photochemical reactions. Furthermore, it is produced as a by-product of incomplete combustions, being present in automobile exhaust gases, tobacco smoke, etc. [22,32,33]. Other emission sources occur in the course of its production and use. Natural emissions are without practical importance [34,35].

As a highly reactive compound, formaldehyde has a half-life in the atmosphere of a few hours. There is no accumulation in the environment, since the compound is rapidly oxidized or biodegraded.

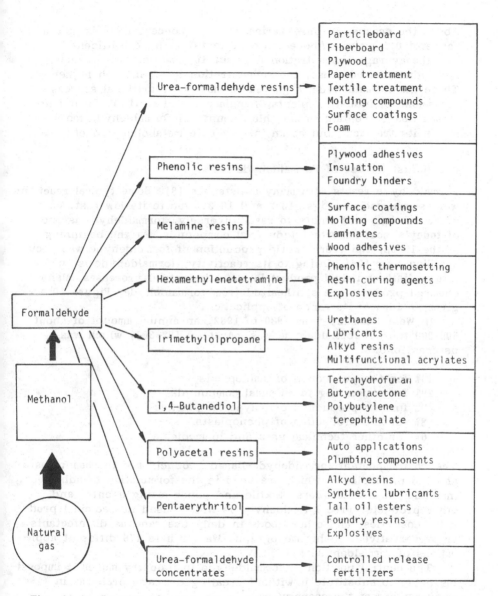

Fig. 10.4 Survey of products in which formaldehyde is a component.

Above the oceans, in pure marine air, the concentration is in the range of 0.1 $\mu g/m^3$. Above the continents, in pure continental air, the average concentration is about 10 times higher. In urban and industrialized areas, the concentration is often much higher. There the formaldehyde is generated mostly by artificial sources (emissions of industrial plants, fireplaces, and traffic). Plant and animal tissues contain measurable amounts of formaldehyde too, not in its free form but as an intermediate metabolic product.

D. Industrial Use of Formaldehyde

Formaldehyde reacts with many substances [8]. Some typical reactions were summarized in Figs. 10.2 and 10.3. Due to its low cost, ease of production, and ability to react diversely, formaldehyde is one of today's most important industrial compounds. At the beginning of the 1980s, the yearly world production of formaldehyde was about 3 million tons [36]. Owing to its reactivity, formaldehyde is an intermediate and final product in many industrial processes. Many chemical products are synthesized from formaldehyde. Figure 10.4 gives an idea of its range of application.

In West Germany from 1980 to 1982, an annual amount of about 500,000 tons of formaldehyde was marketed [22,36], which was used as follows:

48% for the production of aminoplasts
29% for processing to chemical compounds
9% for the production of polyacetal plastics
8% for the production of phenoplasts
6% for other technical uses and in medicine

The most important formaldehyde-based products are the aminoplasts and the phenoplasts, which are used in the formulation of adhesives, insulation foams, lacquers, textile and paper sizing agents, and other products. Free formaldehyde is widely used in medicinal products, cosmetics, and other goods in daily use, and as disinfectants or preservatives. In his monograph, Walker lists 176 different applications of formaldehyde [8].

The central position of formaldehyde in industry makes it impossible today to substitute it without creating dramatic problems in many sectors of the economy.

III.   FORMALDEHYDE: LIMITS, REGULATIONS, AND GUIDELINES

Control and limitation of formaldehyde emissions are necessary to protect the population against risks and harm by exposure to this gas. It is obvious that, in principle, each compound that presents a risk needs to be banned from the human habitat. It is obvious too that an absolute ban of all such compounds is unrealistic. In the case of formaldehyde, an absolute ban is impossible to attain because formaldehyde is also a natural compound of our environment. Protection of the population against hazardous formaldehyde levels can, however, be regulated by control of its concentration in air or by limiting the extent of formaldehyde emissions at the source. Both approaches are in use.

The first step in avoiding danger is to define a concentration level that is harmless. Usually, regulations for limited concentrations are different for workplaces and for living space environments. Contact with the harmful gaseous compound is nearly inevitable in the workplace if the compound or products liberating the compound are used in the industry or service. Concentration limits should minimize the health risks of an adult working person. Odor should be taken into consideration but cannot be excluded in each case.

Limits for rooms in dwellings, day-care facilities, and schools must be defined with more consideration. Here, not only healthy adults but also children and elderly or sick persons may suffer under the influence of a hazardous compound. A further argument to consider is that the time of personal contact with the compound may be longer in such rooms than at a place of work.

A. Hygenic Exposure Limits in the Workplace

At the beginning of the 1970s, exposure limits for formaldehyde at places of work were fixed in most European countries in the range of 5 ppm. During the following years, these limits were tightened. In 1985 in Denmark, Finland, Holland, Italy, Norway, and West Germany, the exposure limit was 1.0 ppm, in Belgium and the United Kingdom 2.0 ppm, and in Sweden 0.8 ppm [5,37]. In some countries, the limits were reduced even further. Since October 1987, in West Germany, the exposure concentration at places of work has been limited to 0.5 ppm [38]. It is probable that during the coming years, most countries in Europe will establish comparable regulatory values.

In the United States the limits are still higher than in Europe. In 1986 Meyer reported current values of 5.0 ppm (threshold limit value: ceiling) and 3.0 ppm (MAC: maximum acceptable concentration) and a proposed value of 1.0 or 1.5 ppm [39].

B. Indoor Air Levels of Formaldehyde

In 1977 the West German Federal Health Board invited a committee of experts to discuss a specification for indoor air levels of formaldehyde. The committee recommended a value of 0.1 ppm for dwellings

[40]. The value was recommended as a guideline but not as a limit specification. Later, values very similar to this guideline were accepted by Denmark, Finland, Holland, and Italy [22,37]. Only Sweden established a much higher indoor ambient level of 0.4 ppm. In some states in the United States, values of 0.4 ppm or higher were adopted [39]. Indeed such organizations as NASA, the U.S. Navy, and the U.S. Air Force introduced a lower level of 0.1 ppm.

All the concentration values mentioned above are only indicative, since this particular area—specifications—is still in rapid flux. The World Health Organization (WHO) has recommended an indoor level of 100 $\mu g/m^3$, corresponding to 0.08 ppm [41].

C. Wood Product Limits

In 1980 two regulations governing formaldehyde release for particleboards being used in buildings were established in West Germany [42-46]. As a consequence of these regulations, the formaldehyde release of particleboards was significantly limited. Based on a large-chamber test, no board should emit more than 0.1 ppm. Boards with higher emission potential must be coated to fulfill this requirement. Three emission classes were established (Table 10.2). Later, other Western European countries also adopted such a regulatory limit or established their own specifications with requirements very similar to the West German one.

After spirited public debate about formaldehyde hazards, a new wider-reaching specification was adopted in West Germany [47]. This "Hazardous Compounds Act" also concerns all wood products. The goal is to reduce the emission for all products to 0.1 ppm [48-50]. Furthermore, the act rules that furniture must be manufactured only with wood products that fulfill the 0.1 ppm requirement.

Table 10.2 Emission Classes of the West German Regulation for Formaldehyde Release of Particleboards Used in Buildings [42]

Emission class	Formaldehyde equilibrium concentration (ppm)[a]	Perforator value EN 120 (mg/100 g)
3	> 1.0-2.3	> 30-60
2	> 0.1-1.0	> 10-30
1	≤ 0.1	≤ 10

[a]Determined in a 40 m^3 test chamber (see Table 10.5).

In Europe, it is expected that most countries will follow West Germany and establish comparable regulatory specifications.

To accelerate the development toward wood products emitting very low amounts of formaldehyde, a voluntary mark called "Blauer Engel" ("blue angel") was established in West Germany by the Federal Board of the Environment. This mark is used for consumer products constructed with formaldehyde-containing boards but with an emission potential of not more than 0.05 ppm [51]. The declared intention of this mark is the promotion of products with minimal risk to the consumer. In North America, the situation is more moderate with respect to formaldehyde-release limits. In a voluntary act, U.S. wood producers have committed themselves to the manufacture of wood products of reduced formaldehyde release [39,52]. The U.S. formaldehyde-release limits are higher than the European ones, however.

IV. REASONS FOR THE RELEASE OF FORMALDEHYDE BY WOOD PRODUCTS

Statements on formaldehyde problems sometimes include the assertion that even pure wood contains formaldehyde. This assertion is incorrect. Wood contains no free formaldehyde, but it does contain substances able to liberate formaldehyde at elevated temperatures. Polysaccharides, as well as lignin, can partially degrade, liberating formaldehyde and other substances [53]. Usually these reactions become significant at temperatures higher than 100°C. At room temperature, formaldehyde release of wood or wood products is normally connected with the presence of a synthetic product. Only if wood or wood products are contaminated with formaldehyde (e.g., by adsorption from a formaldehyde-containing atmosphere), is a mostly insignificant and temporary, limited release at times measurable.

In wood and wood products, the formaldehyde-liberating compounds can be adhesives, lacquers, finishes, and/or coatings. The cause and the extent of formaldehyde release depends on many factors. However, significant emission potentials are mostly due to the presence of aminoplastic resins, such as urea-formaldehyde resins. In this connection, "significance" means that product emission takes place at values above the usual indoor air concentration of 0.02-0.03 ppm and that the emission lasts for years.

A. Survey of Wood Adhesives

The main adhesives currently in use to bond wood in a wide range of application are listed in Table 10.3.

Table 10.3   Survey of Wood Adhesives

Group	Adhesives	Application
Phenolics	Phenol-formaldehyde resins (PF)	Plywood, particleboards, fiberboards, coatings
	Resorcinol-formaldehyde resins (RF)	Special purposes, glulam
	Tannin-formaldehyde resins (TF)	Plywood, particleboards, glulam
Aminoplasts	Urea-formaldehyde resins (UF)	Particleboards, plywood, MDF, coatings
	Melamine-formaldehyde resins (MF)	Coatings
	Melamine-urea-formaldehyde resins (MUF)	Plywood, particleboards
Naturals	Protein glues	Special purposes
	Sulfite liquor resins	Special purposes
Others	Methane-4,4'-diphenyl-diisocyanate (MDI)	Particleboard
	Polyvinyl acetate (PVAc)	Plywood, special purposes
Mineral binders	Gypsum, cement, magnesia	Particleboards, fiberboards

UF resins are the most important adhesives with respect to amount and application. UF resins are used as binders for particleboards, plywood, medium density fiberboards (MDF), glulam, laminated beams, wood veneers, and wood foils. Melamine is the second amino compound of significance in aminoplastic resin production. Pure melamine-formaldehyde (MF) resins are not used as wood adhesives but have found application as impregnating agents of paper and foils. Usually, mixtures of melamine and urea are condensed with formaldehyde to obtain melamine-urea-formaldehyde (MUF) resins. MUF resins are used for the same products in which UF resins are used, but they give bonds of much higher resistance to moisture attack.

The most commonly used water-resistant adhesives are phenol-formaldehyde (PF) resins, which can be prepared in two ways.

The first involves the reaction of phenol with formaldehyde in the presence of an alkali catalyst, giving a concentrated "resol solution." The final curing reaction takes place after gluing, at elevated temperature, in the presence of alkali. The second type of synthesis is obtained by reacting phenol and hexamethylenetetramine (as a formaldehyde producer) in the presence of an acid catalyst. The resulting novolak-type resins are cured at elevated temperatures with an excess of formaldehyde, usually paraformaldehyde.

Resol-type PF resins are the dominant phenolic resins generally used in the production of particleboards and plywood. Novolak-type PF resins are used for the manufacture of special products, particularly in the production of waferboards.

Another type of phenolic resin is based on resorcinol, a phenolic substance having a reactivity much higher than that of phenol. Resorcinol-formaldehyde (RF) resins can be cured at normal room temperatures. Their main application is in the manufacture of glulam and of some other special products.

The last group of phenolics, the tannin-formaldehyde (TF) resins, are based on natural polyphenolic compounds extracted from plant tissues and barks. The reactivity of tannins is closer to that of resorcinol than to that of phenol. Curing takes place so quickly that TF resins can be used only under neutral, weakly alkaline or weakly acid conditions. Most TF resin formulations require the addition of a fortifying synthetic resin. The application of TF resins is universal, but in practice it is restricted to some tannin-producing countries in the southern hemisphere, especially South Africa and Zimbabwe [54,55].

Diphenylmethanediisocyanate (MDI) is also a synthetic adhesive for wood. It is a highly reactive substance giving water-resistant bonds. MDI is used for the production of particleboards.

Spent sulfite liquor resins (SSL) are also usable for the bonding of particleboards. Polymerization processes with pure SSL are possible but uneconomical. SSL is mostly used as a cheap diluting additive for other resins.

The rest of the organic glues listed in Table 10.3 are used for general joinery purposes in bonding wood to wood. Polyvinyl acetate (PVAc) can be used in the manufacture of plywood and in veneering or foiling of panels.

Mineral binders (cement, gypsum, magnesia) are also used for the production of special particleboards and fiberboards. Among the large groups of potential adhesives for wood products, only a few contain formaldehyde, and these belong almost exclusively to the aminoplast and phenoplast categories. Adhesives without formaldehyde are the mineral binders, MDI, and the elastomeric and thermoplastic adhesives. Only PVAc can sometimes contain small amounts of free formaldehyde, added as a preservative agent. SSL can be used without formaldehyde but can also be used with formaldehyde as a crosslinking agent.

B. Formaldehyde Emission Potential of Adhesives

Wood adhesives belonging to the aminoplast (UF, MUF) and phenoplast (PF, RF, TF) groups always contain formaldehyde. In spite of this similarity, there are great differences between these resin groups. In aminoplasts, formaldehyde reacts with the amino group to form bonds of different types (Fig. 10.5). In general, most of the bonding reactions occurring in the formation of UF and MUF resins are reversible; there are, however, enormous differences in stability. Urea reacts with formaldehyde easily forming hydrolyzable bonds. The methylene bonds are more stable than the methylene-ether bonds. The methylol group is easily split. In the presence of moisture, UF resins are slightly hydrolyzed, and the hydrolysis is enhanced under acid conditions and at elevated temperatures.

Due to melamine's poor solubility in cold water, the bonds formed by melamine and formaldehyde are much more resistant to hydrolysis. Pure MF resins are then much more resistant and can be hydrolyzed only under drastic physical and chemical conditions.

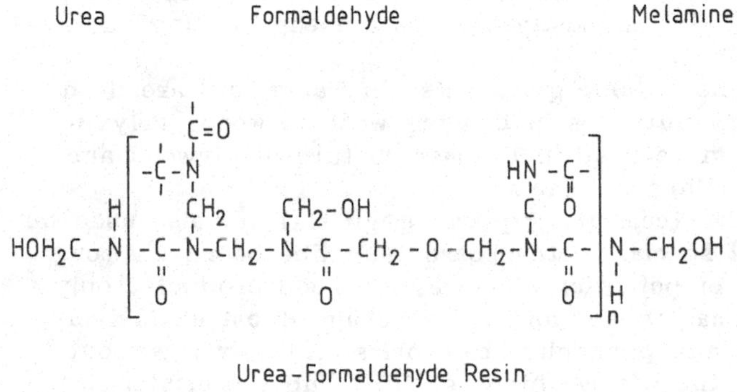

Urea   Formaldehyde   Melamine

Urea-Formaldehyde Resin

Fig. 10.5 Components for aminoplasts and structure of a UF resin.

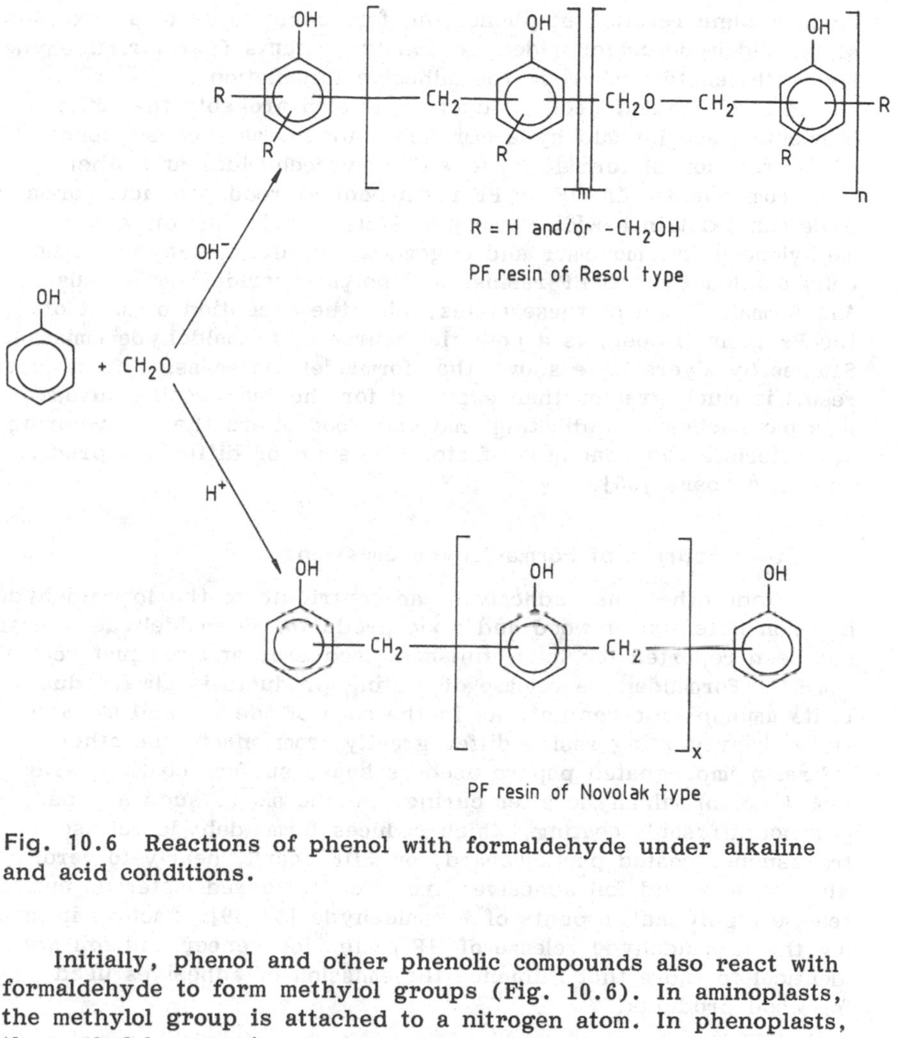
OH
R
CH2
CH2O
R
m
n
R = H and/or -CH2OH
PF resin of Resol type
OH⁻
+ CH2O
H⁺
x
PF resin of Novolak type

Fig. 10.6 Reactions of phenol with formaldehyde under alkaline and acid conditions.

Initially, phenol and other phenolic compounds also react with formaldehyde to form methylol groups (Fig. 10.6). In aminoplasts, the methylol group is attached to a nitrogen atom. In phenoplasts, the methylol group is attached to one of the carbons of the aromatic ring. C—C bonds are stable to hydrolysis, although new data reported by Myers may not exclude the possibility that even phenolic methylols may be a source of formaldehyde emissions source [56]. Further reactions of phenols form methylene, ethylene, and methylene-ether bridges. In phenoplasts, these reactions are irreversible. Thus, formaldehyde added in stoichiometric amounts to the resin formulation can no longer be released after resin curing, provided the curing reaction has been carried to completion. Under these conditions, formaldehyde release by a phenolic resin must always be due to

an incomplete reaction of phenol and formaldehyde or to an excess
of formaldehyde or formaldehyde-donating agents (paraformaldehyde,
hexamethylenetetramine) in the adhesive formulation.

In a PF-bonded wood product, it is also probable that other
compounds are formaldehyde emission sources, as a consequence
of the reaction of formaldehyde with polysaccharides and other
wood components. In UF or PF resin-bonded wood product, formaldehyde can exist in a wide variety of states, including dissolved
methylene glycol monomer and oligomers, paraformaldehyde, chemically bonded UF and PF resins, and polysaccharide hemiformals
and formals. Each of these states, with the exception of most of
the PF resin proper, is a potential source of formaldehyde emission.
Studies by Myers have shown that formaldehyde emission from cured
resins is much greater than expected for the same resins cured
in a particleboard, indicating that the wood alters the resin-curing
characteristics or that other factors, like pH or diffusion, predomi-
nate in a board [56].

C. Other Sources of Formaldehyde Emissions

Compounds other than adhesives can contribute to the formaldehyde
emission potential of wood and wood products. Formaldehyde release
has been reported for foils, finishes, lacquers, and parquet sealants
[56-63]. Formaldehyde release of coating products is always due
to its aminoplastic content. As in the case of the UF and MF adhesives, impregnating resins differ greatly from one to the other.
MF resin-impregnated papers used as board surface coatings emit
almost no formaldehyde after curing. Furthermore, such a paper
is a nondiffusable coating, which reduces formaldehyde release
from such a coated particleboard, or MDF board, nearly to zero.
Also, veneer and foil adhesives are often UF-based materials and
release significant amounts of formaldehyde [57-59]. Factors influencing the formaldehyde release of UF resins for veneer and foil are
identical to those that influence the emission of adhesives used
for wood products.

D. Possibilities for the Substitution of
Formaldehyde-Releasing Adhesives

Each wood adhesive group has advantages and disadvantages when
compared with the other groups. These differences concern applicability, properties, availability, and cost. Substitution of other adhesives for formaldehyde-releasing aminoplastic resins is only partially
possible, even when costs are overlooked. For example, MDI gives

a neutral, light-colored glue line very similar to those of UF or MUF but with the benefits of no formaldehyde content and high water resistance.

Nevertheless, MDI cannot be used for plywood or veneer gluing because of its low viscosity. Moreover, because MDI is a very reactive compound, it is very difficult to increase its viscosity by adding extenders. This makes MDI an interesting adhesive, but its utility is substantially restricted to particleboard applications [64].

Since UF resins are so difficult to substitute, great efforts have been made during the past 20 years in both industry and research institutes to solve the formaldehyde release problem. All in all, these efforts have brought much benefit to the consumer. The formaldehyde release of most wood products has been reduced by 80-90%. Nevertheless, an end to these efforts is not yet in sight. It appears that changes in consumers' convictions and living environments, and stricter limitations set by state and health authorities, will force industry to reduce formaldehyde release of wood products to an emission level that is yet unknown in most countries.

V. ANALYSIS OF ADHESIVES WITH RESPECT TO FORMALDEHYDE

For the production of wood products of low formaldehyde emission, the use of adhesives of particular characteristics is inevitable. In a resin, formaldehyde can exist in several chemical forms that are in a dynamic equilibrium, the proportions of which are sensitive to many factors. Nevertheless, control of the adhesives can bring valuable information and benefits to the user. To control the manufacture of an adhesive, accurate analytical data are necessary. Analytical chemistry of adhesives offers some powerful methods to determine the quality and characteristics of an adhesive. Notwithstanding the progress in the analytical chemistry of adhesives, the ultimate decision about the utility of an adhesive can be given only upon the manufacture and analysis of the final bonded wood product.

When talking about particleboard with low formaldehyde content, manufacturers often use the expression "E1 UF resin." This expression is, of course, incorrect. An adhesive is never E1, but it may be able to give an E1 board by application under certain production conditions. Note that an adhesive, especially a UF resin, is simply an intermediate product that can be significantly influenced, with regard to the formaldehyde release of the board manufactured with it, by production conditions, additives, and posttreatments.

A. Determination of Free Formaldehyde
   and Methylol Group Contents

The free formaldehyde and methylol group contents of a UF resin
supply a great deal of information about the composition of the
resin and its curing properties.

Since free formaldehyde cannot be separated from the resin
monomers and oligomers without significantly disturbing the equilibrium
between the different chemical species present, the formaldehyde
content must be determined directly in the adhesive. The analytical
procedure must always be carried out rapidly to avoid hydrolysis,
and at low temperatures to avoid changes in the composition of
the adhesive.

Accurate methods for the determination of free formaldehyde
and methylol groups content were developed by De Jong and De Jong
[67,68]. The first method involves the determination of free formaldehyde and methylol groups, simultaneously, by reaction with an
alkaline hypoiodite. In the second method, the free formaldehyde
present in the adhesive reacts with an excess of sulfite by forming
a one-to-one adduce. The reaction is carried out under mild alkaline
conditions. The unreacted excess of formaldehyde is back-titrated
with iodine under mild acid conditions. The methylol group content
is then calculated by correcting for the free formaldehyde content
of the resin. Figure 10.7 presents the reactions just described,
as well as two others.

Another method of analysis is based on reaction of formaldehyde
with potassium cyanide, which is added in excess [69]. The unreacted
excess of cyanide is back-titrated with mercury(II) nitrate. This
method, however, is relatively complicated and requires the use
of two highly poisonous substances.

Käsbauer et al. developed a modified method for the determination
of formaldehyde and N-methylolformaldehyde in UF condensates [70].

An indirect method for the determination of free formaldehyde
in UF resins was also developed by Roffael et al. [71]. This determination is based on the formaldehyde release of the adhesive, tested
in the uncured state.

In nonhydrolyzable adhesives (e.g., PF resins), the free formaldehyde can be separated by water vapor distillation without influencing the chemical species equilibrium [72]. An advantage of the
separation of resin and free formaldehyde is that the dark color
of a PF resin cannot disturb the analysis. Determination of the
separated formaldehyde can be carried out by a wide variety of
methods.

B. Determination of Molar Ratio

The molar ratio of a UF resin is a factor that greatly influences
formaldehyde release. The molar ratio can be determined by a total

(a)

$$CH_2O + NH_2OH \cdot HCl \longrightarrow CH_2NOH + H_2O + HCl$$

Sodium Sulfite Method

$$CH_2O + Na_2SO_3 + H_2O \longrightarrow CH_2(NaSO_3)OH + NaOH$$

Potassium Cyanide Method

$$CH_2O + KCN + H_2O \longrightarrow CNCH_2OH + KOH$$

(b)

$$CH_2O + I_2 + 3NaOH \longrightarrow NaHCOO + 2NaI + 2H_2O$$

$$2Na_2S_2O_3 + I_2 \longrightarrow Na_2S_4O_6 + 2NaI$$

Fig. 10.7 Some reactions used in the titration analysis of formaldehyde.

analysis of the resin. The UF resin is hydrolyzed totally under acid conditions, and the liberated formaldehyde is separated by distillation and determined [73,74]. The urea in the resin can be determined directly after hydrolysis or calculated from the nitrogen content of the condensate, provided no other nitrogen-containing compound is present in the adhesive. Urea determination methods are based on enzymatic reaction or on gravimetry, after coupling with benzylamine to form dibenzyl urea [73].

In an unmodified UF resin, the molar ratio of formaldehyde to urea can be calculated directly from the nitrogen-to-carbon ratio (Fig. 10.8). This procedure is also suitable for melamine-modified UF resins, if the melamine content is known.

C. Determination of Formaldehyde Release of Cured Resins

The formaldehyde release of aminoplasts is connected with the hydrolytic stability of a resin. The stability of UF and MUF resins is mostly determined by measuring the increase of formaldehyde concentration in a slurry of water and ground resin powder [75-77]. These methods have several drawbacks. One problem is the solubility of

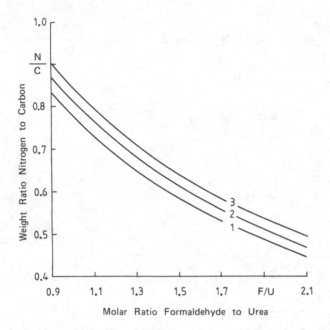

Fig. 10.8 Ratio of nitrogen to carbon as a function of the molar ratio of a UF resin.

polymer fragments in water, another the extraction of compounds like catalysts that, under normal usage conditions, influence the resin degradation rate. It is more practical to determine the formaldehyde release of cured resins in the humid atmosphere of a closed and defined system. Some procedures reported in the literature allow its determination under various temperatures and values of relative humidity [78-80]. The formaldehyde release rate can be measured simply at approximately 100% relative humidity [78,79]. A less reliable method used sulfuric acid solutions to control humidity [80].

Formaldehyde release values of cured resins can give valuable information, but their meaningfulness is relative and is restricted to a limited resin/procedure system. The advantage of such a simple resin test is that many variables affecting resin stability can be studied rapidly. However, measuring formaldehyde release for production control still requires the determination of "correlation" factors, since the formaldehyde emission behavior of pure resins differs significantly from that of a resin included in the complex structure of a wood product.

D. Other Analytical Data

Important analytical data giving information about formaldehyde release properties of UF resins and other adhesives can be determined by chromatography or spectroscopy. Melamine may be determined by ultraviolet absorption measurement at 237 nm after hydrolysis of the resin [73]. More details about useful analytical methods are to be found in the specialist literature [36,73,74,80–82].

VI. ANALYTICAL METHODS FOR DETECTING AND QUANTIFYING FORMALDEHYDE

Methods to detect and quantify formaldehyde are needed to monitor environmental or product emissions and to prepare and assess adhesive resins. There are many methods for formaldehyde analysis, classical as well as recently developed. Detection methods are either direct or indirect, but direct methods such as chromatography or polarography are seldom used. More common is the use of indirect methods, in which formaldehyde is first reacted with another compound, and the resulting derivative or its by-product is then quantified.

A. Sampling of Formaldehyde

If formaldehyde is a component of air, the first step of the analysis is to sample and enrich its content. Its quantification in aqueous mixtures is easier.

Formaldehyde sampling from air can be carried out with simple sampling systems based on gas washing bottles (Fig. 10.9). Different standard methods are described in American and German standard specifications and guidelines [6,32,36,84–88]. In the sampling procedure, a measured volume of gas is brought through a gas washing bottle or a sampling tube, filled with an absorber [84]. The simplest absorber is pure water or an aqueous solution of a reagent. Sampling tubes filled with specially prepared solid adsorbers are also used. Before analysis, adsorbed formaldehyde must be desorbed quantitatively from the solid.

The analytical method selected for quantification of the sampled formaldehyde should reflect the requirements of the sample. If the sample contains other substances, these should not interfere with the analytical method. Most analyses are done with colorimetric or chromatographic methods. Some of these tests require extreme care to obtain reliable results. Others are simple, safe, and also give high reliability.

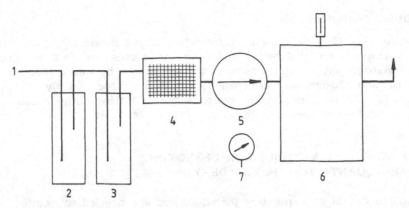

Fig. 10.9 Sampling system for formaldehyde analysis in air: 1, sampling tube; 2, gas washing bottle; 3, safety bottle; 4, moisture absorber; 5, gas pump; 6, gas meter, and 7, manometer.

Detector tubes specific to formaldehyde are also usable for air analysis. Detector tubes of types formerly used, which detected formaldehyde at higher concentrations (0.2 ppm), are usable in factories and other places of work. The more modern detector tubes with specially developed ancillary tubes have lower detection limits (0.04 ppm) and can be used in dwellings, where the formaldehyde concentration often lies between 0.05 and 0.15 ppm [89,90]. Disadvantages of detector tubes are their lack of accuracy and specificity. The tubes have an accuracy of only approximately ±25%, which is much lower than the accuracy of a sampling system like that shown in Fig. 10.9 (±5%).

Detector tubes are also not specific to formaldehyde. Many other organic compounds will also react with the indicator in the tubes. Advantages of detector tubes are simplicity of handling and fast applicability.

Other simple test methods are based on dosimeters ("passive samplers"). Dosimeters collect formaldehyde by passive diffusion into a liquid or solid sorbent medium. Dosimeters do not detect actual formaldehyde concentration but summarize the concentration level over a longer period of time, thus monitoring the average formaldehyde concentration [91-93]. Dosimeter-based methods are simple and inexpensive but have the disadvantage of uncontrolled sampling conditions and lack of reliability.

B.   Classical Methods of Formaldehyde Determination

Gravimetry and titrimetry are classical methods in analytical chemistry. Even today they are still used for quantification of stock solutions.

1. Dimedone Method

The dimedone method is the only formaldehyde determination based on gravimetry that is still used [8,36]. Formaldehyde reacts specifically with dimedone (dimethylcyclohexanedione) to form a dimeric derivative (Fig. 10.10). In cold aqueous solution the derivative precipitates as a crystalline solid, which can be weighed after filtering and drying. The method is formaldehyde-specific and can be used for solutions containing enough formaldehyde to form an amount of precipitate that can be measured by weight. Suitable titrimetric methods are based on hydroxylamine hydrochloride, sodium sulfite, and oxidation with iodine (Fig. 10.7).

2. Hydroxylamine Hydrochloride Method

The reaction of formaldehyde with hydroxylamine hydrochloride liberates one mole of hydrochloric acid (HCl) for each mole of formaldehyde in solution. The liberated HCl is back-titrated with standard sodium hydroxide solution, so that the moles of sodium hydroxide added equal the moles of formaldehyde in the initial solution. By adding a color indicator (bromophenol blue), the titration can be controlled colorimetrically even in opaque solutions. Using a pH meter combined with the hydroxylamine hydrochloride method, formaldehyde can be determined in the presence of phenolics as well as of amino resins. This procedure is particularly useful for the analysis of wood adhesives.

3. Sodium Sulfite Method

Another important titrimetric method is based on the reaction of formaldehyde with sodium sulfite, which liberates one mole of sodium hydroxide (NaOH) for each mole of formaldehyde in solution. The liberated NaOH is back-titrated with a standard acid solution using either a color indicator or a pH meter. The method is suitable for analysis of pure formaldehyde solutions but will cause problems

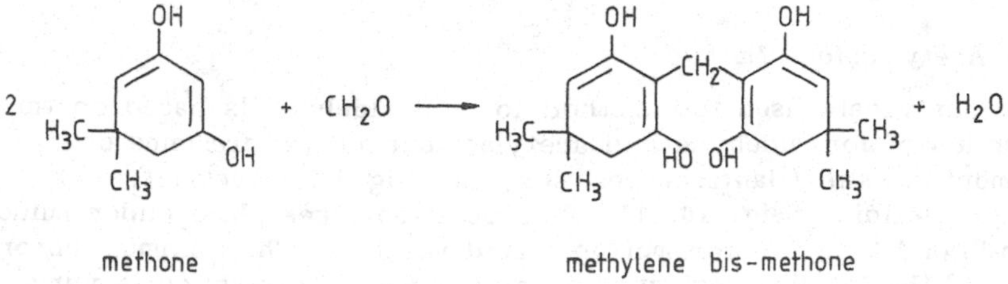

Fig. 10.10 Dimedone reaction used for the gravimetric determination of formaldehyde in solutions.

with most wood adhesives because the sodium sulfite reacts with
certain structures (e.g., methylol groups) in the resin molecule.

4. Iodometric Determination

Another commonly used method is oxidation with iodine. Under
alkaline conditions, formaldehyde is oxidized by iodine. After acidification, the iodine that has not been reduced is back-titrated with
a standard solution of sodium sulfite. To control the titration reaction,
a color indicator for free iodine (starch) is added. The reaction
can also be controlled potentiographically with a platinum electrode.
The iodine oxidation method is suitable for analysis of pure solutions
of formaldehyde. The method gives values that are too high if other
oxidation compounds are present in the solution.

C. Colorimetric Methods

Colorimetric methods are the most commonly used methods in analytical
laboratories for quantifying formaldehyde [6,36,86,94-98]. Some
methods are highly specific and sensitive, requiring only simple
and inexpensive equipment and chemicals. Three methods have found
wide practical application: the chromotropic acid and acetylacetone
methods (Fig. 10.11), and the pararosaniline method (Fig. 10.12,
below).

1. Chromotropic Acid Method

In concentrated sulfuric acid, formaldehyde reacts with chromotropic
acid to form a colored compound (Fig. 10.11). The derivative has
a broad absorption spectrum with a maximum at 570 nm. The chromotropic acid system has been standardized for analysis in air and
is established as the formaldehyde determination reaction in large-
chamber methods and in the desiccator method [85,99,100]. Therefore,
it is predominantly used in laboratories in English-speaking countries,
although it presents some disadvantages [101]. Other aldehydes,
especially acrolein, interfere with the reaction. This method will
fail in the presence of phenols or tannins.

2. Acetylacetone Method

Another widely used test method for formaldehyde is based on the
reaction of formaldehyde and acetylacetone in the presence of
ammonium salts (Hantzsch reaction) forming 3:5-diacetyl-1:4-
dihydrolutidine (Fig. 10.11). This reaction takes place under milder
conditions than the chromotropic acid reaction. The maximum absorption of the lutidine derivative is at 412 nm. The reaction is very
specific but with lower sensitivity than the chromotropic acid reaction

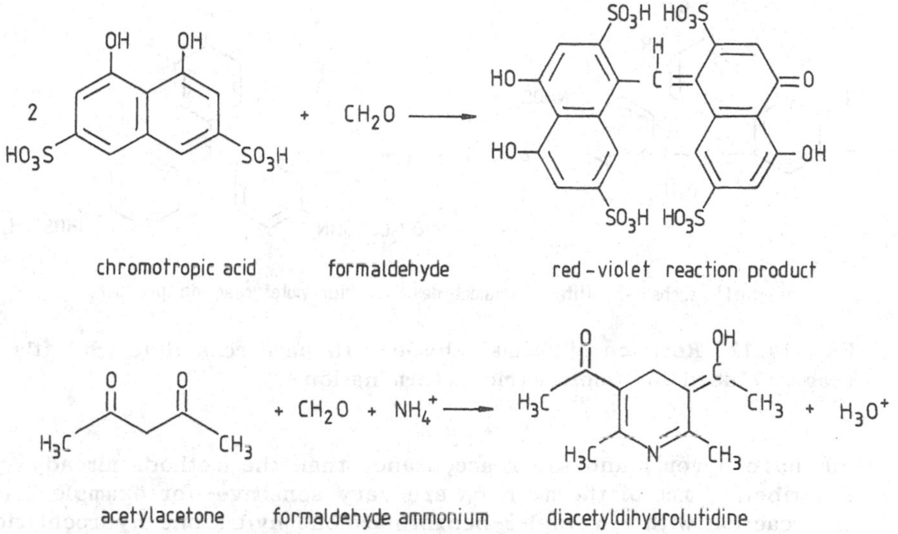

chromotropic acid   formaldehyde   red-violet reaction product

acetylacetone   formaldehyde ammonium   diacetyldihydrolutidine

Fig. 10.11 Reactions of formaldehyde with chromotropic acid and acetylacetone used for photometric determination.

[102]. Sensitivity is enhanced by a factor of about 100 when fluorimetry, instead of photometric detection, is used [104]. The acetylacetone method is unsuitable in the presence of oxidizing agents and of compounds absorbing light at 412 nm. The high specificity and the simple handling afforded by the acetylacetone method renders it very suitable for the analysis of air and solutions containing pure formaldehyde [87].

3. Pararosaniline Method

Formaldehyde reacts with pararosaniline reduced with sulfur dioxide to a purple derivative (Fig. 10.12). The specificity of the test depends on the pH value. Below pH 1 the reaction is specific for formaldehyde [101]. A disadvantage of the pararosaniline method is the instability of the reagent [88,105]. The sensitivity of this method is of the same order as the chromotropic acid method [106-110]. The method is standard in Germany and the Netherlands [86,111].

4. Other Colorimetric Methods

Many other analytical methods for the colorimetric determination of formaldehyde have been developed and described in the literature

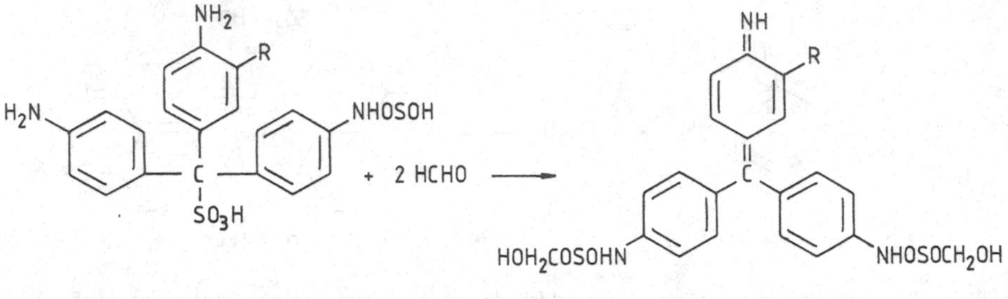

Schiff's fuchsin-bisulfite   formaldehyde   blue-violet reaction product

Fig. 10.12   Reaction of formaldehyde with para-rosaniline (Schiff's reagent) used for photometric determination.

but have never found wider acceptance than the methods already described. Some of the methods are very sensitive—for example, the reaction with 3-methyl-2-benzothiazolone hydrazone hydrochloride (MBTH)—but are not specific to formaldehyde [114,115]. Table 10.4 lists the colorimetric test methods, as well as one fluorimetric method.

D.   Chromatographic Methods

The direct determination of formaldehyde in solution by gas or liquid chromatography is possible only at higher concentrations. The use of chromatographic methods is more convenient for derivatives formed by reaction of formaldehyde with different reagents. A powerful method is the derivatization of formaldehyde with 2,4-dinitrophenyl-hydrazine (DNPH) to a formylhydrazone (Fig. 10.13). The method

Table 10.4   Survey of Some Photometric and Fluorimetric Methods for the Determination of Formaldehyde

Method	Specificity	Ref.
Acetylacetone	High	102–104
Fluoral-P	Low	116
Chromotropic acid	High	100
3-Methyl-2-benzothioazolone hydrazone (MBTH)	Low	114,115
Pararosaniline	High	105–111

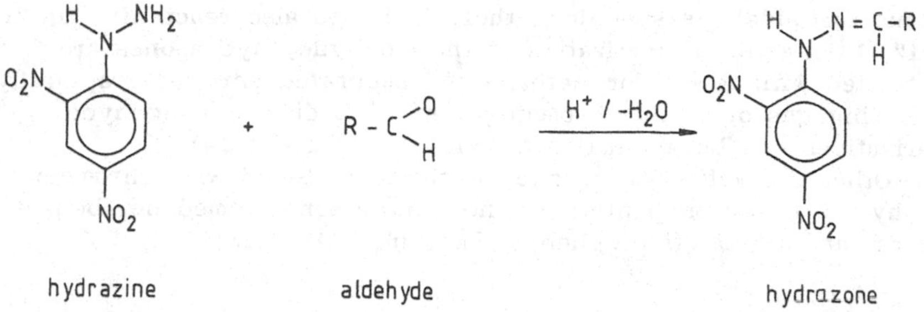

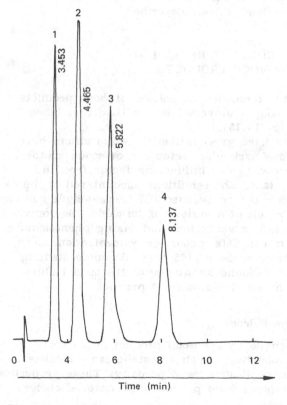

Fig. 10.13 Reaction scheme of 2,4-dinitrophenylhydrazine (DNPH) and aldehydes forming hydrazones.

Fig. 10.14 High permeation liquid chromatogram of a DNPH standard, showing four peaks: 1, formaldehyde; 2, acetaldehyde; 3, propionaldehyde; and 4, butylaldehyde.

is not formaldehyde-specific. Other aldehydes also react with DNPH [112,113]. After derivatization of the aldehyde, hydrazones are extracted with tetrachloromethane and separated chromatographically. By using gas or liquid chromatography, the different aldehyde derivatives can be quantified individually (Fig. 10.14).

Other derivatization reaction methods combined with chromatography are based on pentafluorophenylhydrazone, dimedone, benzyloxime, and o-pentafluorophenyloxime [36,94,116-118].

E. Other Determination Methods

Other methods to determine formaldehyde are based on polarography, atomic absorption spectrometry, and enzymatic reactions [119-124]. In air, formaldehyde is detectable by infrared spectrometry or fluorimetry. The significance of these methods in practice is low when compared with the other methods described.

VII. FACTORS INFLUENCING THE RELEASE OF FORMALDEHYDE IN WOOD PRODUCTS

Many factors influence the formaldehyde release of wood products, making it difficult to develop a universal test method that takes them all into account (Fig. 10.15).

Some of these factors have great influence, while others have negligible effect. Thus one particular factor can override another, creating difficulties in isolating one influencing factor from the other when performing a test. The conditions encountered in indoor environments are more complicated because the formaldehyde concentration is the combined result of a variety of formaldehyde sources and formaldehyde sinks and of ventilation and mixing phenomena. Models to describe and to calculate indoor air concentrations have been developed by different authors [125-132]. An understanding of the models requires a profound knowledge of the main factors influencing the formaldehyde release of wood products.

A. Properties of the Wood Product

The emission of formaldehyde from a wood product is primarily influenced by its own properties, which are structure, thickness, specific weight, surface constitution, and porosity. These properties are determined by the type of wood product, the material choice, and the production process.

B. Variable Properties

Variable properties of wood products are the age, the storage conditions, the moisture content, the room loading, and the size. The

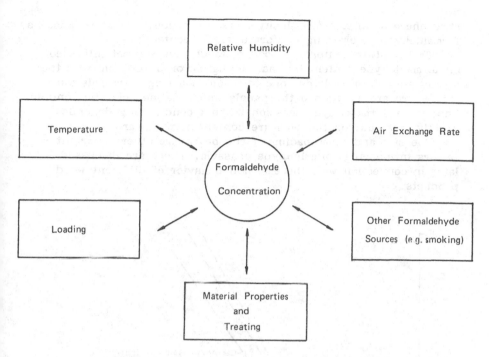

Fig. 10.15 Factors influencing the formaldehyde concentration in a room and in a test chamber.

age of a board influences the emission value, which decreases with time. This decrease is, however, dependent on external factors such as relative humidity, temperature, and ventilation. Often the effect of aging is overvalued. A significant decrease with time may occur during the first days and weeks after production when an excess of free formaldehyde is released. Later, the emission is reduced very slightly. Tests of particleboards that have been produced with a perforator value of about 100 mg (higher than emission class E3) show a loss within 10-20 years of about 80% of their extractable formaldehyde, but no material has reached a value corresponding to emission class E1 (i.e., having a perforator value < 10 mg; the perforator method is discussed in Subsection C of Section VIII, below).

Storage conditions also influence the moisture content of the board, thus changing the formaldehyde content in an indirect way. The formaldehyde content is directly influenced, during storage, by an atmosphere of formaldehyde concentration higher or lower than the equilibrium concentration of the board. In an atmosphere of lower concentration, the aging of the board is accelerated, provided the possibility exists for free formaldehyde exchange. In an atmosphere of higher formaldehyde concentration, the board absorbs formaldehyde, increasing its formaldehyde content.

The moisture content of wood products has a great influence on formaldehyde content but has negligible or no influence on the steady state formaldehyde concentration in a long-term emission test where are maintained the panels under defined climatic conditions. Short-term formaldehyde emission without conditioning depends greatly, instead, on the moisture content of the board.

The size and room loading of the board are other important factors influencing formaldehyde emission. These are discussed later in connection with the emission behavior of different wood products.

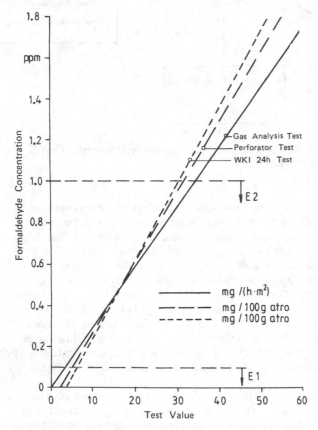

Fig. 10.16 Correlation between formaldehyde values of particleboards and concentration of formaldehyde in the chamber test.

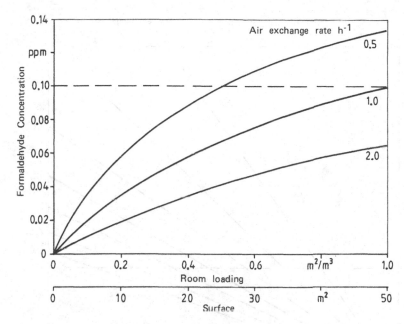

Fig. 10.17 Formaldehyde concentration in the chamber test as a function of the room loading/surface of samples and the air exchange rate under standard test conditions.

C. External Factors

Formaldehyde release is influenced, both in practice and during testing, by many external factors. Important factors are temperature, relative humidity, air exchange rate, and air velocity.

With increasing temperature and relative humidity, the formaldehyde release increases. The correlation factors for the influence of temperature and humidity have been determined both for large-chamber tests and for practical conditions [128,131,133].

Scandinavian scientists, in particular, have evaluated mathematical models describing the formaldehyde release from wood products in test chambers or in normal rooms [125-130].

Based on the Anderson equation, Mehlhorn determined the detailed correlation for the large-chamber test [130]. Some correlations showing the effects of various parameters at the large-chamber, steady state formaldehyde concentration are presented in Figs. 10.16 to 10.18.

Some discrepancies between the values reported by different authors in the literature notwithstanding, it can be concluded from

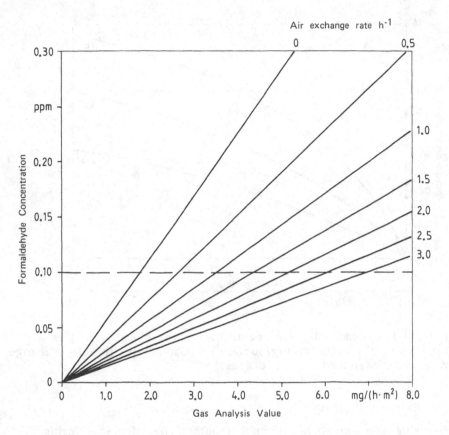

Fig. 10.18 Formaldehyde concentration in the chamber test as a function of the air exchange rate and the material test value of the particleboard under standard test conditions.

the data that under normal climatic conditions an increase in temperature of 5-8°C doubles the formaldehyde concentration; an increase in relative humidity of 10% enhances the formaldehyde concentration level by 15-20%.

The air exchange rate or ventilation also plays an important role as far as formaldehyde concentration is concerned.

Close correlation exists between the steady state equilibrium concentration in a large-chamber test under defined test conditions and the air exchange rate. Because the formaldehyde emission kinetics of formaldehyde-containing wood products are different in a real room, where different emitting sources are present, the influence of the air exchange rate can only be roughly estimated.

The influence of air pressure on formaldehyde emission is instead low.

D. Other Factors

An emission source interacts with the formaldehyde concentration in the air and with other emission sources. At low formaldehyde concentration in ventilated environments, the wood emission source releases its formaldehyde in larger quantities, thus aging faster. At high formaldehyde concentrations, the source may absorb formaldehyde, thus increasing its formaldehyde content. In a room with different emission sources, interaction takes place. Studies, especially by Sundin, have shown that the emission values from different formaldehyde sources in a room are not additive [134]. The equilibrium concentration always adjusts itself at a level that lies between the values of the highest and the lowest emitters at equal loads.

VIII. METHODS FOR DETERMINATION OF FORMALDEHYDE RELEASE IN WOOD PRODUCTS

Consumer protection requires the definition of a maximum specific emission level that fulfills the required conditions under specific conditions of temperature, humidity, loading, and air exchange rate. The formaldehyde emission potential of wood products is a characteristic that is measurable by physical and chemical tests. Hence, exact test methods are required to determine formaldehyde content or formaldehyde emission potential. Only the possession of reliable data on the rate of formaldehyde release of a material allows decisions to be taken about its usefulness with respect to a desired emission level of the product [8].

During the past 20 years, many different test methods have been developed. In this chapter, only methods used in Western Europe and in North America are surveyed. Full-scale testing of final products under actual use conditions is without doubt the best product test, but it is also unrealistic. The next best approach is to test the emitting products in large chambers under standardized conditions. This test method is of great value but too expensive and time-consuming for production control. Thus, great efforts have been made over the past 25 years to develop rapid and reliable product test methods for production control. Of the many published methods, a few have found wide application in laboratories or industry. Some methods have been standardized at a national or international level. Most methods have found limited interest and are largely forgotten.

Formaldehyde release of a wood product is a complex process influenced by a multitude of factors. Hence no simple test can fulfill each specific requirement. As pointed out by Hoetjer and Koerts [132], most of the methods have in common the representation of formaldehyde emission through only one characteristic parameter. On the basis of one characteristic it is difficult to comprehensively judge a wood product. The necessity to classify wood products rapidly and inexpensively for formaldehyde release requires the use of simple methods, especially in quality control. In Europe as well as in North America, large-chamber tests are the basis of formaldehyde classification for wood products. Methods used for quality control are emission tests in smaller chambers, in bottles, or in desiccators. An outstanding quality control test is the perforator test, which determines the formaldehyde content of a panel product by extraction with toluene.

A. Large-Chamber Tests

The determination of the specific emission of formaldehyde from wood products at specified values of temperature, humidity, loading, and air exchange rate requires an environment in which these parameters can be exactly controlled [135-139]. Such an environment is a confined space, which usually is called a chamber or exposure chamber.

Large chambers are needed for testing large wood products. Large-chamber tests have been developed and standardized in West Germany and in the United States.

Large-chamber tests measure dynamic formaldehyde emission from wood products under conditions that simulate those of ambient indoor air. Emission potentials are determined by observing airborne formaldehyde chamber concentrations under defined test conditions.

A large-chamber prototype was developed in 1976 in the Fraunhofer-Institute for Wood Research in Braunschweig, West Germany. The chamber volume was 34 m^3 and the total volume including ventilation system was 38 m^3. From the experience gained with this first chamber, chamber dimensions and construction were optimized. This resulted in a standard test chamber with a chamber volume of 43 m^3 and a total volume of 48 m^3 (Fig. 10.19).

The chamber test method became the reference method in the 1980 German formaldehyde regulation [42] and the basis for classification of particle boards in emission classes (Table 10.2). The classification test was carried out under defined test conditions (Table 10.5). The chamber test was also adopted in the West German Hazardous Compound Act of 1986 [47] to check the emission of all wood panel products.

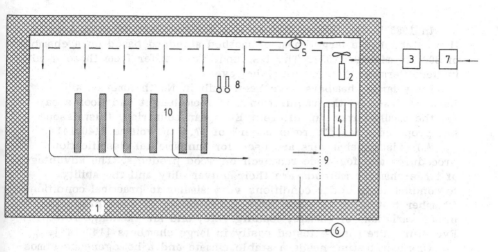

Fig. 10.19 Construction of the 40 m³ chamber of the Fraunhofer Institute for Wood Research (WKI chamber): 1, 40 m³ chamber constructed of stainless steel with PUR foam insulation; 2, ventilation pump in the circulation channel; 3, fresh air flowmeter; 4, heat exchanger; 5, air outlet; 6, sampling system; 7, air inlet; 8, control instruments; 9, circulation channel; and 10, samples.

Table 10.5 Comparison of Test Conditions of Large-Chamber Tests in the United States and West Germany

Test condition	FTM 2-1985[a]	ETB 1980[b]	HCA 1988[c]
Volume (m³)	> 22	ca. 40	> 12
Loading (m²/m³)	0.43	1.00	1.00
Air exchange per hour	1.0	1.0	1.0
Temperature (°C)	25	23	23
Relative humidity (%)	50	45	45
Conditioning time (days)	7	None	None
Time in chamber (hours)		240	96-240 (600)
Air velocity (m/s)	—	—	0.3

[a]Formaldehyde Test Method 2 [85].
[b]ETB Regulation [42].
[c]Proposal of a test procedure for the Hazardous Compound Act [47].

In 1985 in the United States, the National Particleboard Association established a large-chamber method standard based on a chamber of 800 ft^3 minimum [85]. The test conditions differ from those used in West Germany (Table 10.5), however.

Many large chambers have been built in North America and Europe. Based on the requirements of room height and floor area for the smallest room in different European countries, Gustafsson has proposed a test "Euro chamber" of 17.4 m^3 volume [140,141].

All existing chambers are used for fundamental classification procedures and for basic research on wood products. The advantages of large-chamber methods are their universality and the ability to conduct tests under conditions very similar to practical conditions. Chamber test facilities allow large-scale testing of wood products under variations of climate, loading rate, and air exchange rate. Even furniture can be tested easily in large chambers [142-144].

Chamber testing needs a stable climate and a homogeneous atmosphere [145]. To guarantee stable and exact climate conditions in all parts of the test chamber, proper mixing of the chamber atmosphere is necessary to avoid concentration gradients in the chamber and at the surface of the test samples. Emission from a surface leads to the buildup of an air layer of increased formaldehyde concentration immediately above the test surface unless there is sufficient air mixing velocity to break up this surface layer [132]. Increased formaldehyde concentration above the emitting surface, however, increases the resistance against formaldehyde evaporation, hence decreases the emission rate in an uncontrolled way. Therefore, air velocity in the chamber must be high enough to avoid buildup of a surface layer that decreases formaldehyde concentration. In the literature the theoretical velocity needed to break down the surface layer is reported to lie between 0.06 and 0.1 m/s [132,146]. In practice a velocity of 0.2-0.3 m/s in the center of the chamber is necessary for sufficient mixing of the atmosphere in all parts of the chamber [145,147].

The mixing of the chamber atmosphere can be achieved in two ways:

By circulation of the atmosphere with a strong ventilator (favorable for a chamber with external climatization)

By circulation of the total chamber atmosphere through a ventilator in the climatization channel (favorable for a chamber with internal climatization)

The velocity must be checked in the center of the chamber before each test. The maximum velocity should be in the range of 0.2-0.3 m/s.

Table 10.6 Formaldehyde Equilibrium Concentration in the 40 m^3 Test Chamber Determined with Particleboards of Different Thickness

Emission class	Thickness (mm)	Perforator value (mg/100 g)		Equilibrium concentration (ppm)[a]	
		Iodometry	Photometry	Measured value	Calculated value
E1	3	7.7	7.0	0.25	0.16
	16	7.3	6.5	0.15	0.14
	38	7.3	6.6	0.08	0.15
E2	5	21.4	19.0	0.48	0.41
	16	23.6	18.6	0.35	0.39
	38	19.4	15.3	0.18	0.33

[a]Calculated by correlation curve for particleboards of medium thickness.

It must be pointed out that while air circulation is necessary for testing, it also influences the emission potential of the specimen tested. Reduction of the equilibrium concentration of formaldehyde on reduction of air velocity is one effect (Table 10.6). Another effect is that particleboards with the same formaldehyde content but differing in thickness and density will give different emission levels (Tables 10.6-10.8). These effects show clearly that large chambers are sensitive instruments for testing the emission potential of boards and other materials.

The disadvantages of large chambers are the cost of the chamber itself and the cost of the test procedure. It is clearly impossible to

Table 10.7 Formaldehyde Equilibrium Concentration of Particleboards in the 1-m^3 Chamber With and Without Air Circulation

No.	Formaldehyde equilibrium concentration (ppm)	
	With air circulation[a]	Without air circulation[b]
1	0.35 (100%)	0.17 (49%)
2	0.15 (100%)	0.09 (60%)
3	0.13 (100%)	0.08 (62%)

[a]Air velocity in the middle: ~ 0.3-0.4 m/s.
[b]Air velocity in the middle: < 0.01 m/s.

Table 10.8 Formaldehyde Equilibrium Concentration from Particleboards of Different Density in the 1-m^3 Chamber With and Without
Air Circulation

Property	Board 1	Board 2
Thickness (mm)	10	10
Density (g/cm^3)	0.5	0.8
Moisture content (%)	5.03	4.60
Perforator value (mg/100 g)		
iodometric	3.8	3.8
photometric	3.2	2.9
Equilibrium concentration (ppm)		
with air circulation	0.13 (100%)	0.10 (100%)
without air circulation	0.07 (54%)	0.08 (80%)

use a chamber for routine testing and for everyday quality control
tests on a production line. Another disadvantage of chamber tests
is the time required to reach the equilibrium concentration. This
depends on different factors such as the formaldehyde emission
potential itself. The higher the emission potential, the longer the
time needed to reach a steady state concentration. This behavior
appears to depend partially on the state of curing of the resin
binder. Systems in which the resin is undercured, like UF foams
or UF lacquers, show a long-term decrease of emission rates.

Another factor influencing the emission rate is the type of board
tested. Uncoated particleboards and MDF boards usually reach the
equilibrium concentration faster than coated boards or plywood.
The different behavior of wood products depends on diffusibility
of the board and of its coatings (Fig. 10.20). The minimum duration
of a large-chamber test is 3 days; but it can take much longer,
up to 6 weeks, to reach the steady state concentration.

B. Small-Chamber Methods

The large-chamber test is relatively slow and expensive to operate.
To reduce costs and shorten the procedure, it is favorable to reduce
chamber volumes. This can be done in different ways. In Sweden
and in West Germany, chambers with a volume of 1 m^3 have been
developed as standard test methods for the determination of formaldehyde emission [148-153]. The construction diagram of the Fraunhofer-
Institute's 1-m^3 chamber is shown in Figure 10.21. In East Germany,

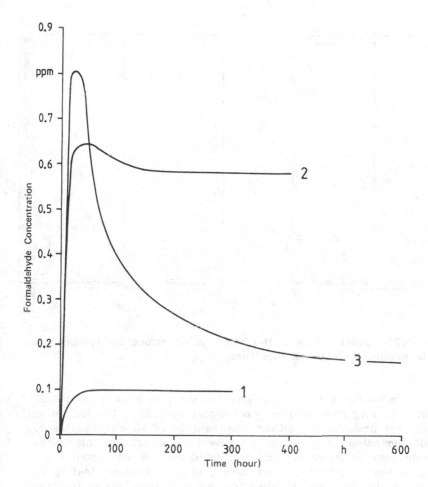

Fig. 10.20 Formaldehyde concentration in the 40 m³ chamber as a function of time: 1, E1 particleboard; 2, E2 particleboard; and 3, eleven-layer plywood.

a small-chamber method based on a test facility with a volume of 585 liters was standardized in 1985 [142]. The Danish standard test is based on a chamber with a volume of 225 liters, while chambers with a volume of 120 liters are used in Finland [143,149].

In 1986 the Scandinavian countries proposed the Nordtest method [150], based on a chamber with a volume of 250 liters, as a universal testing facility for many different products.

Smaller chambers are cheaper to construct than large chambers. For an exact determination of steady state concentration, however,

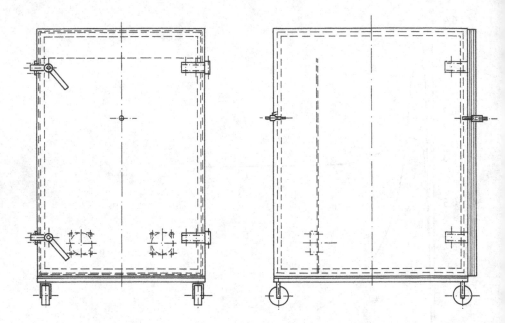

Fig. 10.21   Construction of the WKI 1-m^3 chamber for testing composite products and small furniture.

it is as indispensable, as in large chambers, to have a controlled
climate, loading rate, and air exchange rate [161]. The long duration
of the test procedure, another disadvantage of all chamber tests,
is not eliminated by reducing the chamber volume. Testing times
of conditioned samples in small ventilated chambers is about 4-5
days. If the conditioning is performed in the chamber, testing times
as long as in the large-chamber test are required, which increase
to up to 6 weeks in special cases.

C.   Perforator Method

The formaldehyde content of a board is much easier to determine
than the emission rate. Therefore, methods to determine the formaldehyde content attracted early interest in industry. The European
particleboard association, FESYP, developed in the late 1970s a
test procedure called the perforator method, a simple method that
was established in 1984 as European standard EN 120 [154].
    The perforator method is a procedure for the extraction with
toluene of small particleboard samples in a perforator apparatus
(Fig. 10.22). The extracted formaldehyde is sampled in water and

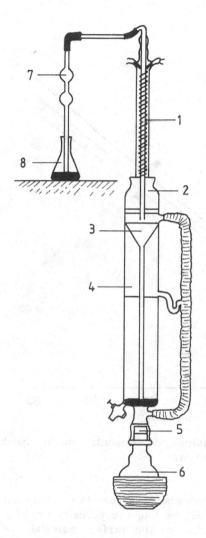

Fig. 10.22 Perforator apparatus for the determination of the formaldehyde content of composite products by toluene extraction.

determined by the iodine method. The formaldehyde content is expressed in milligrams of formaldehyde per 100 g of dry board. It is a simple test, and the total time to run it is about 3 hours. The test is universally applicable on uncoated boards of all types.

The perforator method also has some disadvantages. Although the emission of formaldehyde from a wood product depends on its

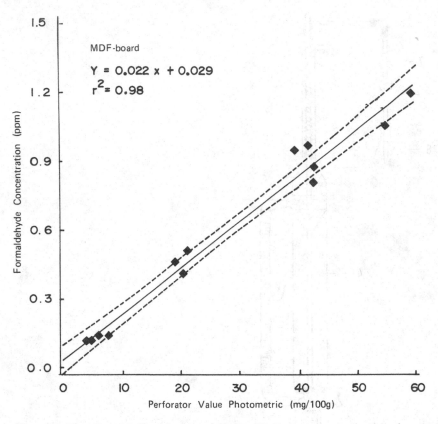

Fig. 10.23 Correlation between formaldehyde concentration in chamber test and perforator value of MDF boards.

formaldehyde content, there is not necessarily a correlation between the content and the emission rate. Factors like the porosity or the density of the board, the properties of the surface materials and surface coatings, and the size of the specimens influence the actual emission rate. For uncoated boards of the same type (e.g., particleboards), the interactions are less complicated. Particleboards or MDF boards show sufficient correlation between emission values determined in large-chamber tests and formaldehyde content to allow the use of the content value as a basis for their classification (Fig. 10.16 and 10.23). Based on the correlation between large-chamber values and perforator values, the perforator method was the second method accepted to determine the emission class of particleboards in West Germany [42].

A definite disadvantage of the standardized test procedure is the nonspecificity of the iodine oxidation. Roffael and others have detected that other oxidizable compounds in board composites are extracted with toluene too, often causing a relatively high error [155-160]. This error, which depends on different factors, makes it impossible to determine exactly the formaldehyde content of a wood product by iodometric determination. The influence of such a method's error increases with decreasing total formaldehyde content. Therefore, a reliable classification of wood products with low formaldehyde contents requires the use of more accurate methods of analysis. A more accurate method is the colorimetric determination of formaldehyde—for example, by the acetylacetone method. Control tests with the acetylacetone method and the DNPH method showed that photometric determination gives the exact formaldehyde content. The new German regulations for particleboards take these discrepancies into consideration and allow only for photometric analysis.

Another disadvantage of the perforator test is the dependence of the perforator value from the moisture content (Fig. 10.24).

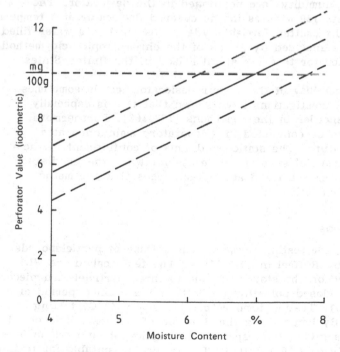

Fig. 10.24 Perforator value of particleboards as a function of moisture content.

Furthermore, a moisture-based increase in formaldehyde content
does not necessarily cause an increase in formaldehyde release.
Sundin has shown that the perforator value of unwrapped particleboard samples can rise during storage, while the emission values,
in contrast, will decrease [158,159]. These analytical problems
notwithstanding, the European wood products industry expects
that the perforator method will be retained as a quality control
test for its simplicity, its quickness, and its wide distribution.

D. Desiccator Method

The 2-hour desiccator test and its modifications are the most widely
used methods in North America, Australia, and Japan to determine
formaldehyde emission from particleboards. Astonishingly, the desiccator test is nearly unknown in Europe. The 2-hour dessicator
test has been adapted from a 24-hour, closed-har method, developed
in the middle 1970s in Japan. The method uses a common glass
desiccator with a volume of 10.5 liters. Eight test specimens, cut
7.0 cm × 12.7 cm from a board and conditioned for 7 days at 23°C
and 50% relative humidity, are positioned in the desiccator. The
emission test lasts 120 minutes in the covered desiccator at a temperature of 23°C. The emitted formaldehyde is absorbed in a water-filled
petri dish and is analyzed by means of the chromotropic acid method.
The 2-hour desiccator test was standardized in the United States
in 1983 [161].

The reproducibility of the 2-hour desiccator test is sometimes
unsatisfactory. Investigations have shown that this is especially
due to inhomogeneities in the specimens [162,163]. Furthermore,
the tests have to be conducted in a laboratory maintained at a
constant temperature. The static conditions of confinement in the
desiccator test restrict spontaneous emission, while the inevitable
time lag between emission and absorption means that the method
cannot be truly quantitative.

E. Bottle Methods

A simple method for testing formaldehyde release of particleboards
was developed by Roffael in 1975 [164]. The test, called the WKI
method, is based on the storage of one to three particleboard pieces
over water in a closed polyethylene bottle for a defined period of
time (Fig. 10.25). Testing temperature is usually 40°C, testing
time 24 hours. By lengthening the duration of the test, it is possible
to obtain curves giving more specific information about small differences between boards [6,165,166]. The method is suitable for testing
of uncoated boards. A disadvantage of the WKI method is the small

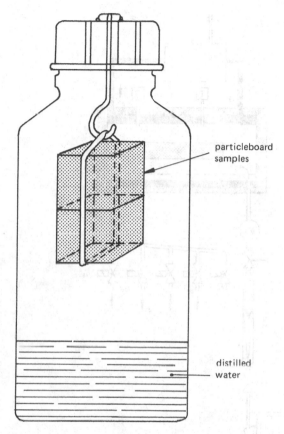

Fig. 10.25 The WKI test apparatus for the determination of the formaldehyde release of composite products.

size of the sample (2.5 cm × 2.5 cm), which makes it difficult to test nonhomogeneous or coated boards.

Later the WKI method was modified in the Netherlands and described as the Roffael method [6,86]. In East Germany a bottle test based on a large bottle with a volume of 4 liters was established as a standard in 1985 [167]. Called the "Hermetik behälter-Methode," this method is suitable for testing larger specimens, even of coated boards [168,169].

In Sweden a fast bottle test was developed named the REM test. Within 3 hours the formaldehyde release of particleboards is determinable. The REM test shows sufficient correlation with other test methods [170,171].

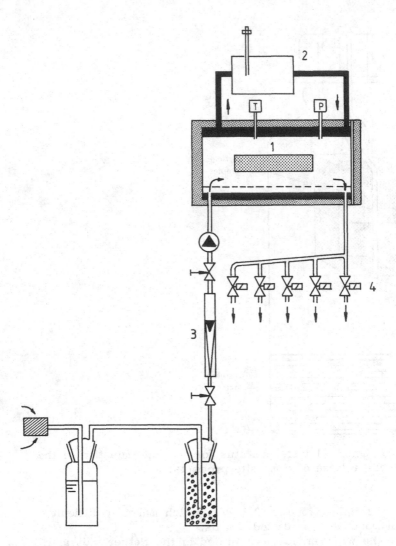

Fig. 10.26 Semiautomatic gas analysis apparatus for the determination of the formaldehyde release of composite products: 1, test chamber with sample; 2, heating system; 3, air inlet and cleaning system; 4, sampling systems with five automatically controlled valves.

A major benefit of bottle methods is that they enable a large
number of samples to be tested simultaneously. Bottle methods are
simple and inexpensive, but it takes one or more days to get the
results. To overcome this time lag, the WKI method was modified
to obtain results quickly by reducing testing time up to 3 hours.
This WKI quick test is to be proposed as a Scandinavian standard
test [171].

After years of stagnation, there is now increasing interest
in bottle tests.

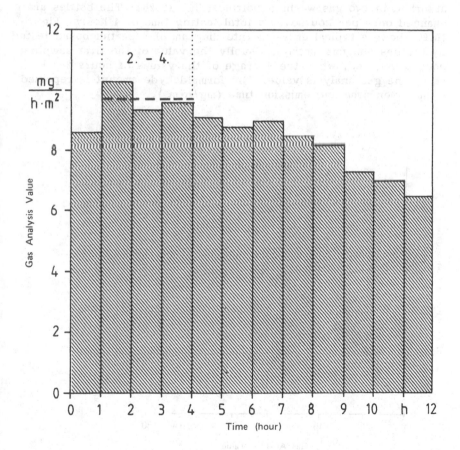

Fig. 10.27 Formaldehyde release of a particleboard in gas analysis
test as a function of time.

F.   Other Methods

A popular test in West Germany is the gas analysis method. Like
the perforator test, the gas analysis method was originally developed
by FESYP. Later it was optimized to a semiautomatic test by Mehlhorn
[6,86]. By 1985, the gas analysis method was established as a German
standard [173]. The gas analysis test requires a specimen of 40
cm × 5 cm × the thickness of the board. The sample is placed into
a test tube at a controlled temperature of 60°C. A gas stream of
cleaned and dry air is blown through the test tube at a rate of
1 liter per minute. The formaldehyde emitted by the specimen is
absorbed in two gas washing bottles (Fig. 10.26). The bottles are
changed once per hour over a total testing time of 4 hours. Figure
10.27 shows a typical emission rate diagram of a particleboard tested
by the gas analysis method. Usually the value of the first sampling
hour is rejected, while the average of the values of hours 2-4 is
called the gas analysis value. The formaldehyde amount is reported
to specimen area and emission time (mg/h·m^2).

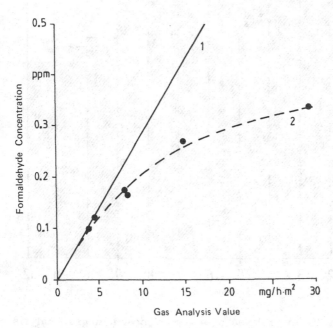

Fig. 10.28   Formaldehyde concentration in a chamber test and gas
analysis value of particleboards (1) and of wood panel products
coated with an acid hardening lacquer (2).

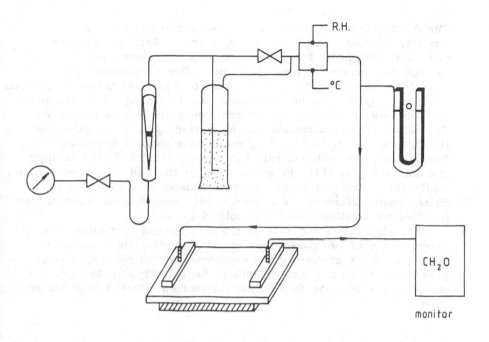

$$n = 200 - 1500 \ h^{-1} \qquad A = 0.015 \ m^2$$

$$V = 0.000075 \ m^3 \qquad a = A/V = 200 \ m^2/m^3$$

Fig. 10.29 Schematic flow system of the microchamber test system (MKS) developed by Hoetjer et al. [132,174].

The gas analysis method is an accurate test to determine the formaldehyde emission of uncoated particleboards or MDF boards. The correlation between gas analysis value and large-chamber value has been well established (Fig. 10.16). This method is also suitable for coated boards or for plywood, provided the edges of the specimen are sealed. It is unsuitable for materials that are not completely cured (e.g., wood samples coated with an acid hardening lacquer). These materials show precuring at 60°C, resulting in a higher equivalent emission rate that that expected from a test at room temperature (Fig. 10.28).

The high test temperature was the main reason for the failure of some European countries to accept the gas analysis method for European standardization.

Another interesting emission method was developed in the Netherlands. First called the MCN test, it is now named microchamber

(MKS) test [175]. In this test a particleboard sample is placed, airtight, between two plastic boards (Fig. 10.29). Between the panel and the plastic boards is a channel 5 mm wide. Air is passed through the channel at high flow rate. After a relatively short time (15-30 minutes) the formaldehyde concentration becomes constant. The test samples must be conditioned before testing. By varying the flow rate, a data set of different concentrations is obtainable. The values of concentrations can be plotted against the values of the flow rate (Fig. 10.30). The equilibrium concentrations as a function of the ventilation rate can be calculated from the straight line obtained [132,174]. From the slope of the plot, the mass transfer coefficient of the test board can be calculated too. Provided the characteristic parameters are known, the formaldehyde concentration in board combinations can be calculated as well.

The microchamber method is the only method that allows the determination of two parameters within a relative short period of testing time. By comparing the confidence of the plot, it is also possible to check the test reliability. Regrettably, up to now, the microchamber test has found little interest outside the MCN laboratories.

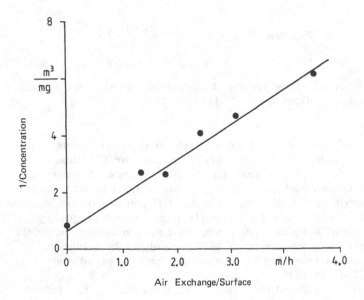

Fig. 10.30 Hoetjer plot of a particleboard tested by the MKS system giving information on formaldehyde release and mass transfer coefficient [132].

Many other methods have been developed [176-184]. Today some methods are more of historic interest, such as the microdiffusion method [177], or have never found interest outside the laboratories in which they were developed. The original literature or specific surveys give more information about these methods [6,36,86,172].

IX. DIFFERENCES IN EMISSION BEHAVIOR AMONG WOOD PRODUCTS

The large differences reported for the emission behavior of various wood products are due to their differing structures. It is to be expected that plywood will have emission characteristics other than those of particleboard. Test results with the most common wood products have shown, in accordance with these assumptions, that the emission behavior of particleboards is similar to that of MDF boards, while plywood emission characteristics diverge significantly. The differences between wood products can be explained by the different structures of the materials that determined the emission path of a formaldehyde molecule (Fig. 10.31). Therefore, the structure influences the test duration of the large chamber test. A particleboard or MDF board normally reaches steady state concentration within 3-10 days, while plywood needs up to 6 weeks, depending on its thickness and the number of plies (Fig. 10.20).

The structure-dependent differences among panel products have a significant influence on the correlation between formaldehyde content and large-chamber emission rate. Particleboards and MDF boards are porous materials containing no significant diffusion barriers. In plywood, on the other hand, the inner layers are encircled by the outer layers and the glue line film acts as an inner barrier comparable to a coated surface. Therefore, large plywood panels can contain 1.5-2.5 times as much extractable formaldehyde as particleboard or MDF board while still giving comparable steady state equilibrium concentration in the large-chamber test.

In the perforator method, the formaldehyde content of wood products is determined by extraction with toluene. The standard analysis is carried out with iodine titration, which is not specific to formaldehyde but detects other oxidizable compounds also. Perforator test results for plywood and MDF boards are comparable, while values for particleboards differ. MDF boards and plywood contain relatively pure wood. The wood fibers and wood veneers are dried under moderate conditions. Particleboard, instead, contains degraded wood, biomass or bark, as well. The particles are, in general, dried more drastically than fibers or veneers. Hard drying of wood causes partial pyrolysis, building up the amount of oxidizable

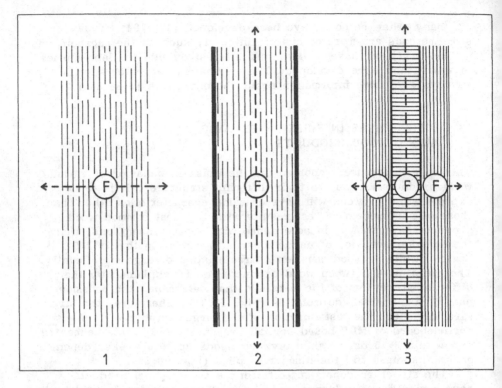

Fig. 10.31 Diffusion path of a formaldehyde molecule in a particleboard or fiberboard (1), a coated board (2), and a three-layer
plywood (3).

compound. This gives higher values for particleboard in the perforator test than for plywood or MDF board (Figs. 10.32-10.34). The
mean variation between iodometric and photometric perforator values
is also relatively greater, especially at low values, in the case of
particleboards.

The size and the room loading of the board are important factors
influencing formaldehyde emission. At a given emission per area,
the formaldehyde concentration in a room with constant climate is
proportional to the room loading. Without any ventilation, the steady
state concentration equals the equilibrium concentration, thus being
"nearly" independent of size and area of the emission source. The
ratio of edges to surface depends on the size of the emitting panel.
Different investigations have shown that the edges emit 3-6 times
more formaldehyde than the surfaces [57,186]. In plywood, even
larger differences in emission rate of edges and surfaces are to be

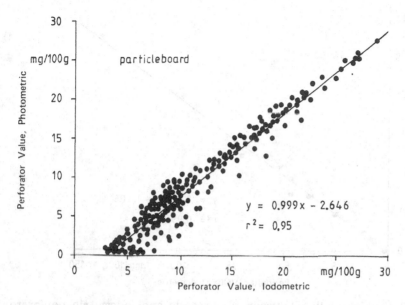

Fig. 10.32 Correlation between photometric and iodometric perforator values for particleboards.

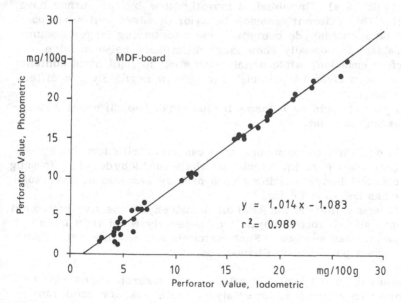

Fig. 10.33 Correlation between photometric and iodometric perforator values for MDF boards.

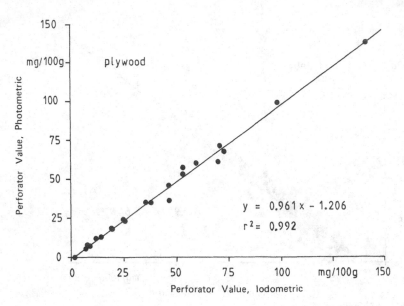

Fig. 10.34 Correlation between photometric and iodometric perforator values for plywood.

found (Table 10.9). Unpublished investigations by the author have shown that the different emission behavior of edges and surfaces depends on formaldehyde content. Panels containing larger amounts for formaldehyde normally show large differences between edge and surface emission, while panels with small formaldehyde contents (e.g., emission class E1 particleboards) show negligibly low differences or none.

   If a product acts as a formaldehyde sink, two different adsorption possibilities exist.

1. It does include compound that can react with formaldehyde (e.g., a UF resin, which absorbs formaldehyde when forming chemical bonds). Being also a primary emission source, such sinks act as a buffer.
2. Materials that do not contain formaldehyde-reactive compounds may absorb formaldehyde too, especially when they have large polar surfaces. Such materials are cement- and gypsum-bonded particle- or fiberboards.

In a cement-bonded board, formaldehyde is disproportionated by the Cannizarro reaction route catalyzed under alkaline conditions by calcium ions. Gypsum contains calcium ions as cement but, unlike

Table 10.9 Influence of Edges on Formaldehyde Release of Different Wood Composite Products

Wood composite product	Thick-ness	Formaldehyde release in $mg/(h \cdot m^2)$[a]	
		Unsealed edges	Sealed edges
Particleboard E1	19 mm	19,5	14,6
Particleboard E2	19 mm	3,3	2,7
MDF Board "E2"	19 mm	23,5	18,4
MDF Board "E1"	19 mm	4,2	4,0
Plywood "E2"	12 mm	24,1	2,9
Plywood "E1"	12 mm	4,4	0,3

[a]Determined by Gasanalysis method DIN 52 368.
E1 = Emission class with low formaldehyde content.
E2 = Emission class with medium formaldehyde content.

Table 10.10 Formaldehyde Release of Gypsum-Bonded Particleboard After Storage in a Formaldehyde Atmosphere or in Contact to Formaldehyde Containing Particleboards ("Memory Effect")

Test	Time after storage d	Formaldehyde release	
		$mg/(h \cdot m^2)$[a]	mg/m^3 [b]
A	—	0,3	0,02
B	1	0,2	0,02
C	1	2,9	0,55
	5	1,3	0,26
	10	0,9	0,12
	15	0,6	0,08
D	1	5,0	0,92
	5	2,3	0,42
	15	1,0	0,21
	15	0,7	0,11

[a]Determined by Gasanalysis method DIN 52 368.
[b]Determined in an $1m^3$-chamber at 23°C without air exchange.
A = just after production of gypsum-bonded boards.
B = after 14 d storage in an atmosphere with a formaldehyde content below 0,02 mg/m^3.
C = after 14 d storage between emission class E2 particleboards.
D = after 14 d storage in an atmosphere with a formaldehyde content of about 1,5 mg/m^3.

cement, gypsum is a neutral material with an enormous inner surface
due to its microcrystalline structure. Hence, gypsum-bonded boards
can absorb relatively large amounts of formaldehyde. After removal
of the primary emission source, those absorbers, which are manufactured without any formaldehyde themselves, act as secondary emission
sources, releasing the absorbed formaldehyde only over a relatively
long period of time (Table 10.10). The "memory" effect of materials
like gypsum-bonded boards introduces an added difficulty to the
study of real formaldehyde emission complaints [89].

X. REDUCTION OF FORMALDEHYDE RELEASE
IN WOOD PRODUCTS

The release of formaldehyde in wood products can be minimized
during the manufacturing process, or by posttreatment of the board
and by surface treatment. These include the addition of wax scavengers or urea to the wood finish, treatment of the composite boards
with ammonia gas or ammonia salts, and surface treatment of boards
with paints, lacquers, veneers, and papers. The surface treatment
can be both a physical and a chemical minimization of formaldehyde
diffusibility. The most efficient route is the modification of the
adhesives chosen. The use of phenolics and isocyanates or of UF
formulations with lower molar ratios are powerful concepts used
to reduce the formaldehyde emission potential of wood products,
in some case to nearly zero.

The influences of the pressing time, press temperature, and
some more production-dependent factors were investigated intensively
in the 1960s and 1970s [177, 188-192]. Increasing the press temperature and lengthening the press time generally serve to reduce the
formaldehyde release of composite products. The moisture content
of the wood particles and of the finished board, the hardener content
and the pH of the wood particles, and the glue formulation are
factors influencing the formaldehyde release of the final product
too.

In the initial years of use of formaldehyde-reducing procedures,
the control of production factors was an easy and safe way to influence the extent of a board's formaldehyde release. Nowadays, the
situation is somewhat more complicated. Modern resins with new
emission characteristics are more sensitive to the influence of the
production process, wood particle properties, and handling procedures
than the resins used in the past. In the late 1980s, the production
of composite products makes it absolutely necessary to take into
consideration the different behavior of low formaldehyde resins.

A. UF Resins

The molar ratio of formaldehyde to urea (F/U) is the most important factor affecting formaldehyde release in the production of boards. Another effective way to reduce formaldehyde release is the addition of formaldehyde-binding substances ("scavengers") to the resin or to the wood particles. Last but not least, the combination of UF resins with formaldehyde-free adhesive systems may also be a way to reduce formaldehyde emission. Posttreatment methods to decrease and minimize formaldehyde release are based on compounds like ammonia, ammonium salts, or urea. Many reports, patents, and publications describe these different possibilities in more detail [6,7,194-208].

B. Influence of Molar Ratio

Different methods can be used to produce a UF-bonded particleboard that releases low amounts of formaldehyde. The most common one is to reduce the formaldehyde content of the UF resin. Formaldehyde release from particleboards was decreased with the successive lowering of the molar ratio of formaldehyde to urea of the UF resin. Figure 10.35 shows the decline of formaldehyde release as a function of the molar ratio. Unfortunately, the decrease in the molar ratio has a negative influence on other adhesive characteristics.

Over a period of 15 years, the formaldehyde release of particleboard adhesives was reduced step by step, and particleboard release decreased by more than 90%. At the beginning of the 1970s, the average formaldehyde content of particleboards was between 100 and 120 mg per 100 g board dry weight (perforator value). In 1985 about 40% of the West German particleboard production had releases lower than 10 mg/100 g (emission class E1), and boards with formaldehyde contents above 30 mg/100 g were no longer produced (Fig. 10.36). Nowadays, more than 90% of West German particleboards belong to emission class E1.

To reach the level of class E1 board (perforator value < 10 mg) without any more modifications, the molar ratio F/U of the UF resin must be lower than 1.2 in the absence of phenolic compound [198]. Modern E1-type UF resins often are fortified with about 4-8% melamine on resin, dry weight. The modification requires sensitive and accurate control of the resin manufacturing process.

The relative reactivity of UF resins is defined by the resin gel time. In a literature critique Myers illustrated the reduction in reactivity of lower F/U resins [200]. Resin gel time, using ammonium salt catalyst, increases twofold or more as F/U decreases from 2.0 to 1.2. The loss of reactivity is also influenced by the

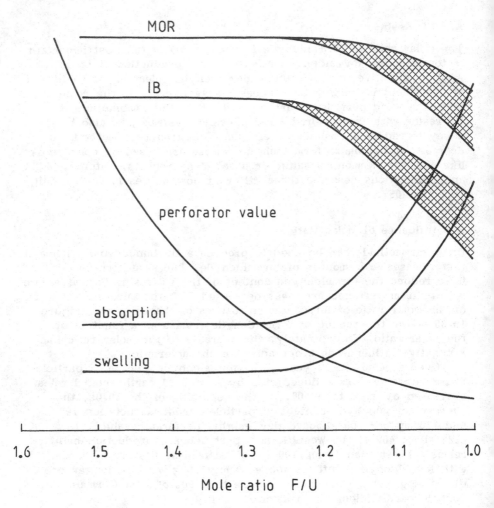

Fig. 10.35 Some characteristics of UF-bonded particleboards as a function of the mole ratio [158,159].

free formaldehyde content of the resin, which decreases with the decrease in molar ratio. The free formaldehyde content of a resin does not decrease linearly with the molar ratio. It falls rapidly between F/U 2 and 1.5 but only slightly below 1.3. The decline in reactivity of UF resins can be overcome by using more hardener or by the selection of a more reactive hardening system.

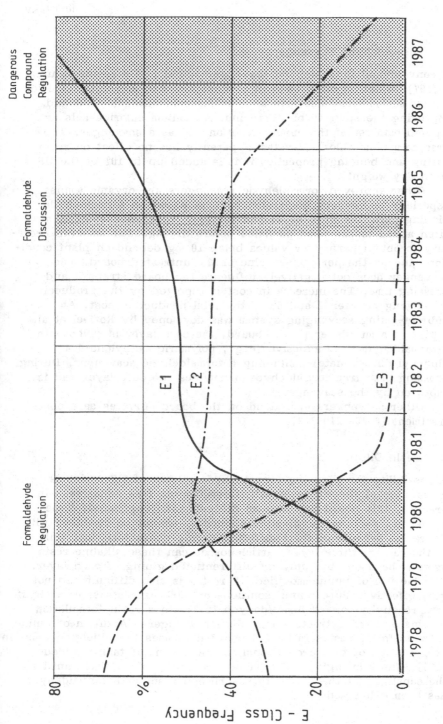

Fig. 10.36 Formaldehyde emission class frequency of particleboards in West Germany between 1978 and 1987 tested in quality control.

C. Formaldehyde Scavengers

A conventional formaldehyde scavenger added to UF resins is urea
[6,197], which is often used in combination with ammonium chloride
(about 20:1). Urea also acts as a buffer controlling the pH and
improving the stability of UF resins. Ammonium chloride acts as
an acid catalyst of the curing reaction and as a scavenger. The
urea/ammonium chloride system apparently has no effect on resin
curing and bonding properties if it is added up to 10% of the UF
resin dry weight.

Other approved formaldehyde scavengers are organic amines.
Especially in Sweden, dispersion systems based on formaldehyde-
binding paraffins were developed [5,159]. These wax systems are
often added to the wood particles before drying. The amounts necessary to obtain perforator values below 10 mg depend on plant conditions and on the perforator value of the untreated board. The
scavenger does not negatively influence the board strength and
pressing time. The increase in cost is reported by the producer
as varying between 1 and 5% of the total production cost. An
indirect-acting scavenging system was developed by Roffael et al.
[194]. In a multilayer particleboard, the core layer is glued with
a formaldehyde-free adhesive (e.g., polymeric methylene-4,4'-
diphenyl-diisocyanate) containing a formaldehyde scavenger. During
pressing, the free formaldehyde penetrates the core layer and is
captured by the scavenger.

Other scavengers are added on the board surfaces as a posttreatment [6,206,210-212].

D. Resin Dilution

A possibility for reducing formaldehyde release of UF resins is
by dilution of the UF resin with formaldehyde-free adhesives. In
practice, this possibility is more difficult than suggested. Some
adhesives, such as the alkaline phenolics, cannot be combined with
UF resins because of their different curing characteristics. Only
in the case of three-layer particleboards can these alkaline resin
systems be used, but only by differentiated gluing of each layer.

The use of tannin-modified UF resins is also difficult and not
very effective. Pure tannin solutions or bark extracts show a significant reactivity toward formaldehyde [213]. In a resin formulation,
tannins or bark extracts act as weak scavengers. Addition of tannin
extracts to UF resins in particleboard decreases formaldehyde emission
over periods of time proportional to the amount of tannin added
[215]. The scavenging effect lasts a relatively short time, until
the capability of the tannin to absorb and react with formaldehyde
has been exhausted.

The addition of tannin-formaldehyde resins may be another way to influence formaldehyde release of UF-bonded composite products. Tannin resins are cured normally at neutral or weak acid conditions; thus, in principle, their combination with UF resins should pose no difficulties. Unfortunately, tannin resins require an excess of formaldehyde to obtain sufficient hardening. This makes it impossible to combine tannin adhesives with UF resins in the production of boards of very low formaldehyde release. The background to this failure is discussed in more detail in Subsection F ("Phenolic Resins") below.

A formaldehyde-free adhesive is polymeric methylene-4,4'-diphenyl-diisocyanate (PMDI). PMDI is very reactive and can be combined easily with UF resins. Astonishingly, boards produced with combinations of MDI and UF resin often release more formaldehyde than those produced with pure UF resin. This effect appears to be due to the great reactivity of the isocyanate group toward amino groups. The isocyanate groups compete effectively with the formaldehyde carbonyl groups or with the UF resin methylol groups.

The addition of spent sulfite liquor (SSL) as a scavenger has also been suggested. Laboratory test results showed that more than 50% of the UF resin may be replaced by SSL, the formaldehyde emission being reduced proportionally to the UF replacement [209].

A more sophisticated application of this principle was the use of the ammonium salt of SSL copolymerized with a liquid phenolic resin. For the combination tests, conventional UF resins with a relatively high molar ratio have often been used. UF resins of lower molar ratio are much more sensitive to modification. From this standpoint, it is doubtful that low formaldehyde resins will be able to tolerate glue mixing similarly to conventional UF resins. From the data reported, it is difficult to predict whether SSL can be combined successfully with UF resins for the production of boards of emission class E1 and lower values.

E. Posttreatment Procedures

The benefit to a wood product manufacturer of posttreatment is the possibility of using UF resins with a higher molar ratio, which have better handling and gluing characteristics than lower formaldehyde resins. Two procedures to reduce formaldehyde release of boards by posttreatment are based on ammonia gas: the RYAB and the Verkor FD-EX methods [6,197].

The FD-EX method is a two-step procedure that is done in a three-chambered plant. In the first chamber, the board is treated with an excess of ammonia gas at temperatures between 30 and 35°C, thus reducing the formaldehyde by the formation of hexamethylene-

tetramine (Fig. 10.2). In the second chamber, most of the excess
ammonia is desorbed. In the third chamber, the rest of the ammonia
reacts with formic acid and other acidic wood compounds to be fixed
as ammonium salt in the board. As a consequence of the last reaction
chamber, no ammonia odor should be present to lessen the suitability
of the finished boards.

The ammonia treatment not only binds the free formaldehyde
in the board but also influences the acidity by changing the pH
value from abour 3.5 to about 6 [196]. This increase in the pH
value also positively affects the hydrolyzability of the UF resin.
The FD-EX method requires further investigation with exposure
chambers. The relatively high cost of this method has hindered
its wider distribution in industry.

In the RYAB method, the particleboards are treated with ammonia
gas in a single step. The treatment takes place in a special chamber
using a pressure gradient (Fig. 10.37). The reaction is completed
during storate of the treated board. The reaction is the same as
in the FD-EX method.

A disadvantage of the RYAB method is the ammonia odor of
the boards, which can last for weeks. Other disadvantages are

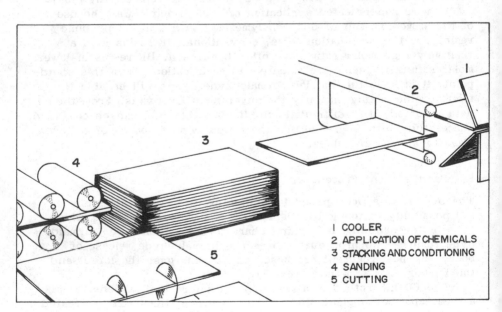

1 COOLER
2 APPLICATION OF CHEMICALS
3 STACKING AND CONDITIONING
4 SANDING
5 CUTTING

Fig. 10.37   The posttreatment of particleboards by the RYAB
process.

the difficulty in handling the gas and problems in maintaining evenness of treatment.

Another posttreatment is the ASSI method, which uses such ammonia salts as ammonium carbonate, ammonium bicarbonate or ammonium carbamate as scavenging agents (196,197). The active scavenger is ammonia, which is slowly liberated from the salts. The salts can be added to the adhesive or sprayed as solids or solutions on the surface of the boards.

Ammonia reacts with many wood substances, darkening the color of the wood. Hence much care must be taken in determining the exact dosage of ammonia-based scavengers that will prevent darkening of the surfaces after veneering of the treated boards, especially with light-colored veneers. To avoid the unwelcome darkening effect, it is also possible to use urea solutions instead of ammonia or ammonium salts. Sprayed on the hot surfaces of particleboards just after pressing, urea solutions can reduce formaldehyde release under optimized conditions by more than 50% [6,196]. Urea reacts in two ways. On the one hand it is itself a scavenger; on the other hand it decomposes slightly to ammonia, especially under acid conditions:

$$CO(NH_2)_2 + H_2O + 2H^+ \rightarrow 2NH_4^+ + CO_2$$

Posttreatment procedures have elicited moderate interest in industry.

If regulations like the West German Hazardous Compound Act are found to enforce the production of boards with perforator values much lower than 10 mg, the combination of UF resins with low formaldehyde content and posttreatment procedures may become of more interest. Therefore, the long-term reliability of these posttreatments must be more accurately verified.

F. Phenolic Resins

Alkaline phenol resins contain, even in their uncured state, very little residual free formaldehyde. On the one hand, the molar ratio of formaldehyde to phenol is low in PF resins, while on the other hand nearly all formaldehyde reacts with the phenol, forming nonhydrolyzable bonds.

The cured resin is stable and does not break down; hence it does not release formaldehyde, even under extreme conditions. The few formaldehyde molecules remaining entrapped in the cured resin are slowly converted to methanol and formic acid by the Cannizzaro reaction. In contrast to UF resins, formaldehyde loses its chemical identity in PF resins. Testing of PF-bonded particleboard

and plywood usually gives values in the range of the experimental
error of the method or only a very little above.

Formaldehyde levels associated with wood products bonded with
alkaline-cured phenol-formaldehyde adhesives are in most cases
extremely low. Moreover, their formaldehyde release approaches
zero as the small amount of formaldehyde initially present in the
panels, is released. There are some indications that wood products
bonded with newly developed PF resins of low alkalinity emit significantly more formaldehyde than wood products bonded with conventional PF resins of higher alkaline content. The situation thus changes
if instead of an alkaline PF resin, a resin type is used that is cured
under neutral or weak acid conditions. Such a resin may be a tannin-
formaldehyde resin, a resorcinol-formaldehyde resin, or a novolak-
type PF resin.

The reason for the different behavior of neutral and acid PF
resins is usually an unreacted excess of formaldehyde. An excess
of formaldehyde or paraformaldehyde is necessary to cure resins
such as tannins or PF novolaks. During hot-pressing, only part
of the formaldehyde excess is emitted. The residual formaldehyde
may be unable to disproportionate in the Cannizzaro reaction. Wood
products containing neutral or acid PF resins can then release significant amounts of formaldehyde. Even this emission potential will,
however, decrease after release of the unreacted formaldehyde or
formaldehyde polymers. To reduce the initial emission potential
of wood products bonded with these adhesives, it is necessary
to carefully control the pH and the formaldehyde excess in the
adhesive formulation.

The reactivity of some phenols also depends on pH. In the
neutral or weak acid range, the reactivity of the nucleophilic centers
of the aromatic ring is much lower than in the alkaline range. Precuring or other polymerizing reactions take place very slowly. The
influence of formaldehyde excess is easily detectable with tannin
resins. When the paraformaldehyde content is varied in a tannin
adhesive formulation with a pH value of about 4-4.5, the formaldehyde
release of the cured resin increases significantly, especially above
6-8% paraformaldehyde addition (Fig. 10.38). These results corre-
spond with stoichiometric data.

It can be calculated that approximately 10.5% of formaldehyde
or paraformaldehyde could react with a tannin extract of 100% polyphenolic content. The polyphenolic content of the tannin extracts
used for the experiments lay between 65 and 80%. It is also for
this reason that some commercial tannin adhesive formulations for
particleboard are used at a pH of ±7 or higher. The pH value of
the adhesive formulation has also a dominant influence on the formaldehyde release of the cured tannin resin. Further experiments with

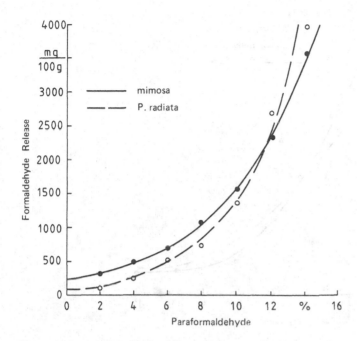

Fig. 10.38 Influence of paraformaldehyde content on formaldchyde release of cured tannin formaldehyde resins determined by the WKI method (24 hours).

tannin-formaldehyde formulations containing 10% paraformaldehyde show that by increasing the pH value of the adhesive, one causes the formaldehyde release of the cured resin to decrease drastically (Fig. 10.39). At pH 8-9, formaldehyde release becomes minimal and is no longer influenced significantly by a further increase of pH.

G. Coating and Laminating

The application of coatings and of laminates is very effective in reducing the formaldehyde release of a UF-bonded wood composite product. Laminating a particleboard with a melamine-formaldehyde impregnated paper will reduce the formaldehyde emission to nearly zero, independently of its formaldehyde content [144,145]. The primary effect of coatings and laminates is their action as diffusion barriers, which reduce the mass transfer coefficient of the boards [102]. Care must be taken when the overlay itself contains urea-formaldehyde resins, thus increasing, instead of decreasing, the

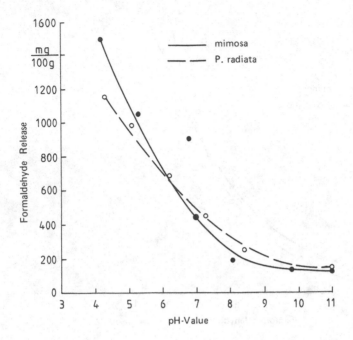

Fig. 10.39 Influence of pH value on formaldehyde release of cured tannin formaldehyde resins determined by the WKI method (24 hours).

emission potential of the treated boards. On top of the physical effect of coatings and laminates, scavenging compounds can further enhance their release-reducing potential. More information about the usefulness of coatings and laminates can be found in the specialist literature [6,103,144,206].

XI. FUTURE DEVELOPMENTS

During the past 25 years, the issue of formaldehyde release has received much attention by almost everyone involved in the production, use, and regulation of materials containing formaldehyde. Thus it has been of extreme concern to those in the wood products industry, which depends heavily on the use of UF resins. Much effort has been directed toward developing better resin systems, posttreatment procedures, and other techniques to reduce the emission potential of wood products. Research institutes have supported these efforts by developing test methods suitable for the accurate determination of formaldehyde release. As a result of these efforts, a significant reduction in formaldehyde emissions from wood products has been achieved. The emission potential of most UF-bonded wood products has been reduced by 80-90%. The decrease of emission values was accompanied by an increase in the demand for products of lower emission potential. In some countries, politicians and environmental pressure groups have called for products with an effective formaldehyde release of zero.

The important question in fulfilling such a requirement is not the higher costs involved. If consumers are willing to pay a higher price for formaldehyde-free products, industry will, without any doubt, deliver such products. The important question nobody can presently answer is whether an equivalent alternative in adhesives or to adhesives exists.

REFERENCES

1. Butlerov, A. M. (1859). "Uber einige Derivate des Jodmethylens." *Justus Liebigs Ann. Chem.*, 111:242-252.
2. Hofmann, A. W. (1868). "Zur Kenntnis des Methylaldehyds." *Justus Liebigs Ann. Chem.*, 145:357-361.
3. Baekeland, L. H., U.S. Patents 939,966; 942,699; 942,809. German patents 233,803; 237,790; 281,454.
4. Wittmann, O. (1962). "Die nachträgliche Formaldehydabspaltung bei Spanplatten." *Holz Roh- Werkst.*, 20:221-224.
5. Sundin, B. (1985). "The formaldehyde situation in Europe." *Proceedings of the 19th Particleboard Symposium at Washington State University*, Pullman, pp. 255-276.
6. Roffael, E. (1982). *Die Formaldehyd-Abgabe von Spanplatten und anderen Werkstoffen.* Stuttgart, DRW-Verlag.
7. Meyer, B., Andrews, B. A. K., and Reinhardt, R. M. (1986). "Formaldehyde Release from Wood Products." ACS Symposium Series No. 316. Washington, DC, American Chemical Society.
8. Walker, J. F. (1964). *Formaldehyde.* London, Reinhold Publishing.
9. Swenberg, J. A., Kerns, W. D., Mitchell, R. I., Gralla, E., and Pavkov, K. L. (1980). "Induction of Squamous Cell Carcinomas of the Rat Nasal Cavity by Inhalation Exposure to Formaldehyde Vapour." *Cancer Res.*, 40:3398-3407.
10. Kerns, W. D. (1980). "Long-term Inhalation and Carcinogenity Studies of Formaldehyde in Rats and Mice." CIIT Conference on Formaldehyde Toxicity at Raleigh, NC.
11. Boreiko, C. J., Couch, D. B., and Swenberg, J. A. (1982). "Mutagenic and Carcinogenic Effects of Formaldehyde." *Environ. Sci. Res.*, 25:353-367.

12. Professional Consultants in Occupational Health, Inc. (1980).
    "A Report on the Human Health Effects from Low Dose Exposures
    to Formaldehyde Gas." Bethesda, MD.
13. European Chemical Industry Ecology & Toxicology Center.
    (1981). "Assessment of Dates on the Effects of Formaldehyde
    on Humans." Technical Report No. 1, Brussels.
14. European Chemical Industry Ecology & Toxicology Center. (1981).
    "The Mutagenic and Carcinogenic Potential of Formaldehyde."
    Technical Report No. 2, Brussels.
15. Uehleke, H. (1981). "Wo liegen Sicherheit und Gefahren von
    Formaldehyd?" (Part 1), *Umweltmedizin*, 3:48-51.
16. Uehleke, H. (1983). "Wo liegen Sicherheit und Gefahren von
    Formaldehyd?" (Part 2), *Umweltmedizin*, 4:72-73.
17. Baumann, H. (1981). "Stellungnahme zur Debatte über mut-
    massliche Gesundheitsgefahren durch Formaldehyd." *Kunststoffe*,
    71(11):2-6.
18. Andersen, I., and Molhave, L. (1982). "Controlled Human Studies
    with Formaldehyde." In *Formaldehyde Toxicity*, J. E. Gibson,
    Ed. Washington, Hemisphere, pp. 154-165.
19. Goh, K.-O., Cestero, R. V. M. (1982). "Health Hazards of
    Formaldehyde." *JAMA*, 247:2778.
20. Heilmann, B. (1982). "Formaldehyde." *Environ. Sci. Technol.*,
    16:543-547.
21. Griesshammer, R., Vahrenholt, F., and Claus, F. (1984).
    "Formaldehyd—Eine Nation wird geleimt." Rowolt Taschenbuch
    No. 5543: Hamburg.
22. Bundesminister für Jugend, Familie, und Gesundheit (Ed.).
    (1985). *Formaldehyd—Gemeinsamer Bericht des Bundesgesundheits-
    amtes, der Bundesanstalt für Arbeitsschutz und des Umwelt-
    bundesamtes*. Schriftenreihe des Bundesministers für Jugend,
    Familie, und Gesundheit, No. 148. Bonn-Bad Godesberg.
23. Brunner, P., and Warich, U. (1985). "Todesursachen bei
    Pathologen—Zum möglichen Einfluss des Formaldehyds." *Patho-
    logie* (6):43-45.
24. Fischer, M. (1986). "Gesundheitliche Gefahren von Baustoffen—
    Erfahrungen und Erkenntnisse." *Dtsch. Archit. Blatt*, 6/86:753-
    758.
25. Petri, N. K. (1986). "Formaldehyde, A Brief Survey." Paper
    presented at 14th International Congress on Occupational Health
    in the Chemical Industry, Ludwigshafen.
26. Liebling, T., Rosenmann, K. D., Pastides, H., Griffith, R. G.,
    and Lemeshow, S. (1984). "Cancer Mortality Among Workers
    Exposed to Formaldehyde." *Am. J. Ind. Med.*, 5:423-428.
27. Blair, A., Stewart, P., O'Berg, M., Gaffey, W., Walrath, J.,
    Ward, J., Bales, R., Kaplan, S., and Cubit, D. (1986).

"Mortality Among Industrial Workers Exposed to Formaldehyde." *J. Natl. Cancer Inst.*, 76:1071-1084.

28. Loewe, O. C. (1972). "The Tropospheric Distribution of Formaldehyde." Report No. 1756: KFA Jülich, Jülich.

29. Ehhalt, O. M. (1974). "The Atmospheric Cycle of Methane." *Tellus*, 26:58-70.

30. Fushimi, K., and Miyake, Y. (1980). "Contents of Formaldehyde in the Air Above the Surface of the Ocean." *J. Geophys. Res.*, 85:75233-75236.

31. Seiler, W. (1982). "Der atmosphärische Kreislauf des Formaldehyds." In *Arbeitsbericht für den Zeitraum 1979-1981 des DFG*. Sonderforschungsbereichs 73, Frankfurt Main.

32. Lahmann, E., and Jander, K. (1968). "Formaldehydbestimmungen in Strassenluft." *Gesundheits-Ingenieur*, 18(1):18-21.

33. Grosjean, D. (1982). "Formaldehyde and Other Carbonyls in Los Angeles Ambient Air." *Environ. Sci. Technol.*, 16:254-262.

34. Nantke, H.-J. and Lohrer, W. (1986). "Umweltschadstoff Formaldehyd: Belastungssituation und Regelungen." *Adhäsion*, 32(11): 17-28.

35. Lohrer, W., Nantke, H.-J., and Schaaf, R. (1985). "Formaldehyd in der Umwelt." *Staub - Reinhaltung Luft.*, 45:239-247.

36. Petersen, H., and Petri, N. (1985). "Formaldehyd—Allgemeine Situation, Nachweismethoden, Einsatz in der Textilhochveredelung." *Melliand Textilber.*, 66:217-222, 285-295, 363-368.

37. Coutrot, D. (1986). "European Formaldehyde Regulations: A French View." ACS Symposium Series No. 316. Washington, DC, American Chemical Society, pp. 209-216.

38. Senatskommission der Deutschen Forschungsgemeinschaft. (1987). *Maximale Arbeitsplatzkonzentration und Biologische Arbeitsstofftoleranzwerte*, XVI. Verlag Chemie, Weinheim.

39. Meyer, B. (1986). "Occupational and Indoor Air Formaldehyde Exposure: Regulations and Guidelines." ACS Symposium Series No. 316. Washington, DC, American Chemical Society, pp. 217-229.

40. Bundesgesundheitsamt. (1977). Bewertungsmasstab für Formaldehyd in der Raumluft. BGA-Pressedienst 19/77, Berlin.

41. Meyer, B. (1988). "Formaldehyde Control and Regulation in the USA." WKI Report No. 19, Fraunhofer-Institute for Wood Research, Braunschweig.

42. ETB-Richtlinie über die Verwendung von Spanplatten hinsichtlich der Vermeidung unzumutbarer Formaldehydkonzentrationen in der Raumluft—Fassung April 1980. (1980). Herausgegeben vom Ausschuss für Einheitliche Technische Baubestimmungen (ETB). Berlin, Beuth Verlag.

43. ETB-Richtlinie über die Klassifizierung von Spanplatten bezüglich der Formaldehydabgabe. (1980). Appendix to Reference 42.
44. Verband der Deutschen Holzwerstoffindustrie (Ed.). (1981). *Spanplatten und Formaldehyd: Erläuterungen zu den "Formaldehyd-Richtlinien."* Giessen.
45. Verband der Deutschen Holzwerstoffindustrie (Ed.). (1981). *Spanplatten und Formaldehyd: Anwendungstechnische Empfehlungen.* Giessen.
46. Deppe, H.-J. (1982). "Emissionen von organischen Substanzen aus Spanplatten." In *Luftqualität in Innenräumen*, K. Aurand, B. Seifert, and J. Wegner (Eds.). Stuttgart, Fischer Verlag, pp. 91–128.
47. Verordnung über gefährliche Stoffe (Gefahrstoffverordnung) vom 26. August 1986. (1986). Bundesgesetzblatt Part 1: 1470–1487.
48. Deppe, H.-J. (1985). "Holzindustrielle Fertigung und Formaldehydemission." *Holz- Kunststoffverarb.*, 20(7/8):12–15.
49. Deppe, H.-J. (1986). "Zum Stand der Erarbeitung von "Formaldehyd-Regelungen" bei Holzwerkstoffen und Möbeln." *Holz- Kunststoffverarb.*, 21(7/8):12–18.
50. Marutzky, R. (1984). Holzkebstoffe und Formaldehyd." *Adhäsion*, 28(12):10–15.
51. RAL-UZ 38. (1986). *Grundlage für Umweltzeichenvergabe—Formaldehydarme Produkte aus Holz/Holzwerkstoffen.* Berlin, Beuth Verlag.
52. Groah, W. G. (1986). "Formaldehyde Emissions: Hardwood Plywood and Certain Wood-Based Panel Products." ACS Symposium Series No. 316. Washington, DC, American Chemical Society, pp. 17–25.
53. Marutzky, R., and Roffael, E. (1977). "Über die Abspaltung von Formaldehyd bei der thermischen Behandlung von Holzspänen." *Holzforschung*, 31:8–12.
54. Pizzi, A. (1980). "Tannin-Based Adhesives." *J. Macromol. Sci. Rev. Macromol. Chem. C*, 18:247–315.
55. Pizzi, A. (1983). "Tannin-Based Wood Adhesives." In *Wood Adhesives*, Vol. I, A. Pizzi, Ed. New York, Dekker, pp. 177–246.
56. Myers, G. E. (1986). "Mechanisms of Formaldehyde Release from Bonded Wood Products." In ACS Symposium Series No. 316. Washington, DC, American Chemical Society, pp. 87–106.
57. Marutzky, R., Mehlhorn, L., and Menzel, L. (1981). "Verminderung der Formaldehyd-Emission von Möbeln." *Holz Roh- Werkst.*, 39:7–10.
58. Tinkelenberg, A., and Versloot, D. (1982). "Measurement of Formaldehyde Emissions from Paint Films." *Proceedings of the 16th Fatipec Congress*, Liege, pp. 131–153.

59. Marutzky, R., and Flentge, A. (1985). "Neuere Erkenntnisse zur Formaldehydabgabe von Möbeln." *Holz Kunststoffverarb.*, 20(1):38-45.

60. Wittmann, O. (1985). "Formaldehydarme Flächenverleimung mit Aminoplastharzen." *Holz Roh- Werkst.*, 43:187-191.

61. Scheithauer, M., Boehme, P., Kehr, E., Riehl, G., and Rinkefeil, R. (1985). "Formaldehydabgabe oberflächenbeschichteter Bauteile für Möbel." *Holztechnologie*, 26:188-192.

62. Boehme, P., Dammer, S., (1986). "Umweltfreundliche Dekorfolien." *Holz- Kunstoffverarb.*, 21(12):26-29.

63. Marutzky, R. (1987). "Prüfmethoden als Grundlagen der Formaldehydabgabeminderung bei Möbeln." *Holztechnologie*, 28:301-304.

64. Marutzky, R., Pittkaenen, H., and Flentge, A. (1988). "Zur Formaldehydabgabe von Parkett und Parkettsiegeln." *Holz Kunststoffverarb.*, 23(1):86-88.

65. Dix, B. (1987). "Untersuchungen zur Verleimung von Furniersperrholz mit modifizierten Diisocyanat-Klebstoffen." *Holz Roh-Werkst.*, 45:487-494.

66. Marutzky, R., and Dix, B. (1985). "Formaldehyde Release of Tannin-Bonded Particleboards." *J. Appl. Polym. Sci., Appl. Polym. Symp.*, 40:49-57.

67. De Jong, J. I., and De Jonge, J. (1952). "The Reaction of Urea with Formaldehyde." *Rec. Trav. Chim. Pays-Bas*, 71:643-667.

68. De Jong, J. I. (1953). "A Determination of Methylol Groups in Condensates of Urea and Formaldehyde." *Rec. Trav. Chim. Pays-Bas*, 72:653-654.

69. Pfeil, E., and Schroth, G. (1953). "Mercurimetrische Formaldehydbestimmung." *Z. Anal. Chem.*, 134:333-334.

70. Käsbauer, F., Merkel, D., and Wittmann, O. (1976). "Bestimmung von freiem Formaldehyd und N-Methylol-Formaldehyd in Harnstoff-Formaldehyd-Kondensaten. *Z. Anal. Chem.*, 281:17-21.

71. Roffael, E., and Schriever, E. (1985). "Formaldehydabgabe von definierten UF-Präpolymeren nach der WKI- und der Perforator-Methods." *Holz Roh- Werkst.*, 43:110.

72. DIN 16 916 Part 2: Kunststoffe-Reaktionsharze-Phenolharze-Prüfverfahren. Berlin, Beuth Verlag.

73. Morath, J. C., and Woods, J. T. (1958). "Analysis of Amino-Formaldehyde Resins." *Anal. Chem.*, 30:1437-1440.

74. Meyer, B. (1979). *Urea-Formaldehyde Resins.* Reading, MA, Addison-Wesley.

75. Freeman, H. G., and Kreibich, R. E. (1968). "Estimating Durability of Wood Adhesives in Vitro." *Forest Prod. J.*, 18(7):39-43.

76. Troghton, G. E. (1969). "Accelerated Aging of Glue-Wood Bonds." *Wood Sci.*, 1:172-176.

77. Myers, G. E., and Nagaoka, M. (1983). "Hydrolytic Stability of Cured UF Resins." *Wood Sci.*, 15:127-138.

78. Marutzky, R., Roffael, E., and Ranta, L., "Untersuchungen über den Zusammenhang zwischen dem Molverhältnis und der Formaldehydabgabe bei Harnstoff-Formaldehyd-Leimharzen." *Holz Roh- Werkst.*, 37:303-307.

79. Marutzky, R., and Ranta, L. (1979). "Die Eigenschaften formaldehydarmer HF-Leimharze und daraus hergestellter Holzspanplatten. 1. Einfluss des Molverhältnisses auf die Eigenschaften der Rohleimharze." *Holz Roh- Werkst.*, 37:389-393.

80. Myers, G. E., and Koutsky, J. A. (1987). "Procedure for Measuring Formaldehyde Liberation from Formaldehyde-based Resins." *Forest Prod. J.*, 37(9):56-60.

81. Pizzi, A. (Ed.). (1973). *Wood Adhesives—Chemistry and Technology*. New York, Dekker.

82. Dankelman, W., and De Wit, J. (1976). "Moderne Methoden zur Analyse von Harnstoff-Formaldehyd-Harzen." *Holz Roh-Werkst.*, 34:131-134.

83. Kumlin, K., and Simonson, R. (1978). "Urea-Formaldehyde Resins. 1. Separation of Low Molecular Weight Components in Urea-Formaldehyde Resins by Means of Liquid Chromatography." *Angew. Makromol. Chem.*, 68:175-184.

84. Matthews, T. G., and Howell, T. C. (19). "Solid Sorbent for Formaldehyde Monitoring." *Anal. Chem.*, 54:1495-1498.

85. National Particleboard Association (1985). Formaldehyde Test Method 2: Large-Scale Test Method for Determining Formaldehyde Emissions from Wood Products—Large-Chamber Method FTM2—1985. Gaithersburg, MD.

86. VDI-Richtlinie 3484: Messen von Aldehyden—Bestimmen der Formaldehyd-Konzentration nach dem Sulfit-Pararosanilin-Verfahren. Düsseldorf, VDI-Verlag, 1979.

87. Menzel, W., Marutzky, R., and Mehlhorn, L. (1981). "Formaldehyd-Messmethoden." WKI Report No. 13. Braunschweig, Fraunhofer-Institute for Wood Research.

88. Henschler, D. (Ed.). (1976). *Luftanalysen*, Vol. 1. Weinheim, Verlag Chemie.

89. Flentge, A. (1988). "Einfache Messmethoden für Formaldehyd." WKI Report No. 19. Braunschweig, Fraunhofer-Institute for Wood Research.

90. Marutzky, R. (1986). "Formaldehyd in Innenräumen: Zur Problematik von Beschwerdefällen." *Gesundheits-Ingenieur*, 107:327-334.

91. Prescher, K. E., and Schöndube, M. (1983). "Die Bestimmung von Formaldehyd in Innen- und Aussenluft mit Passivsammlern." *Gesundheits-Ingenieur*, 104:198-200.

92. Bisgaard, P., Molhave, L., Rietz, B., and Wilhardt, P. (1984). "A Method for Personal Sampling and Analysis of Nanogram Amounts of Formaldehyde in Air." *Am. Ind. Hyg. Assoc. J.*, 45:425-429.

93. Prescher, K. E., and Seifert, B. (1988). "Erfahrungen bei der Formaldehydbestimmung in Innenräumen." WKI Report No. 19. Braunschweig, Fraunhofer-Institute for Wood Research.
94. Sawicki, E., and Sawicki, C. R. (1965). *Aldehydes—Photometric Analysis*, Vol. 1. London, Academic Press.
95. Johansson, E., and Sundin, B. (1982). "Verfahren zur Bestimmung von Formaldehyd in der Raumluft." *Proceedings of the FESYP Conference*, Giessen, pp. 17-38.
96. Gollob, L., and Wellons, J. D. (1980). "Analytical Methods for Formaldehyde—A Review." *Forest Prod. J.*, 30(6):27-34.
97. Baumbach, G. (1983). "Messverfahren für Aldehyd-Emissionen in Verbrennungsabgasen." *Staub- Reinhaltung Luft*, 43:95-101.
98. Godish, T. (1985). "Residential Formaldehyde Sampling—Current and Recommended Practices." *Am. Ind. Hyg. Assoc. J.*, 46: 105-110.
99. Clermont, L. P., and Carroll, M. N. (1976). "Improved Chromotropic Acid Determination of Formaldehyde Evolution from Wood Composites." *Forest Prod. J.*, 26(8):35-37.
100. Schlüter, E. (1985). "Messung von Formaldehyd in der Raumluft." *Gesundheits-Ingenieur*, 106:200-204.
101. van der Waal, J. F., Korf, C., Kuypers, A. T. J. M., and Neele, J. (1988). "Interference by Chemicals in the Determination of Formaldehyde." *Environ. Int.*, in press.
102. Bisgaard, P., Molhave, L., Rietz, B., and Wilhardt, P. (1983). "Quantitative Determination of Formaldehyde in Air Using the Acetylacetone Method." *Anal. Lett.*, 16:1457-1468.
103. DIN 55666: Prüfung von Formaldehyd emittierenden Werkstoffen-Bestimmung der Ausgleichskonzentration au Formaldehyd in einem Kleinen Prüfraum (Norm-Entwurf). Berlin, Beuth Verlag, 1988.
104. Belman, S. (1963). "The Fluorimetric Determination of Formaldehyde." *Anal. Chim. Acta*, 29:120-126.
105. Kuijpers, A. T. J. M., and Neele, J. (1983). "Reagent Stability in the Modified Pararosaniline Method for the Determination of Formaldehyde." *Anal. Chem.*, 55:390-391.
106. Lahmann, E., and Jander, K. (1968). "Formaldehyd-Bestimmungen in Strassenluft." *Gesundheits-Ingenieur*, 89:18-21.
107. Rayner, A. C., and Jephcott, C. M. (1961). "Microdetermination of Formaldehyde in Air." *Anal. Chem.*, 33:627-630.
108. Lyles, G. R., Dowling, F. B., and Blanchard, V. J. (1965). "Quantitative Determination of Formaldehyde in the Parts per Hundred Million Concentration Level." *J. Air Pollut. Control Assoc.*, 15:106-108.
109. Ahonen, I., Priha, E., and Äijälä, M.-L. (1984). "Specificity of Analytical Methods to Determine the Concentration of Formaldehyde in Workroom Air." *Chemosphere*, 13:521-525.

110. Lowe, D. C., Schmidt, U., Ehhalt, D. E., Frischkorn, C. G. B., and Nürnberg, H. W. (19). "Determination of Formaldehyde in Clean Air." *Environ. Sci. Technol.*, 15:819-823.

111. Nederlands Norm NEN 2795; Bepaling van de concentratie aan methanal (formaldehyde)—Fotometrische methode met pararosaniline. Nederlands Normalisatie-instituut, Delft, 1985.

112. Tuss, H., Neizert, V., Seiler, W., and Neeb, R. (1982). "Method for Determination of Formaldehyde in the ppt-Range by HPLC After Extraction as 2,4-Dinitrophenyl-Hydrazone." *Fresenius Z. Anal. Chem.*, 312:613-617.

113. Wentrup, G. J., and Wenzel, M. (1988). "Messtechnische Erfassung von Formaldehyd nach der DNPH-Methode am Arbeitsplatz mit Beispielen aus der Praxis." WKI Report No. 19. Braunschweig, Fraunhofer-Institute for Wood Research.

114. Sawicki, E., Hauser, T. R., Stanley, T. W., Elbert, W. (1961). "The 3-Methyl-2-benzothiazolone Hydrazone Test." *Anal. Chem.*, 33:93-95.

115. VDI-Richtlinie 3862 (Vorentwurf): Messen aliphatischer Aldehyde nach dem MBTH-Verfahren. Düsseldorf, VDI-Verlag, 1987.

116. Compton, B. J., and Purdy, W. C. (1980). "Fluoral-P, a Member of a Selective Family of Reagents for Aldehydes." *Anal. Chim. Acta*, 119:349-357.

117. Bombaugh, K. J., and Bull, W. C. (1982). "Gas of Formaldehyde in Solution and High Purity Gas." *Anal. Chem.*, 34:1237-1239.

118. Stahovec, W. L., and Mopper, K. (1984). "Trace Analysis of Aldehydes by Pre-Column Fluorogenic Labeling with 1,3-Cyclohexanedione and Reversed-Phase High-Performance Liquid Chromatography." *J. Chromatogr.*, 298:399-406.

119. Dumas, T. (1982). Determination of Formaldehyde in Air by Gas Chromatography." *J. Chromatogr.*, 19:289-295.

120. Le Botlan, D. J., Mechin, B. G., and Martin, G. J. (1983). "Proton and Carbon-13 Nuclear Magnetic Resonance Spectrometry of Formaldehyde in Water." *Anal. Chem.*, 55:587-591.

121. Bombaugh, K. J., and Bull, W. C. (1962). "Gas Chromatographic Determination of Formaldehyde in Solution and High Purity Gas." *Anal. Chem.*, 34:1237-1241.

122. Marutzky, R., and Heindl, H. D. (1977). "Formaldehydbestimmung mit Hilfe einer enzymatischen Methode." WKI Short Report No. 1/77. Braunschweig, Fraunhofer-Institute for Wood Research.

123. Guilbault, G. G. (1983). "Determination of Formaldehyde with an Enzyme-Coated Piezoelectric Detector." *Anal. Chem.*, 55:1682-1684.

124. Ho, M. H., and Weng, J. L. (1986). "Enzymatic Methods for Determining Formaldehyde Release from Wood Products." ACS Symposium Series No. 316. Washington, DC, American Chemical Society, pp. 116-124.

125. Andersen, I., Lundqvist, G. R., and Molhave, L. (1974). "Formaldehydafgivelse fra spanplader—En matematisk model." *Ugeskr. Laeg.*, 136:2145-2150.
126. Hanetho, P. (1979). "The Influence of Climate Parameters on the Release of Formaldehyde from Particleboard in a Climate Chamber." Paper presented at the joint DSM/MCN/CASCO/DYNO meeting in Stockholm.
127. Berge, A., Mellegaard, B., Hanetho, P., Ormstad, E. B. (1980). "Formaldehyde Release from Particleboard—Evaluation of a Mathematical Model." *Holz Roh- Werkst.*, 38:251-255.
128. Molhave, L. (1982). "The Model Room Method for Measurement of Formaldehyde Emission from Particle Boards." *Holzforsch. Holzverwert.*, 34:24-27.
129. Molhave, L., Bisgaard, P., and Dueholm, S. (1983). "A Mathematical Model of Indoor Air Pollution Due to Formaldehyde from Urea-Formaldehyde Glued Particleboards." *Atmos. Environ.*, 17:2105-2108.
130. Mehlhorn, L. (1986). "Normierungsverfahren für die Formaldehydabgabe von Spanplatten." *Adhäsion*, 29(6):27-33.
131. Gustafsson, H. N. O., "Measurement of Formaldehyde Release from Building Materials in a Ventilated Test Chamber." ACS Symposium Series No. 316. Washington, DC, American Chemical Society, pp. 145-153.
132. Hoetjer, J. J., and Koerts, F. (1986). "Experiences with a Model for the Formaldehyde Release of Particle Boards. ACS Symposium Series No. 316. Washington, DC, American Chemical Society, pp. 125-144.
133. Lehmann, W. F. (1987). "Effect of Ventilation and Loading Rates in Large-Chamber Testing of Formaldehyde Emissions from Composite Panels." *Forest Prod. J.*, 37(4):31-37.
134. Sundin, B. (1987). "Formaldehyde Source Interaction Studies. *Proceedings of FESYP Technical Conference*, Giessen, pp. 30-47.
135. Andersen, I., Lundqvist, G. R., and Molhave, L. (1974). "Formaldehydafspaltning fra spanplader i klimakammer. " *Ugeskr. Laeg.*, 136:2140-2145.
136. Myers, G. E., and Nagaoka, M. (1981). "Emission of Formaldehyde by Particleboard: Effect of Ventilation Rate and Loading on Air-Contamination Levels." *Forest Prod. J.*, 31(7):39-44.
137. Matthews, T. G., Reed, T. J., Daffron, C. R., and Hawthorne, A. R. (1984). "Environmental Dependence of Formaldehyde Emission from Pressed-Wood Products." *Proceedings of the 3rd International Conference on Indoor Air Quality and Climate*, Stockholm, Vol. 3, pp. 121-126.
138. Myers, G. E. (1981). "Formaldehyde Emission: Methods of Measurement and Effects of Several Particleboard Variables." *Wood Sci.*, 13:140-150.

139. Myers, G. E. (1984). "Effect of Ventilation Rate and Board Loading on Formaldehyde Concentration: A Critical Review of the Literature." *Forest Prod. J.*, 34(10):59-68.

140. Gustafsson, H. N. O. (1987). "Proposal of a Euro Chamber for Testing Formaldehyde and Other Indoor Air Pollutants." *Proceedings of the 4th International Conference on Indoor Air Quality and Climate*, Berlin, Vol. 1, pp. 45-53.

141. Gustafsson, H. N. O. (1988). "Vorschlag zu einem Europäischen Referenzprüfraum für Emissionprüfungen." WKI Report No. 19. Braunschweig, Fraunhofer-Institute for Wood Research.

142. Merker, O. (Ed.). (1985). "Formaldehydemission von Möbeln— Möglichkeiten der Prüfung in der DDR." Dresden, Zentrale Prüfstelle für Möbel im WTZ.

143. Dueholm, S. (1985). "Formaldehydregler for byggeri og moebler— Historie, erfasringer og udviklingstendenser." Limspecialisten 1. Kemi Casco, Denmark.

144. Marutzky, R. (1988). "Zur Herstellung formaldehydarmer Möbel aus furnierten Holzwerkstoffen." *HK Int. - Holz- Möbelind.*, 23:272-276.

145. Marutzky, R., Mehlhorn, L., Roffael, E., and Flentge, A. (1988). "Prüfverfahren nach Gefahrstoffverordnung als Grundlage für formaldehydarme Holzwerkstoffe." *Holz Roh- Werkst.*, 46:253-258.

146. Kazakevics, A. A. R., and Spedding, D. J. (1979). "The Rate of Formaldehyde Emission from Chipboard." *Holzforschung*, 33:155-158.

147. Marutzky, R. A. (1988). "Die Prufraum-Methode als Referenzverfahren nach der Gefahrstoffverordnung." WKI Report No. 19. Braunschweig, Fraunhofer-Institute for Wood Research.

148. Gustafsson, H. N. O. (1984). "A Test Method for Determination of Pollutants in Indoor Air Due to Building Materials." *Proceedings of the 3rd International Conference on Indoor Air Quality and Climate*, Stockholm, Vol. 3, 81-83.

149. Gustafsson, H. N. O., Isaksson, I., and Muameleci, E. (1985). "Formaldehyde Till Inomhusluft." Teknisk Rapport No. 29, Statens Proveningsanstalt, Boras, Sweden.

150. Nordtest Draft: Nordtest Method 481-84: Building Materials— Emission of Gases and Vapors. Danish Building Research Institute, Division of Building Physics, Horsholm, 1986.

151. Stetter, K., Ackva, W., and Tröger, J. (1986). "Bestimmung der Formaldehydabgabe von Holzwerkstoffen und Möbeln in einer 1 m^3 Kammer." *Holz-Zentralbl.*, 112:2024.

152. Marutzky, R., Flentge, A., and Mehlhorn, L. (1987). "Zur Messung der Formaldehydabgabe von Holzwerkstoffen, Baustoffen und Möbeln mittels der 1 m^3-Kammer-Methode." *Holz Roh-Werkst.*, 45:1-5.

153. Jann, O., Deppe, H.-J. (1988). "Formaldehydabgabemessungen mit einer 1 m³-Kammer." WKI Report No. 19. Braunschweig, Fraunhofer-Institute for Wood Research.

154. DIN EN 120: Spanplatten—Bestimmung des Formaldehydgehalts—Extraktionsverfahren genannt Perforatormethode. Berlin, Beuth Verlag, 1984.

155. Roffael, E., Greubel, D., and Mehlhorn, L. (1978). "Uber die Bestimmung der Formaldehydabgabe von Spanplatten nach dem Perforator—Verfahren und nach der WKI-Methode." *Holz-Zentralbl.*, 104:396-397.

156. Roffael, E., and Mehlhorn, L. (1980). "Einfluss der Randbedingungen bei der Bestimmung des extrahierbaren Formaldehyds in Holzspanplatten nach der Perforatormethode." *Holz Roh- Werkst.*, 38:85-88.

157. Sundin, B. (1984). "The Perforator Method—Sources of Error and Background Values." Paper presented at Dynobel's Particleboard Meeting at Roeros, Norway.

158. Sundin, B. (1986). "The E1 Norm—Analytical Problems and Other Consequences for the Particleboard Industry." Paper presented at the Dynobel Particleboard Symposium in Sundsvall, Sweden.

159. Sundin, B., Mansson, B., and Endrody, E. (1987). "Particleboard with Different Contents of Releasable Formaldehyde. A Comparison on the Board Properties, Including Results from Different Formaldehyde Tests." *Proceedings of the 21st Particleboard Symposium*, Washington State University, Pullman, pp. 139-186.

160. Romeis, M. (1988). "Influence of Time and Storage on Perforators." WKI Report No. 19. Braunschweig, Fraunhofer-Institute for Wood Research.

161. National Particleboard Association. (1983). Formaldehyde Test Method 1: Small-Scale Test Method for Determining Formaldehyde Emissions from Wood Products—Two-Hour Desiccator Test FTM 1—1983. Gaitersburg, MD.

162. Rybicky, J., Horst, K., and Kambanis, S. M. (1985). "Assessment of the 2-Hour Desiccator Test for Formaldehyde Release from Particleboard." *Forest Prod. J.*, 33(9):50-54.

163. Rybicky, J. (1985). "Kinetics for Desiccator, Jar and Alike Tests for Formaldehyde Release from Particleboards." *Wood and Fiber Sci.*, 17(1):29-35.

164. Roffael, E. (1975). "Messung der Formaldehydabgabe—Praxisnahe Methode zur Ermittlung der Formaldehydabgabe harnstoffharzgebundener Spanplatten für das Bauwesen." *Holz-Zentralbl.*, 101:1403-1404.

165. Roffael, E., and Mehlhorn, L. (1976). "Erfahrungen mit einer einfachen Methode zur Bestimmung der Formaldehydabgabe von Spanplatten." *Holz-Zentralbl.*, 102:2202.

166. Roffael, E. (1980). "Progress in the Elimination of Formaldehyde Liberation from Particleboards." *Proceedings of the 14th Washington State University Symposium on Particleboard*, Pullman, pp. 19-31.

167. WIHS 313: Bestimmung des Gehaltes an abgebbarem Formaldehyd—Hermetikbehälter-Methode. VEB WTZ der holzverarbeitenden Industrie, Dresden, 1985.

168. Scheithauer, M., Böhme, P., Kehr, E., Riehl, G., and Rinkefeil, R. (1985). "Formaldehydabgabe oberflächenbeschichteter Bauteile für Möbel." *Holztechnologie*, 26:188-190.

169. Böhme, P., and Merker, O. (1988). "Schadstoffemissionen von Dekorfolien und Dekorfoliebeschichteten Holzwerkstoffen." WKI Report No. 19. Braunschweig, Fraunhofer Institute for Wood Research.

170. Roffael, E. (1988). "Bestimmung der Formaldehydabgabe von Holzspanplatten und anderen Werkstoffen nach der WKI-Flaschen-Methode und hiervon abgeleiteten Verfahren." WKI Report No. 19. Braunschweig, Fraunhofer-Institute for Wood Research.

171. Sundin, B. (1988). "Can the Perforator Method Be Replaced? Presentation of an Alternative Method." WKI Report No. 19. Braunschweig, Fraunhofer-Institute for Wood Research.

172. Roffael, E., and Mehlhorn, L. (1977). "Methoden zur Beurteilung der Formaldehydabgabe von Spanplatten." *Holz- Kunststoffverarb.*, 12(10):770-777.

173. DIN 52 368: Prüfung von Spanplatten—Bestimmung der Formaldehydabgabe durch Gasanalyse. Berlin, Beuth Verlag, 1984.

174. Hoetjer, J. J., and Koerts, F. (1981). "Verfahren zur Bestimmung der Formaldehydabgabe aus Spanplatten unter Berücksichtigung der Raumluft-Konzentration." *Holz Roh- Werkst.*, 39:391-393.

175. Koerts, F., and Koster, R. (1988). "Erfahrungen mit dem Mikrokammer-Schnelltest." WKI Report No. 19. Braunschweig, Fraunhofer-Institute for Wood Research.

176. Klug, L. (1988). "Bestimmung der Formaldehyd aus Gleichskonzentration, die sich durch vorhandene Holzwerkstoffe, Lacke, und sonstige Formaldehyd-emittierende Werkstoffe bei Raumtemperatur einstellt." WKI Report No. 19. Braunschweig, Fraunhofer-Institute for Wood Research.

177. Plath, L. (1966). "Bestimmung der Formaldehydabspaltung aus Spanplatten nach der Mikrodiffusionsmethode." *Holz Roh-Werkst.*, 24:312-318.

178. Kratzl, K., and Silbernagel, H. (1968). "Eine Mikromethode zur Bestimmung des Formaldehydgehaltes von Harnstoffharzen in gehärtetem und ungehärtetem Zustand." *Holzforsch. Holzverwert.*, 20:131-132.

179. Silbernagel, H., and Kratzl, K. (1968). "Zur quantitativen Bestimmung der Formaldehydabgabe von Spanplatten." *Holzforsch. Holzverwert.*, 20:113-116.

180. Mehlhorn, L., Roffael, E., and Miertzsch, H. (1978). "Erfahrungen mit den vom FIHH-Karlsruhe vorgeschlagenen Prüfmethoden zur Bestimmung des Formaldehyds." *Holz-Zentralbl.*, 104:345-346.

181. Mohl, H. R. (1978). "Saug- und Spaltmethode zur Bestimmung der Formaldehydabgabe von Holzwerkstoffen und Leimen sowie zur allgemeinen Luftanalyse." *Holz Roh- Werkst.*, 36:69-75.

182. Christensen, R., Robitschek, P., and Stone, J. (1981). "Formaldehyde Emission from Particleboard." *Holz Roh- Werkst.*, 39:231-234.

183. Matthews, T. G., Hawthorne, A. R., Daffron, C. R., Corey, M. D., Reed, T. J., and Schrimsher, J. M. (1984). "Formaldehyde Surface Emission Monitor." *Anal. Chem.*, 56:448-454.

184. Hall, G. (1987). "Presentation of Different National Test Methods and Possibilities of Simplification with Special References to Formaldehyde." *Proceedings of FESYP Technical Conference*, Giessen, pp. 30-37.

185. Myers, G. E. (1982). "Formaldehyde Dynamic Air Contamination by Hardwood Plywood: Effects of Several Variables and Board Treatments." *Forest Prod. J.*, 32(4):20-25.

186. Grigoriou, A. (1987). "Formaldehydabgabe aus den Schmal- und Breitflächen verschiedener Holzwerkstoffe." *Holz Roh-Werkst.*, 45:63-67.

187. Matthews, T. G., Hawthorne, A. R., Daffron, C. R., Reed, T. J., and Corey, M. D. (1983). "Formaldehyde Release from Pressed-Wood Products." *Proceedings 17th Particleboard Symposium*, Washington State University, Pullman, pp. 179-202.

188. Plath, L. (1967). "Bestimmung der Formaldehydabspaltung aus Spanplatten nach der Mikrodiffusionsmethode—Zweite Mitteilung: Einfluss von Presszeit und Presstemperatur auf die Formaldehyd-Abspaltung." *Holz Roh- Werkst.*, 25:63-68.

189. Plath, L. (1967). "Bestimmung der Formaldehydabspaltung aus Spanplatten nach der Mikrodiffusionsmethode—Dritte Mitteilung: Einfluss der Härterzusammensetzung auf die Formaldehyd-Abspaltung." *Holz Roh- Werkst.*, 25:169-173.

190. Plath, L. (1967). "Bestimmung der Formaldehydabspaltung aus Spanplatten nach der Mikrodiffusionsmethode—Vierte Mitteilung: Einfluss der Feuchtigkeit im Spanvlies auf die Formaldehyd-Abspaltung." *Holz Roh- Werkst.*, 25:231-238.

191. Petersen, H., Reuther, W., Eisele, W., and Wittmann, O.
     (1972). "Zur Formaldehyd-Abspaltung bei der Spanplatten-
     erzeugung mit Harnstoff-Formaldehyd-Bindemitteln: Part I."
     Holz Roh- Werkst., 30:429-436.
192. Petersen, H., Reuther, W., Eisele, W., and Wittmann, O.
     (1973). "Zur Formaldehyd-Abspaltung bei der Spanplatten-
     erzeugung mit Harnstoff-Formaldehyd-Bindemitteln: Part II.
     Holz Roh- Werkst., 31:463-469.
193. Petersen, H., Reuther, W., Eisele, W., and Wittmann, O.
     (1974). "Zur Formaldehyd-Abspaltung bei der Spanplatten-
     erzeugung mit Harnstoff-Formaldehyd-Bindemitteln: Part III."
     Holz Roh- Werkst., 32:402-410.
194. Roffael, E., Greubel, D., and Mehlhorn, L. (1980). "Verfahren
     zur Herstellung von Holzspanplatten mit niedrigem Formalde-
     hydabgabepotential." *Adhäsion*, 24(4):92-94.
195. Marutzky, R., and Ranta, L. (1980). "Die Eigenschaften
     formaldehydarmer HF-Leimharze und daraus hergestellter
     Holzspanplatten. 1. Mitteilung: Einfluss des Molverhältnisses
     auf die Eigenschaften der Holzspanplatten." *Holz Roh- Werkst.*,
     38:217-223.
196. Roffael, E., Miertsch, H., and Menzel, W. (1982). "Nach-
     trägliche Behandlung von Spanplatten zur Verminderung ihres
     Formaldehydabgabepotentials." *Adhäsion*, 26(3):18-23.
197. Ernst, K. (1982). "Die verschiedenen Herstellverfahren für
     Spanplatten der Emissionsklasse E1." *Holz Roh- Werkst.*, 40:249-
     253.
198. Wittmann, O. (1983). "Herstellung von Spanplatten mit
     verstärkten Aminoplast-Leimharzen." *Holz Roh- Werkst.*, 41:
     431-435.
199. Westling, A. "Ein neues Verfahren zur Verminderung der
     Formaldehydabgabe von Holzspanplatten und anderen Holz-
     werkstoffen." *Holz-Zentralbl.*, 108:1802-1803.
200. Myers, G. E. (1984). "How Mole Ratio of UF Resin Affects
     Formaldehyde Emission and Other Properties: A Literature
     Critique." *Forest Prod. J.*, 34(5):35-41.
201. Matthews, T. G., Reed, T. J., Tromber, B. J., Daffron,
     C. R., and Hawthorne, A. R. (1984). "Formaldehyde Emissions
     from Consumer and Construction Products: Potential Impact
     on Indoor Formaldehyde Concentrations." *Proceedings of the
     3rd International Conference on Indoor Air Quality and Climate*,
     Stockholm, Vol. 3, pp. 115-120.
202. Meyer, B., Hermanns, K., and Smith, D. C. (1984). "Formalde-
     hyde Release from Urea-Formaldehyde Bonded Wood Products."
     J. Adhes., 17:297-308.
203. Riehl, G. (1985). "Zur Herstellung von Spanplatten mit formalde-
     hydarmen Harnstoffharzen." *Holztechnologie*, 26:206-209.

204. Wittmann, O. (1985). "Formaldehydarme Flächenverleimung mit Aminoplastharzen." *Holz Roh- Werkst.*, 43:187-191.
205. Dunky, M. (1985). "Harnstoff-Formaldehyd-Leime: Einfluss des Molverhältnisses auf die Eigenschaften der Leime und die erreichbaren Bindefestigkeiten bei Sperrholzverleimungen." *Holzforsch. Holzverwert.*, 37:75-82.
206. Myers, G. E. (1986). "Effects of Post-Manufacture Board Treatments on Formaldehyde Emission: A Literature Review (1960-1984)." *Forest Prod. J.*, 36(6):41-50.
207. Meyer, B., and Hermanns, K. (1986). "Formaldehyde Release from Wood Products—An Overview." ACS Symposium Series No. 316. Washington, DC, American Chemical Society, pp. 1-16.
208. Merker, O., Boehme, P., and Kehr, E. (1987). "Anmerkungen zur Senkung der Formaldehydemission bei Spanplatten." *Holztechnologie*, 28:4-6.
209. Calve, L. R., and Brunette, G. G. (1984). "Reducing Formaldehyde Emission from Particleboard with Urea Salts or Sulfite Liquor." *Adhes. Age*, 27/8:39-43.
210. Groah, W. J., Gramp, G. D., and Trant, M. (1984). "Effect of a Decorative Vinyl Overlay on Formaldehyde Emissions." *Forest Prod. J.*, 34(49):27-29.
211. von Mildenstein, M. (1966). "Formaldehydbindende Anstriche für Spanplatten." *Holz-Zentralbl.*, 91:758.
212. Kubitzky, C. (1972). "Vermeidung von Geruchsbelästigung bei Spanplatten." *Industrie-Anzeiger* No. 53.
213. Roffael, E. (1976). "Einfluss des Formaldehydgehaltes in Harnstoffharzen auf ihre Reaktivität und die Formaldehydabgabe damit gebundener Spanplatten." *Holz Roh- Werkst.*, 34:385-390.
214. Roffael, E. (1976). "Über die Reaktivität von wässrigen Rindenextrakten gegenüber Formaldehyd." *Adhäsion*, 20(11):306-311.
215. Cameron, F. A., and Pizzi, A. (1985). "Tannin-Induced Formaldehyde Release Depression in Urea-Formaldehyde Particleboard." ACS Symposium Series No. 316, Washington, DC, American Chemical Society, pp. 198-201.

11

Expanding Resorcinol Cold-Sets for Gap-Filling Adhesives for Wood

GÉRARD ELBEZ / *Centre Technique du Bois et de L'Ameublement Paris, France*

I. INTRODUCTION

Increasing prices of wood-based semifinished products indicate that in many cases the cost of timber as a raw material has considerable influence on the cost of the semifinished product (often more than half). Consequently any technique aimed at improving the utilization of such a raw material will have a fundamental effect on the final cost of the product. An analysis of manufacturing done in the industrial glulam sector in France has shown that the resale price of timber can reach up to 70% of the cost of the product and that the losses of timber due to the planing of members constituting a laminated beam represent between 13 and 20% of the raw material used. These relative losses are higher when timber of lower thickness is used. Utilization of local timber means in many cases that panels of lower thickness must be used and also that the production of timber panels must be followed by planing to eliminate defects and uneven surfaces. All this is necessary (a) to improve the yield of usable timber from small diameter logs, and (b) to minimize the influence of both deformations and defects on the timber panels produced. It is logical to assume, then, that any technology allowing reduction in planing losses is seen as particularly beneficial to the industry.

The study presented in this chapter was based on timber that was only partially planed. The key needs that lead to the developments outlined were:

The possibility of obtaining glulam made from members of variable thickness with glue spreads comparable to those used for normally planed timber

A comparable production rate
Acceptable manufacturing costs
Acceptable raw materials inventories

Such needs were realized by preparing an expanding resorcinol-based phenolic [phenol-resorcinol-formaldehyde or resorcinol-formaldehyde (PRF or RF)] adhesive that comprises not only the basic resin but also

Blowing agents, such as methane- or ethane-derived fluorocarbons, or their azeotropes, not presenting flammability
or toxicity risks
An amount of blowing agent allowing control of the rate of
adhesive expansion—for instance, the initial glue-mix doubles
in volume with 5% of blowing agent added at the exothermic
increase in temperature of 15°C of the PRF adhesive
A gelling agent able to contain the glue-mix expansion, hence
capable of distributing only where necessary the foam cells
obtained within the glue line and the mass of the adhesive,
without weakening the mechanical characteristics of the joints
A wetting agent to improve glue transfer on unplaned timber

In earlier formulations the "in situ" adhesive expansion necessitated an increase in temperature of about 15°C. Under these conditions satisfactory results were obtained [1,2]. This however meant that under industrial conditions the temperature of the glue lines had to reach nearly 35°C for the system to be effective. Only radiofrequency generators, which allow direct and rapid heating of the glue lines, would then allow the use of such an expanding adhesive. Because, however, radiofrequency generators could be used for less than 50% of glulam manufactured by the French industry, the use of such expanding adhesives would be somewhat limited to straight beams.

II. DEVELOPMENT OF LOW TEMPERATURE FORMULATIONS

Although at present the local timber industry is again becoming interested, for productivity reasons, in radiofrequency curing, it appeared wiser instead to develop formulations based on the initial exothermic condensation reactions within the adhesive itself to achieve total or partial vaporization of the blowing agent [3-5]. To obtain safety of usage, trichlorofluoromethane (CCl_3F) was used as the blowing agent, rather than one of its azeotropes having a lower boiling point. Blowing agents other than fluorocarbons can also be used. Only one commercial PRF or RF adhesive having an initial

TIME	0	30"	1'	1'30	4'30	7'30	10'
TEMPERATURE	21°	22°	22.5°	23°	24°	24.5°	25°

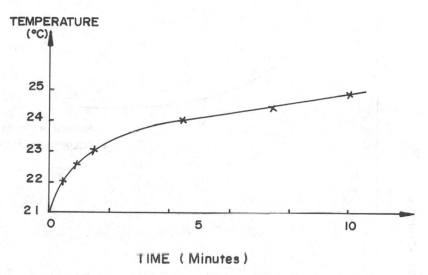

Fig. 11.1 Cascophen RL7051 (DL 7050 hardener): temperature increase as a function of time.

exothermic effect apt for this use was found in France (Cascophen RL7051 with DL7050 hardener). This resin is generally used in the proportions of 1 part liquid hardener to 4 parts (by weight) of liquid resin. Figure 11.1 shows the results obtained on verifying the initial exothermic characteristics of the adhesive after addition of hardener.

Several studies (3,6-10] have shown that partial planing of 38-mm material to only 37 mm thickness can occasionally lead, on local spruce or fir planks of greater width (220 mm), to planing defects up to 1 mm deep and approximately along one-third of the width of the plank. To evaluate the expanding adhesive formulations, a cavity measuring 60 mm × 90 mm × 1 mm and covered by a planed plank was used. The object of the exercise was to obtain a glue line 1 mm thick for a glue spread of 500 g/m^2 and a curing temperature between 20 and 25°C. Several tests were carried out under different conditions and addition of different surfactants to the basic hardener, but the final optimized formula that was finally used for industrial experiments was as follows:

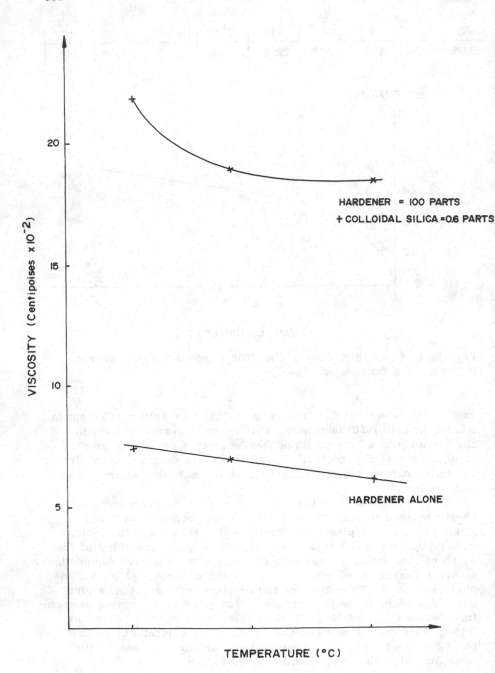

Fig. 11.2 Variation of viscosity of the hardener components as a function of temperature.

Ingredient	Parts by mass
Liquid resorcinol-based resin (Cascophen RL7051)	100
Liquid hardener (DL7050)	25
Colloidal silica	0.15
Surfactant (Zanyl)	1
Trichlorofluoromethane (CCl_3F)	5

The role of the colloidal silica is to retain the microbubbles of trichlorofluoromethane as well as to render the hardener thixotropic. The surfactant improves the incorporation of the blowing agent in the adhesive glue mix. The preapplication temperature of the glue-mix constituents was kept at 16°C or lower, and the timber used was conditioned to 20°C (Fig. 11.2). During gluing the temperature of the glue mix was 18°C or lower while the joint was maintained, after pressure was applied, in a climatic chamber at a temperature between 20 and 25°C. This formulation allowed the filling of the gap that remained after cold pressing and corresponded to approximately 40% more of the initial volume of the glue mix.

The gap filling obtained was satisfactory: it appeared that the diameter of the foam cells formed was approximately constant and regularly distributed within the body of the cured glue line. As regards viscosity and thixotropicity, the Brookfield viscosities at 16°C were 16,800 mPa and 12,700 mPa for 10 and 20 rpm, respectively. Glue-mix densities were 1.14 g/liter at both 14 and 17°C, indicating stability of the glue mix at the latter temperature. The rate of expansion measured in a test tube at 20°C starting with a glue mix at 18°C was measured as being between 20 and 30%.

III. INDUSTRIAL APPLICATION TESTS

Comparative industrial application trials were carried out using the expanding formulation described and the identical adhesive in a normal nonexpanding glue mix. Spruce glulam beams 4.5 m long, composed of timber planks of sections 115 mm × 50 mm, were manufactured at a moisture content of 14 ±2%. Half the timber was properly planed with a thickness decrease of a total of 5 mm for each timber plank to a total thickness of 45 mm. The other half of the timber was only partially planed, to a total decrease of thickness of 0.5 mm and a residual thickness of 49.5 mm of the plank.

To have comparable beams on the bending tests, the timber planks
were divided according to their density, before planing or partial
planing, into three lots, namely:

Lot number	Density (g/cm^3)
1	≤ 0.42; ≤ 0.46 for the central laminae of the beam
2	≤ 0.47; ≤ 0.50 for the intermediate laminae of the glulam beam
3	≤ 0.51; ≤ 0.55 for the outer laminae of the glulam beam

Application of the glue mix to the timber was obtained by using
a glue extruder as shown in Fig. 11.3. Such an application system
is a continuous mixing system. The application of the adhesive
was later attempted, successfully, with "classical" extruders. The
trichlorofluoromethane holding tank was equipped with a cooling
unit (working by circulation of a glycol solution) with a thermostat
setting of 3°C at the beginning of the trial run. After 2 hours
of the industrial run, temperatures on the outside surface of the
holding tank, 20 cm from its base, showed the pattern indicated
in Fig. 11.4.

The surface temperature of the timber during glue application
was 18°C, while after mechanical mixing of the glue-mix ingredients
and before introduction into the extruder head, the formulation

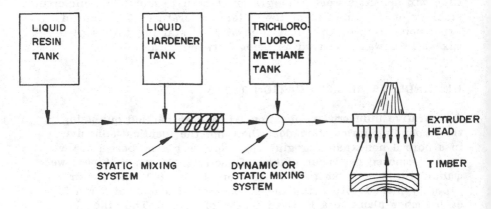

Fig. 11.3 Schematic representation of glue-mix application system.

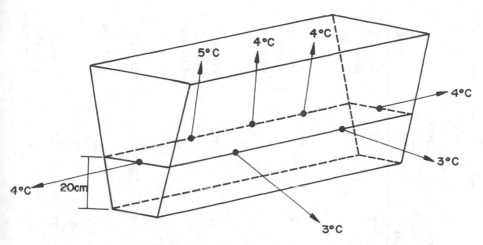

Fig. 11.4 Temperature 20 cm from its base of cooled trichlorofluoromethane holding tank, 2 hours into the industrial run.

had a temperature of 16°C. The rate of linear movement of the
timber under the glue extruder was 60 m/min. Table 11.1 reports
the conditions of assembly and manufacture of the 10 glulam beams
prepared.

The disposition of the 8 beams in the L-clamps is shown in Fig.
11.5. Two beams were clamped separately (beams II and VI), since a
lower ambient temperature was used during curing.

The beams were evaluated globally by a bending strength test,
while the quality of the adhesive bond was evaluated by block-shear
tests and delamination tests.

The bending test was chosen as a comparative mean because
its results depend heavily on the quality of the timber used to
prepare the beam. The test was then used to verify that the expanding adhesive does not weaken the mechanical behavior of long glulam
beams where the principal stress in service is a flexural one (Fig.
11.6). The bending test was carried out by applying a force at
one-third of the span on two beams of each type.

The span used was 4000 mm (4 m; ≥ 18 times the beam thickness), while the beam was 170-220 mm thick and 105-170 mm wide.
The results of the test are shown in Table 11.2.

The results of the tests aimed at verifying the quality of the
glue bond are shown in Table 11.3.

When the results in the three tests are analyzed, it appears
that the particularly high values obtained for the elastic modulus
of the laminated beams reflect the effects of both the orientation

Table 11.1 Composition and Detailed Conditions of Bonding of Industrial Laminated Beams

Beam number	(g/cm³)	Formulation	Glue temperature at application (°C)	Time interval between glue-mixing and application (min)	Assembly time (min)	Ambient temperature during curing (°C)	Planing
I	0.52 0.50 0.46 0.49 0.52	Nonexpanding	19	26	13	28	Lightly planed, 49.5 mm
II	0.52 0.52 0.46 0.47 0.55	Nonexpanding	19	30	9	28	Lightly planed, 49.5 mm
III	0.54 0.50 0.44 0.47 0.52	Expanding	17	6	40	18	Lightly planed, 49.5 mm
IV	0.51 0.45 0.45 0.49 0.50	Expanding	17	8	29	28	Lightly planed, 49.5 mm
V	0.52 0.51 0.45	Expanding	17	10	27	28	Lightly planed, 49.5 mm

	0.49 0.52						
VI	0.57 0.47 0.44 0.46 0.54	Expanding	15	23	23	18	Lightly planed, 49.5 mm
VII	0.54 0.46 0.44 0.47 0.55	Nonexpanding	19	14	25	28	Fully planed, 45 mm
VIII	0.51 0.49 0.45 0.49 0.54	Nonexpanding	19	8	31	28	Fully planed, 45 mm
IX	0.53 0.48 0.44 0.47 0.50	Expanding	15	28	8	28	Fully planed, 45 mm
X	0.53 0.49 0.44 0.47 0.52	Expanding	15	26	11	28	Fully planed, 45 mm

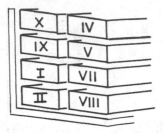

Fig. 11.5   Beam disposition in L-clamps.

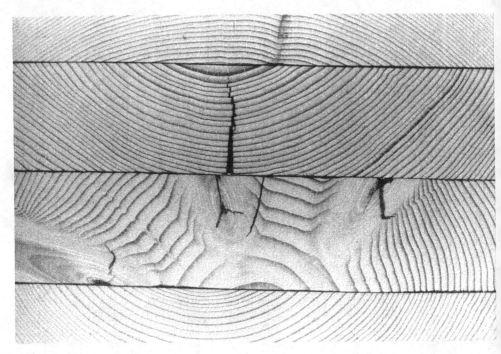

Fig. 11.6   Example of cured expanding adhesive filling cracks in glulam beam section.

Table 11.2 Results of Beams Bending Tests

Beam number	Moisture content (%)	Composition	Mass (kg)	I (cm⁴)	IV (cm³)	MOE (MPa)[a]	MOR (MPa)[a]	Breaking point	Density (g/cm³)
1. I	12	Nonexpanding formulation, lightly planed timber (18°C)	49.400	9573	862	12,600	43.6	2820	0.493
II			50.200	9573	862	12,800	60.6	3920	0.501
						12,700*	52.1*		
2. III	12	Expanding formulation, lightly planed timber (18°C)	50.500	9573	862	12,500	47.9	3100	0.504
IV	13		49.300	9834	878	12,800	56.4	3715	0.487
						12,650*	52.5*		
3 IV	13	Expanding formulation, lightly planed timber, heating at 28°C	48.700	9573	862	11,500	60.6	3920	0.486
V	12		50.200	9573	862	13,100	46.1	2980	0.496
						12,300*	53.4*		
4. VII	12	Nonexpanding formulation, fully planed timber (28°C)	49.400	9573	862	14,100	59.5	3850	0.488
VIII	12		48.900	9573	862	12,000	43.3	2800	0.484
						13,050*	51.4*		
5. IX	12	Expanding formulation, fully planed timber (28°C)	47.700	9573	862	12,200	58.6	3790	0.472
X	12		49.400	9573	862	12,900	53.0	3430	0.488
						12,500*	55.8*		
Average	12.2		49.37			12,650	52.96	3432.5	0.490
S	0.42		0.82			698	7.12	465.2	0.009
G	0.40		0.78			662	6.75	441.3	0.009

[a]Asterisks indicate average value; "G" allowed on the whole assembly = 15.40 MPa.

Table 11.3　Compression Block Shear Strength Test and Delamination Test by Steam Injection/Drying

Beam number	Application conditions	Shear strength test at 12% equilibrium moisture content (strength values ≤ 6.5 MPa and wood failure ≥ 75% not considered)			Adhesion (wood failure)			Shear strength tests at 12% equilibrium moisture content (all results considered)			Delamination test percentage		
		Average (MPa)	Standard deviation (MPa)	Variation (%)	Average (%)	Standard deviation (%)	Variation (%)	Average (MPa)	Standard deviation (MPa)	Variation (%)	1st cycle	2nd cycle	3rd cycle
I + II	Nonexpanding formulation, lightly planed timber, heating at 28°C	9.2	1.4	15.5	94	9	10	8.9	1.8	19.7	0.9	1.6	2.3
III + VI	Expanding formulation, lightly planed timber (18°C)	9.5	1.6	16.5	87	13.2	15	9.3	1.8	18.8	0.3	0.5	1.1
IV + V	Expanding formulation, lightly planed timber (28°C)	8.9	1.2	13.4	81	10.5	13	8.7	1.3	15	0.4	0.6	0.8
VII + VIII	Nonexpanding formulation, fully planed timber (28°C)	9.0	1.2	13.3	87	8.5	9.8	8.7	1.6	18.2	1.3	2.7	3.2
IX + X	Expanding formulation, fully planed timber (28°C)	9.0	1.2	13.7	90	8.5	9.4	9	1.2	13.7	1.0	1.1	1.8

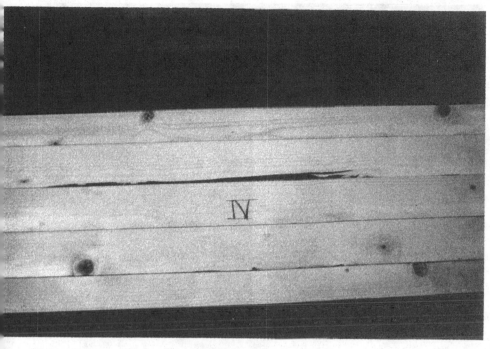

Fig. 11.7 Example of cured expanding adhesive filling gap caused by bad planing of lamina of a glulam beam.

of the beam members and the high volume of the beams. In short, the values obtained were to be expected and are correct in relation to the strength requirements of the beams in use. The values of the modulus of rupture are slightly more favorable for the expanding adhesive system. This is more noticeable in the beams composed of fully planed timber, where the use of the expanding adhesive has elasticized to take advantage of the bonded joints (note the narrower distribution of values). In general the modulus of rupture is adversely affected by bad timber surfaces, due both to the presence of physical defects on the surface and to the presence of joints (Fig. 11.7). All this notwithstanding, and notwithstanding the intentionally poor condition of the timber surfaces, satisfactory results were obtained for the beams made by using the expanding adhesive system (Fig. 11.8).

The results obtained in the block-shear test also satisfy the requirements of the relevant international specifications. Comparison of the results between beams bonded with the two adhesives show that those bonded with the expanding adhesive system give more

Fig. 11.8 Visual example of timber saving when using partially planed timber bonded with expanding adhesive. Note the difference in height between two beams. The left-hand one has been prepared by using timber laminae originally of the same thickness as for the right-hand beam. The laminae of the right-hand beam were only partially planed. The number of laminae is the same in both beams.

favorable results both on partially planed timber (between 7 and 12.5% better) and on fully planed timber (17.5% better). The variation of temperatures examined (18 and 28°C) has an effect of approximately 5% on the results obtained. The delamination tests showed similar trends to the block-shear tests, although the standard deviations obtained were far greater. The expanding adhesive system reduced delamination between 2 and 2.8 times on partially planed timber, and at least twice on properly planed timber. The presence of microcavities in joints bonded with expanding resorcinolic adhesives had to be checked with regard to the influence of these anomalies on possible delamination in the presence of elevated temperatures. After fire tests for periods of up to 20 minutes, no delamination was apparent on any of the beams manufactured with the expanding adhesive system. There also appeared to be no difference in the fire behavior between joints bonded with normal resorcinol and with expanding resorcinol adhesives.

IV. ECONOMIC ASPECTS

From an economic point of view three aspects of the expanding adhesive system must be considered, namely:

1. Gain on investment per cubic meter of laminated timber
2. The investment necessary to introduce the process
3. Net gain (investment is included, but not the costs of research and licensing)

Several savings are generated by the industrial introduction of the process. First, there are savings in the amount of timber otherwise lost or wasted through planing. The results in Table 11.4 are self-explanatory in this respect. It is apparent that partial planing loses more material than proper, normal, planing by as much as a factor of 5. In France glulam manufacturers use mainly material of both 50 and 38 mm nominal thickness: for this reason an average of 9% savings on timber volume can be obtained.

Second, there are savings derived during glulam manufacture as opposed to raw timber alone. To this effect there are two main processes in the manufacture of glulam: (a) the processes involving drying, constraining for drying, and possibly preservation of the raw timber planks, which in France represent approximately 9% of the price of the product; and (b) the processes involving planing, gluing, and pressing, which instead represent approximately 20%

Table 11.4   Timber Saved in the Expanding Adhesive System

Thickness of laminate (mm)	Thickness after planing (mm)	Thickness after partial planing (%)	Loss due to planing (%)	Loss due to partial planing (%)	Timber saved (%)
50	44	48	12	4	8
38	33	37	13	2.6	10.5
32	27	31	15	3.1	12.5
25	20	24	20	4	16

of the cost of the finished product (excluding the cost of the adhesive).

Third, there is the saving on the cost of the adhesive used. In normal cases the cost of the adhesive represents approximately 15% of the cost of the finished product. There is no doubt that the new adhesive system decreases the amount of glue consumed. More important, the addition to the glue mix of trichlorofluoromethane decreases the cost of the glue mix by an average of 10%.

As a consequence of all this, considering the cost of timber delivered to the factory, in France, at approximately \$236 (U.S.) per cubic meter, the total savings obtained, taking into account all the factors considered above, is then:

$$236(1 + 0.09 + 0.20 + 0.15) \times 0.09 + 236 \times 0.15 \times 0.10 = \$34.10/m^3$$

of finished product

which represents a 10% total savings on the cost of the finished product. [In French currency, at the time of writing, the total savings is 195.25 FF/m^3.]

There are of course capital investments necessary to introduce such a process, such as modifications to the extruder/glue spreader (cooling, dosing pump for trichlorofluoromethane) for a total of \$2500-9000; installation of cooling for the trichlorofluoromethane holding tank, for a total of approximately \$1500-\$2000; and a manual timber sorter for thickness and shape at approximately \$35,000. Except for the sorter, at a labor time utilization of 1 hour/10 m^3, all the other investments are accountably recoverable in 5 years.

Table 11.5 Production Distribution of the Glulam Industry in France

Capacity ($m^3 \times 10^3$/year)	Number of factories
0.5-1	20
1-2	7
2-5	14
5-10	5
10-15	1
20	1

Table 11.6   Net Gains for Small Capacity Factories: 500 m^3/year[a]

Lower production case: 200 m^3/year

investment	5000 FF/year = $\dfrac{15,000 + 10,000}{5}$
timber sorting	$\dfrac{8000 \text{ FF} \times 13 \times 20}{220 \times 7.8} \approx 1200$ FF
net gain	195.25 FF × 200 m^3/year - (5000 + 1200)
	= 32,850 FF/year = \$5743/year

Higher production case: 400 m^3/year

investment	5000 FF/year
timber sorting	$\dfrac{8000 \text{ FF} \times 13 \times 40}{220 \times 7.8} \approx 2400$ FF
net gain	195.25 FF × 400 m^3/year - (5000 + 2400)
	= 70,860 FF/year = \$12,388/year

[a]FF = French francs.

Table 11.7   Net Gains for Medium Capacity Factories: 3000 m^3/year

Lower production case: 1200 m^3/year

investment (most pessimistic case)	$\dfrac{50,000 + 200,000 = 10,000}{5} = 52,000$ FF/year
net gain	195.25 FF × 1200 m^3/year - 52,000
	- 182,300 FF/year = \$31,870/year

Higher production case: 2400 m^3/year

investment	52,000 FF/year
net gain	195.25 FF × 2400 m^3/year - 52,000
	= 416,600 FF/year = \$72,832/year

Table 11.8 Net Gains for Higher Capacity Factories: 20,000 m^3/year

---

Lower production case: 8000 m^3/year

   investment   52,000 FF up to 100,000 FF/year
                          (in the most unfavorable case)

   net gain   195.25 FF × 8000 m^3/year − 100,000

                          = 1,462,000 FF/year = \$255,000/year

Higher production case: 16,000 m^3/year

   investment: one should consider the acquisition of:

     a glue spreader at 250,000 FF, equipped with a third holding
     tank (for CCL$_3$F) and cooling

     a timber sorter at 200,000 FF

     various cooling materials at 20,000 FF

   for a total investment of $\dfrac{250,000 + 200,000 + 20,000}{5 \text{ years}}$

                          = 94,000 FF/year

   net gain   195.25 FF × 16,000 m^3 − 94,000

                          = 3,030,000 FF/year = \$530,000/year

---

It is interesting to evaluate the net gains obtained, including the value of the investments. To this effect three scenarios for different levels of timber consumption have been examined, for factories of yearly capacities of 500, 3000 m^3, and 20,000 m^3. This breakdown was dictated by the factory type distribution in France (see Table 11.5).

Furthermore, for each type of factory, two cases can be considered: a lower production case (when 40% of the manufacturing capacity is utilized in producing straight glulam) and a higher production case (when 80% of the manufacturing capacity is used to produce both straight and curved glulam). The monetary amounts involved are shown in Tables 11.6-11.8.

V. CONCLUSIONS

The development of a resorcinol adhesive system expanding at temperatures of 18-25°C appears to be a positive step for the production of exterior-grade glulam. The adhesive produced positive

results on partially planed timber. The bond on properly planed
timber also appears to be improved by the use of such adhesives.
This is an advantage for improving bonding of timber damaged
by planing defects. From an economical point of view it is apparent
that the introduction of such a process presents only minor constraints in relation to the advantages achieved. The level of investment needed is indeed minor in relation to the level of savings
possible. In conclusion, the use of expanding resorcinolic adhesives
appears to be a better and more efficient system for gap filling
of glued joints, especially when the gaps are of a thickness and
a size that cannot be filled effectively by traditional gap-filling
adhesives.

REFERENCES

1. Elbez, G. "Etude Expérimentale du Collage de Planches Épicéa
   Brutes de Sciage par Colle à Moussage Contrôlé." Report 283,
   Centre Technique du Bois et de L'Ameublement, Paris.
2. Elbez, G. "Essais de Flexion et de Résistance sur Poutres en
   Bois Lamellé-Collé 'Colle Expansible' Réalisées Industriellement."
   Report Centre Technique du Bois et de L'Ameublement, Paris.
3. Elbez, G. "Influence de la Distribution des Épaisseurs de Joints
   de Colle sur la Qualité du Collage." Report Centre Technique
   du Bois et de L'Ameublement, Paris.
4. Elbez, G. and De Leeuw, J. (1981). European patent 81/400/769-6.
5. Elbez, G., Schambourg, F., and Vinatier, J.-M. (1988). "Ad-
   hesif Resorcin Expansant à 20-25°C pour Application en Lamelles-
   Colles." Report Centre Technique du Bois et de L'Ameublement,
   Paris, February.
6. "Procédé de Collage Entre Elles de Planches de Bois à Surface
   Brute ou Blanchie, de Placages Tranchés ou Déroulés." Report
   Centre Technique du Bois et de L'Ameublement, Paris.
7. Elbez, G. (1983). "Etablissement d'un Algorithme pour le Con-
   trôle de la Qualité des Pièces Blanchies avant Lamellation."
   Report Centre Technique du Bois et de L'Ameublement, Paris,
   April.
8. Elbez, G. (1984). "Lamellé-Collé avec Adhésiv Expansif. Influence du Taux de Blanchiment sur la Qualité du Collage." Report
   Centre Technique du Bois et de L'Ameublement, Paris, July.
9. Elbez, G. (1984). "Lamellé-Collé avec Colle Moussante. Tests
   de Collage Selon Différentes Configurations d'Etat de Surface
   après Rabotage." Report Centre Technique du Bois et de
   L'Ameublement, Paris, July.

10. Elbez, G. (1988). "Influence des Adhésifs Expansif et de leur
    Condition de Mise en Oeuvre sur la Qualité des Collages, Cas
    de Bois: Présentant des Noeuds, Pelucheux, sans Défaut."
    In preparation, Report Centre Technique du Bois et de
    L'Ameublement, Paris.

Index

Adhesion
  adsorption, 75,76
  to cellulose, 97,98,112
  chemical bonding, 75-92
  contact angles, 76
  Coordinate bonding, 89,90
  covalent bonds, 76-92
  diffusion, 75
  donor-acceptor interactions,
    75,90
  electrostatic interactions/
    forces, 75,97,99
  hydrogen(H)-bonding, 40,
    78,88,89,97,99,102,109,
    110,113-118,129,131,137,
    143
  mechanical entanglement, 75,
    89,91
  molecular entanglement, 75,
    89
  PF/cellulose, 112,118
  secondary forces, 97,98
  specific adhesion, 75,76,91
  surface theory, 112
  theories of, 75
  thermodynamics, 76
  torsional forces, 99
  Van der Waals forces, 78,
    79,97,99,102,109,110,113-
    118
  water/cellulose, 112,11
Animal glues, 26

Blood glues
  air-extended(foamed) PF, 18

[Blood glues]
  aldehyde compounds, 16
  alkaline dispersion, 14
  application, 19
  cold press formulations, 14,
    15
  denaturants, 14-16
  drying, 14
  exterior, 18
  filler, 14
  glue life, 19
  lysine content, 13
  midexterior, 17
  mold-resistant, 14,22
  phenol-formaldehyde, 17,18
  plywood, 15,16,18-21
  pressing times, 19-22
  proteins, 13
  solubility, 14,15
  soybean blends, 22
  viscosity, 14,15,17,19,22
  water resistant, 14,16

Casein glues
  acid precipitation, 22
  alkaline dispersion, 23
  assembly time tolerance, 25
  denaturants, 23,24
  fire resistance, 26
  fungicide, 23
  furniture, 22
  label gluing, 25
  laminating, 25
  lime content, 23
  paper sizing, 25

[Casein glues]
  particle size, 23
  plywood, 25
  single package, 23
  soybean blends, 11,25
  viscosity, 23,24
  water resistance, 22,23,25
Cellulose, 97,98,112-119
  adhesion, 112,113
  reaction with formaldehyde,
    97
  water sorption isotherm, 113
Chemical bonding
  by chemical treatment, 78,
    79,80,91
  by heat treatment, 81,91
  by heat-chemical treatment,
    82-87,91
  by secondary forces, 78,80,
    82,88,89
  coordinate bonding, 89,90
  covalent, 75-77,80,88-92
  glue to wood, 77,78,80
  hemiacetal linkage, 81
  in polymer systems, 77
  isocyanates, 78-80
  noncovalent, 78,88,90
  paper, 79,80,81,82,90
  phenol-formaldehyde, 77,78,
    81,85,91
  urea-formaldehyde, 78
  water resistance, 77-79,89-91
Collagen glues, 1,26
Cresol-formaldehyde resins, 140

3,3'-dichloro-4,4'dihydroxydi-
  phenylmethane, structure
  of, 98
Dimethylolurea, 24

Epoxy resins, 6,145,147

Fast-set adhesives, 229-305
  assembly time dependence,
    287-300
  clamping time dependence,
    282-285
  curing time dependence,
    242,243,245,279-281
  hardeners composition de-
    pendence, 250,257-282
  pH dependence, 231,235,
    236
  planing time dependence,
    244,247
  temperature dependence,
    242,243,255,271,273-282
  testing, 240
  tolerance to quantities varia-
    tions, 237,250,253,255,
    261,264,266-273
  tolerance to viscosity varia-
    tions, 237,250,253,261
  viscosities, 247
  wood density dependence,
    236,241-243,255
Fillers, 97,215-217,220
Finger joints, 192,229-303
Fish skin adhesives, 1,26
Formaldehyde
  adhesives emission potential,
    320-322
  analytical methods, 327-334
  board loading rates of, 310
  bottle methods, 350-353
  Cannizzaro reaction, 142,
    369-373
  dessicator method, 350
  determination of, 324
  effect of air velocity on,
    344
  emission classes, 22,316,335,
    340
  emission differences by prod-
    uct, 357-362
  emission differences by resin,
    362,363

[Formaldehyde]
  emission sources, 312,320-
    322,338,339
  exposure limits, 315
  factors influencing release of,
    334-339
  hazardous emission, 13,307,
    314,340
  health effects, 308,311,312
  indoor air level of, 315,316
  industrial uses of, 314
  influence of molar ratio on
    resins, 363-365
  large chamber test, 338-344
  limit of awareness of, 312
  limits for wood products, 316,
    317,323,340
  low emission of, 317
  methylol content, 324
  perforator method, 346-350,
    369
  properties of, 308-312
  regulatory specifications, 317,
    340
  release/emission, 308,317,318
    323,337,338,339,362-372
  release determination, 325
    326,339-357
  release from diphenylmethane
  diisocyanate (MDI), 319,
    323,367
  release from melamine-
    formaldehyde resins (MF),
    318,322,371
  release from melamine-urea-
    formaldehyde (MUF) resins,
    318,320,325
  release from mineral binders,
    319
  release from phenol-
    formaldehyde (PF) resins,
    308,318,319,321,370-371
  release from PVA, 319
  release from resorcinol-
    formaldehyde (RF) resins,
    319

[Formaldehyde]
  release from spent sulphite
    liquor (SSL) resins, 319
  release from tannin-
    formaldehyde (TF) resins,
    319,367,370-372
  release from urea-formaldehyde
    (UF) resins, 318,320-323
    324-325,362-373
  release from UF foams, 308
  release reduction/low content,
    323,362-373
  resins molar ratio determina-
    tion, 324,325
  scavengers, 363,366,369
  small chamber methods, 344-
    346
  states, 322

Gelatin glues, 22,26
Glulam, 192,229-303,318,319,
    389,393,404,406

Hexamethylenetetramine (hexa,
  hexamine), 6,11,309,319,
  367,368 (*see also* Phenol-
  formaldehyde resins)
"Honeymoon" adhesives, 192,
  229-303 (*see also* Fast-set
  adhesives)
Hotmelt adhesives, 75,211-227
  antioxidants, 217
  application guidelines, 222-
    224,226
  application pressure, 222,
    226
  application weight, 222
  edgebander procedure for,
    226
  edgebanding, 213
  edge veneering, 212,213,
    216,220

[Hotmelt adhesives]
  edging materials, 214,215,
    221
  environmental conditions—
    requirements for, 220
  ethylene-vinylacetate (EVA),
    213,215-220
  fault detection procedures
    for, 225-227
  filler, 215-217,220
  formulations, 216-218
  gravure applicator wheel, 222-
    224
  health/safety, 227
  high performance, 219
  melting time, 222
  polyamide, 213,219,220
  production technique, 218
  resins for, 216
  softening point, 215
  stability, 216
  tackifiers, 216
  techniques of use of, 220-227
  type of laminates for, 213
  veneer splicing, 212
  vinyl acetate content—effect
    of, 216
  viscosity, 215,216,218
  wettability, 215
Hydrogen peroxide, 83,84,86

Isocyanates, 78,79,80

Lignin, 84,127,166-194
  constitution, 155
  demethylation, 163
  lignosulphonates, 155,163
  meta-reaction, 173-186
  phenolic groups pK$_a$, 156
  phenolization, 162,163
  polymerization reactions, 157-
    159

[Lignin]
  quinone methides, 157
  structure, 157-159
Lignin-formaldehyde adhesives/
  resins
  alkali lignin, 156,157,163
  α-carbinol groups, 157,160
  cold-set, 164-166, 184-186
  cross-linking agents for,
    167,168
  dimers, 175
  formaldehyde consumption,
    181-184
  formaldehyde release, 321
  honeymoon fast-sets, 165,166
  from kraft wood lignin, 181-
    186
  with melamine, 168,169
  meta-modification, 173-186
  moisture dependence, 169-
    173
  for particleboard, 169-173
  with PF resins, 155,160-163,
    168-173,186
  pressing time, 169-173
  quinone methides, 157
  resin content, 169
  with resorcinol, 164,185
  from soda AQ wood lignin,
    181-184
  from soda bagasse lignin,
    161,162,164,166-173,181-
    186
  tetramers, 175
  thermosetting, 166-173
  with UF resins, 159,168,169
Lignosulphates, 84,86,87

m-Aminophenol, 229,130,232,
  235,247
Melamine, 308,318,320,325
Melamine-formaldehyde (MF)
  resins, 140,318,322,372

Melamine-urea-formaldehyde
(MUF) resins, 147,148,318,
320-325

Particleboard, 78,79,84,169-173,
212,308,315-323,339-371
Phenol-formaldehyde (PF)
resins
adhesion, 77,78,81,85,91,97,
112-118
assembly times, 146,148,149
analytical techniques, 123,
125-127
bond quality, 145,148,149
branching, 123,125,129-133,
135,136,144-146,148,149,
192-195,197,199
breaking strength, 145
catalysts, 124,129,140
cold-setting, 192
characterization techniques,
121,123-150
curing, 135-137,140,142,144,
145,147,197
dimers, 99-102,109,110,112,
113-118,135,139,140
dynamic light scattering,
126-128
effect on cellulose water-
sorption isotherm, 113
end-use properties, 122,133
extenders, 146,148,149
fast-set, 232
fillers, 148,149
formaldehyde emission, 310,
320-324,371-374
fracture toughness, 148,149
gel permeation chromatography,
123,124-129,131
helices, 110,111
hemiformal structures, 135,
136,139,144
hexamers, 99-102,110,111

[Phenol-formaldehyde (PF)
resins]
hexamethylenetetramine
(hexa, hexamine), 124,127,
136,139,140,141,144
hydrodynamic volume, 130
intrinsic viscosity, 130
with lignin, 155,160-163,168-
173,175,180,186,192
linear, 129,130,139,140,192-
195,197
low angle laser light scattering (LALLS), 128-130
methylene ethers, 138,142
methylene/methylol ratio,
137,144
minimum energy conforma-
tions, 97-112
molar ratio, 124,125,128,129,
137-149,205,207
molecular association, 127-
129,131,133
molecular weight, 130-134,
137,142,143,145-149
molecular weight distribution,
123,124,125-129
natural polyphenolics, 127,
128
NMR, 133,135-144
nonlinear, 124,129
ortho-methylol phenol, 129,
146,148
ortho-orienting additions,
118,139,140,142
ortho-ortho dimers, 98,101,
103,108,109,112,114,116-
118
ortho-para dimers, 98,101,
103,105,113,115,118
para-para dimers, 98,101-
104,112,118
performance, 121,144,150
properties, 121,144-147,150
reaction time, 136,144
resin age, 126,129,136,144

[Phenol-formaldehyde (PF)
    resins]
  resin shrinkage, 145,147,149
  rigidity, 148,149
  rotamers, 102,109
  rotational isomers, 102
  size exclusion chromatography,
    124-127,129,130,133
  solvent association, 124,127-
    129
  structural parameters, 122,
    145-147,150
  structure, 98,112
  substrate impregnation, 113
  synthesis parameters, 122-127,
    133,144,150
  thermal stability, 138,142
  torsional brain apparatus for,
    145,148,149
  trimers, 98-103,106,107,109,
    110,135,139,140
  trimethylolphenol, 196,197
  viscosity, 130-134
  water resistance, 112
Phenol-resorcinol-formaldehyde
    (PRF) adhesives/resins
  acid catalyzed formulations,
    205-208
  assembly time, 202,203
  base catalyzed formulations,
    205-209
  blowing agents, 310
  branched characteristics, 201-
    205
  branching, 192-200
  branching molecules, 195-200
  condensation time, 206,207
  cost, 192
  curing temperature, 201,395
  economic advantages, 301,302,
    403-407
  exothermic effect, 390-392,402
  expanding types, 389,408
  fast-set, 231,232,234 (see
    also Fast-set adhesives)

[Phenol-resorcinol-formaldehyde
    (PRF) adhesives/resins]
  fire tests, 402,403
  fluorocarbons for, 390-393,
    394
  gap-filling, 389,393
  gelling agents, 390
  glue extruders for, 394
  hardener, 191,199,201
  "Honeymoon" types, 203,204,
    205 (see also Fast-set
    adhesives)
  lignin, 192
  linear, 192-195,204
  molecular weight, 195
  pH of application, 201,202
  pot-life, 202,205-209
  reaction time, 202
  reactive sites, 198
  resorcinol, 191-193,197
  resorcinol content, 192,193,
    195,202
  shelf-life, 202,205-208
  solids content, 195,201
  surfactants for, 393
  timber planing losses, 389,
    391,402,403
  trimethylophenol, 196,197
  thixotropicity, 393
  viscosity, 195,201,202,393
Plywood, 223,318,319,323
Polyurethanes, 124,127
Polyvinyl acetate (see PVA)
PVA
  azeotropic copolymerization,
    50
  batch reactors/processes,
    31,51,53,54,60-63
  branching, 48,51
  buffers, 43
  chain transfer, 45-48,57
  coalescents, 43
  colloids, 41,42,46
  continuous reactors/processes,
    31,55-60,61,62,63

[PVA]
copolymers, 33-35,42,61,62
cross-linking agents, 43
electrostatic stabilization, 54
ethylene copolymers, 35,213
fillers, 44
film properties, 65
flammability, 35-37
formulation, 32
freeze-thaw resistance, 35,42
fungicides, 44
gelling agents, 43
general, 212
glass transition temperature,
34,51
health hazards reduction, 35
heat of polymerization, 48,63
heat stability, 42
homogeneous nucleation, 54
hydrolysis, 65,13
hydrolysis protection, 35
inhibitors, 32,34,38,50,61
initiators, 32,36-38,46,47,50,
54-56,60-62
initiators activation energy, 38
initiators stability, 37,38
kinetics of copolymerization,
49,50
kinetics of reactions, 45-48,
53-56,60
mean residence time, 57
mechanical stability, 35
minimum filming temperature,
34
molecular weight, 48,49
multifunctional monomers, 35
nonvolatiles contents, 32,53,
55,64
particles structure, 57
pH adjusters, 43,44
pigments, 44
plasticizers, 43,49,63
polymerization reactions, 32,
45
polyvinyl alcohol, 32,41,42,
54,61

[PVA]
postreaction functionalization,
35
practical production aspects,
62-65
preservatives, 44
radical desorption, 57
rate-determining step, 47,59
reaction temperature, 32,50
remoistenable adhesives, 42
safety factors, 64
shear viscosity, 40
softening, 34,35,43,49
steric stabilization, 54,55
surfactants, 38-40,43,46,51,
54,56
tackifiers, 43
thickeners, 40,43
vinyl acetate monomer properties, 33
viscosity, 41
water resistance, 35,41,42
PVC, 211,213,215,222,223,225,
226

Resorcinol, 136,164,185,232,
319
Resorcinol-formaldehyde
adhesives/resins, 192,212
branching, 145
blowing agents, 390
expanding types, 389-408
flourocarbons for, 390
formaldehyde release, 319
gap-filling, 389-393
gelling agents, 390
moisture content, 143,145
molar ratio, 144,145
molecular weight, 145,147
pot-life, 145
resin shrinkage, 145
surfactants for, 393
thixotropicity, 311

[Resorcinol-formaldehyde
     adhesives/resins]
  timber planing losses, 389,391,
     402
  viscosity, 393

Soybean adhesives
  alkaline solubility, 2
  blood blends, 9,10,22
  briquetting, 5,6
  casein blends, 11,25
  cellulose alkali burn, 3
  cold pressing, 6,8
  crosslinkers, 5,6
  denaturants, 5,6
  dispersion, 3,5,6,9,11,12
  fire performance, 11
  glue spread, 7
  high alkali formulations, 3,4
  humidity and water resistance,
     1,4-6,9,11
  hydrogen bonds, 3
  insoluble proteinates, 4
  low alkali formulations, 3,5
  lumber laminating, 6
  no-clamp process, 8
  nonstaining, 5
  paper laminating, 5,6,9
  plywood, 5-9,11
  preservative, 4,11
  pressing schedules, 5-9
  soft board laminating, 5,6
  staining, 3
  structural uses, 5
  viscosity, 4,6,13
Soybean meal
  alkaline dispersion, 3,5,9,11,
     12
  carbohydrates, 3

[Soybean meal]
  particle size, 3
  protein content, 2,3
  specific surface values, 3
  test method, 3
Surfactants, 38,39,40,43,46,
     51,54,56
  associative monomers, 40
  critical micelle concentration,
     39,56
  hydrophilic-liporhilic balance,
     39
  micelles, 39,53,54
  polymerizable, 40
  surface activity, 41

Tannin, 192,193,229-231,232,
     368
  pine, 231,233,234,236
  wattle, 231,233-236,282
Tannin-formaldehyde adhesives/
     resins, 192,193,212,319,
     367,371-373
Tannin-resorcinol-formaldehyde
     adhesives/resins, 192,193,
     231,232,282

Urea, 310,320,322,327,368,371
Urea-formaldehyde (UF)
  adhesives/resins, 40,78,
     140,159,168,212,308,318
     320-325,362-371

Wood buffering capacity, 78

Printed in the United States
by Baker & Taylor Publisher Services